高等学校土木工程专业国际化人才培养全英文系列教材

Introduction to Geotechnical Engineering

岩土工程概论

王培涛　主　编

中国建筑工业出版社
CHINA ARCHITECTURE & BUILDING PRESS

图书在版编目(CIP)数据

岩土工程概论 = Introduction to Geotechnical Engineering / 王培涛主编. — 北京 : 中国建筑工业出版社, 2024. 3

高等学校土木工程专业国际化人才培养全英文系列教材

ISBN 978-7-112-29401-5

Ⅰ. ①岩… Ⅱ. ①王… Ⅲ. ①岩土工程-高等学校-教材-英文 Ⅳ. ①TU4

中国国家版本馆 CIP 数据核字(2023)第 241186 号

责任编辑: 聂 伟
文字编辑: 卜 煜
责任校对: 赵 力

高等学校土木工程专业国际化人才培养全英文系列教材

Introduction to Geotechnical Engineering

岩土工程概论

王培涛 主 编

*

中国建筑工业出版社出版、发行(北京海淀三里河路 9 号)

各地新华书店、建筑书店经销

北京科地亚盟排版公司制版

建工社(河北)印刷有限公司印刷

*

开本: 787 毫米×1092 毫米 1/16 印张: 16½ 字数: 494 千字

2024 年 3 月第一版 2024 年 3 月第一次印刷

定价: **48.00** 元 (赠教师课件)

ISBN 978-7-112-29401-5

(42162)

Brief Introduction

The book provides a comprehensive overview of various aspects of geotechnical engineering. It is divided into nine chapters that cover different topics related to the field: introduction to geotechnical engineering, the properties of rock materials, the properties of rock mass, the physical properties of soil, constitutive relationships of rocks, slope engineering and stability, underground engineering, numerical methods for geotechnical engineering, and developments of geotechnical engineering. The book also includes a list of academic journals, explanations of technical terms, and key vocabulary related to the field of geotechnical engineering. It provides readers with additional resources to further understand the research trends and professional terminology in the field.

In addition to being used as a textbook for undergraduate students, this book can also be selected as a textbook for graduate students. It can also serve as a reference book for teachers in higher education institutions, researchers in scientific research institutes, and engineering departments, as well as engineering technicians.

本书介绍了岩土工程专业领域的基本常识，分九个章节，各章分别涵盖了与该领域相关的不同内容：岩土工程背景、岩石材料的性质、节理岩体的性质、土壤的物理性质、岩石本构关系、边坡工程与稳定性、地下工程基本知识、岩土工程数值方法以及岩土工程发展。本书还列出了岩土工程专业相关的一些学术期刊、术语解释和主要词汇。这为读者进一步了解该领域的研究趋势提供了帮助。

除作为本科生教材使用外，本书也可用作研究生的教材，还可以作为高等院校相关专业的教师、科研院所和工程部门的科研人员、工程技术人员的参考用书。

为了更好地支持教学，我社向采用本书作为教材的教师提供课件，有需要者可与出版社联系，邮箱 jckj@ cabp. com. cn，电话（010）58337285。

Preface

Geotechnical engineering plays a crucial role in the design, construction, and maintenance of civil infrastructure. It encompasses the study of soil and rock mechanics, as well as their interaction with structures and the environment. As a multidisciplinary field, geotechnical engineering requires a comprehensive understanding of geology, engineering mechanics, materials science, and so on. The development and advancement of geotechnical engineering have led to the emergence of new theories, methods, and technologies. It is essential to document and disseminate this knowledge to students and researchers in the form of textbooks. Geotechnical engineering textbooks serve as a fundamental resource for acquiring theoretical knowledge, practical skills, and problem-solving abilities.

The principal purpose of writing this book is to provide students with a structured and systematic approach to learning the subject. By presenting the fundamental principles, theories, and methodologies, this book lays the foundation for further studies and research. This book also provides a comprehensive overview of the current state of the field, including recent advancements and innovative techniques. It also encompasses a wide range of technical terms and specialized vocabulary. By providing clear and concise explanations of these terms, it helps students learn to master the English knowledge. This book refers to many contents by many world-famous researchers, and I would like to express my sincere gratitude to the researchers whose work has been referenced in this book. Their contributions to the field of geotechnical engineering have greatly influenced and enriched the content of this publication.

I would also like to extend my appreciation and gratitude to all those who have contributed to the creation and development of this book. First and foremost, I would like to express my sincere gratitude to my postdoctoral co-supervisor, Cai Meifeng, whose invaluable expertise, guidance, and support have been instrumental in shaping the content and structure of this book. Furthermore, I would like to thank the students who participated in the pilot testing and provided invaluable feedback. Their constructive criticism and suggestions by Ma Chi, Liu Cao, Qi Zhenwu, Zhang Bo, Liu Qingru, Fu Yilin, Huang Boran, and Huang Hao have significantly enhanced the clarity and effectiveness of the material. I am indebted to Nie Wei and the publishing team at China Architecture & Building Press for their continuous support, meticulous editing, and dedication throughout the production process. Their attention to detail and commitment to excellence have been pivotal in bringing this textbook to fruition.

This book is funded by the Textbook Construction Fund of USTB and fully supported by the Department of Academic Affairs of USTB. And part of the work is also financially supported by the University of Science and Technology Beijing (Planned Textbooks for 2021), and the National Natural Science Foundation of China (52074020). All of these supports are gratefully acknowledged.

Although we have worked hard to make this book as comprehensive as possible, we recognize that a book about such a hot topic will not be up-to-date for very long. If you have comments, questions or suggestions for improvement, we are welcome. We anticipate updating this book in the future, and we look forward to hearing how you have used these materials and how we might improve this book.

Wang Peitao
Beijing, March 2023

前言

岩土工程在土木基础设施的设计、施工和维护中起着至关重要的作用，涵盖了岩石、土力学以及它们与结构、环境的相互作用等诸多内容。作为一个跨学科领域，岩土工程需要全面了解地质学、工程力学和材料科学等。岩土工程的发展和进步导致了新的理论、方法和技术的出现。将这些知识以教科书的形式记录和传播给学生和研究人员是十分重要的。岩土工程教科书是获取相关理论知识和实践技能、提升解决工程问题能力的重要资源。

编写本书的主要目的是为岩土工程专业英语系统学习提供参考材料。通过介绍岩土工程领域的基本原理、理论和方法，为进一步的学习和研究打下基础。本书介绍了岩土工程多个研究方向的现状，包括最新的进展和创新技术，另外还涵盖了岩土工程相关术语解释和专业词汇。通过对这些术语进行清晰简明的解释，可帮助学生学习掌握相关英文知识。本书引用了许多国内外著名科研人员的成果，在此向这些研究人员表示衷心的感谢，他们的工作为岩土工程领域做出了巨大的贡献，并丰富了本书的内容。

编者同时感谢为本书的编写和内容完善提供帮助的所有人。首先，我要衷心地感谢我的博士后合作导师：中国工程院院士蔡美峰教授，蔡先生为教材的内容和结构提供了指导和支持；同时，感谢我的研究生在教材内容编写过程中付出的辛苦劳动，包括马驰博士生、刘操硕士生、齐振武硕士生、张博硕士生、刘庆如硕士生、付翊林硕士生、黄铂然硕士生和黄浩硕士生；还要感谢中国建筑出版传媒有限公司的聂伟老师及出版团队，在整个制作过程中给予我持续的支持、细致入微的编辑和专注的工作，他们对细节的关注在本教材的出版过程中起到了关键作用。

本书获得北京科技大学教材建设经费资助，得到了北京科技大学教务处的全程支持，同时部分工作内容也得到了北京科技大学（2021 年规划教材）以及国家自然科学基金项目（52074020）的支持，对以上支持表示由衷感谢。

书中如有错误或不妥之处，敬请批评、指正。

王培涛

2023 年 3 月于北京

Contents

Chapter 1
Introduction to Geotechnical Engineering

1.1 Background

Civil engineering is to point to use basic subject knowledge, such as mathematics, physics, chemistry, mechanics, materials and other technical and scientific knowledge to planning, design, construction of various buildings and other structures of a discipline.

The scope of civil engineering is very extensive, it includes engineering, bridge engineering, highway and urban road engineering, municipal pipeline engineering, railway engineering, hydraulic engineering, mining engineering, port engineering, marine engineering, etc. Specific forms: houses, bridges, tunnels, dams, health system, transport system, mining system, energy system, power system, communication system of all the fixed part—highway, railway, airport, mine, oil wells and port facilities, power facilities, telecommunication facilities.

Geotechnical engineering is the sub-discipline of civil engineering that involves natural materials found close to the surface of the earth. It is the discipline that deals with the construction on and with earth, soil and rock. It includes the application of the principles of soil mechanics and rock mechanics to the design of foundations, retaining structures, and earth structures (Braja M. D. and Sobhan K., 2012). All involved in the civil engineering part of the soil and rock are included in the scope of geotechnical engineering. Geotechnical engineering mainly includes the following main aspects (Fig. 1-1): soil science, geology (including hydrology), engineering survey, foundation (foundation treatment, foundation engineering), tunnel excavation of foundation, pit excavation, excavation engineering, supporting engineering (foundation pit supporting, slope supporting and debris flow control), engineering detection and monitoring. Actually, the above problems can be summed up in the slope stability, soil and rock pressure and bearing capacity of the foundation. Geotechnical engineering is important, because virtually everything that civil engineers build must be built on or with soil. The performance of constructed facilities most often depends on soil.

Fig. 1-1 The scope of civil engineering (one)

(a) Construction; (b) Railway; (c) 3D urban city; (d) Bridge engineering; (e) Reclamation; (f) City

Fig. 1-1 The scope of civil engineering (two)

(g) Municipal engineering; (h) Pit mine; (i) Sea; (j) Open pit; (k) Subsidence; (l) Deep open pit; (m) Tunnel engineering; (n) Retaining engineering(concrete-bolting); (o) Retaining engineering(Grille reinforcement); (p) Dam engineering; (q) Earthquake; (r) Deep space; (s) 3D printed building; (t) Fractured rock mass; (u) 3D printed specimens

1.2 Historical Perspective

The history of geotechnical engineering can be traced back to ancient times. It is not possible to record the first use of soil or rock in building materials in human's history.

As early as 3000 B. C., people began to dig and pile up soil to form embankments for flood control and irrigation in the Yellow River and Tigris-Euphrates River basins. In ancient China, there were also many famous geotechnical engineers, such as the Yellow Emperor who invented the Sluice Box and the Three Gorges Dam, and the engineer Qin Shi Huang who built the Great Wall.

In modern times, geotechnical engineering has become an important branch of civil engineering. The development of geotechnical engineering is closely related to the development of other branches of civil engineering. With the construction of various civil engineering projects, such as highways, railways, airports, harbors, dams, etc., geotechnical engineers have made continuous efforts and explorations in soil and rock mechanics, foundation engineering, ground treatment, etc., and gradually established a relatively complete geotechnical engineering theory and practice system.

At the same time, with the development of science and technology, new technologies and new materials have also been applied in geotechnical engineering. For example, the application of computer technology in geotechnical engineering can achieve simulation and optimization design of geological structure, and improve the accuracy and efficiency of geotechnical engineering design. In addition, new materials, such as geotextiles, geogrid, and geosynthetics, have also been widely used in geotechnical engineering, which effectively improve the performance of soil and rock and ensure the safety and stability of geotechnical engineering projects.

In short, geotechnical engineering is an important part of modern civil engineering, and its development is closely related to the development of other branches. In the past few decades, with the continuous progress of science and technology, geotechnical engineering has been constantly developing and innovating, and will continue to play an important role in future civil engineering projects.

The understanding of geotechnical engineering as it is known today began early in the 18th century in true engineering terms, as reported by Skempton A. W. (1985). There are several stages for the development of Geotechnical Engineering.

1. Geotechnical Engineering Prior to the 18th Century

Recorded history tells us that ancient civilizations flourished along the banks of rivers, such as the Nile (Egypt), the Tigris and Euphrates (Mesopotamia), the Yellow River (China), and the Indus (India). Dykes, which were built for irrigation purposes, were found all over the world.

Beginning around 2700 B. C., several pyramids were built in Egypt, most of which were built as tombs for the country's Pharaohs and their consorts during the Old and Middle Kingdom periods. The construction of the pyramids posed formidable challenges regarding foundations, stability of slopes, and construction of underground chambers.

One of the most famous examples of problems related to soil-bearing capacity in the construction of structures prior to the 18th century is the Leaning Tower of Pisa in Italy.

After encountering several foundation-related problems during construction over the past centuries, engineers and scientists began to address the properties and behaviors of soils in a more methodical manner starting in the early part of the 18th century.

2. Preclassical Period of Soil Mechanics (1700—1776)

This period concentrated on studies relating to natural slope and unit weights of various types of soils, as well as the semiempirical earth pressure theories. In the preclassical period, practically all theoretical considerations used in calculating lateral earth pressure on retaining walls were based on an arbitrarily based failure surface in soil.

3. Classical Soil Mechanics——Phase Ⅰ (1776—1856)

During this period, most of the developments in the area of geotechnical engineering came from engineers and scientists in France. The end of Phase Ⅰ of the classical soil mechanics period is generally marked by the year (1857) of the first publication by William John Macquorn Rankine (1820—1872), which provided a notable theory on earth pressure and equilibrium of earth masses.

4. Classical Soil Mechanics——Phase Ⅱ (1856—1910)

Several experimental results from laboratory tests on sand appeared in the literature in this phase. One of the earliest and most important publications is one by French engineer Henri Philibert Gaspard Darcy (1803—1858). In 1856, he published a study on the permeability of sand filters. Based on those tests, Darcy defined the term coefficient of permeability (or hydraulic conductivity) of soil, a very useful parameter in geotechnical engineering to this day.

5. Modern Soil Mechanics (1910—1927)

In this period, results of research conducted on clays were published in which the fundamental properties and parameters of clay were established. Albert Mauritz Atterberg (1846—1916) defined clay-size fractions as the percentage by weight of particles smaller than 2 μm in size, realized the important role of clay particles in a soil and the plasticity thereof, explained the consistency of cohesive soils by defining liquid, plastic, and shrinkage limits, and defined the plasticity index as the difference between liquid limit and plastic limit.

Jean Fontard (1884—1962), carried out investigations to determine the cause of the failure of a 17-meter-high earth dam at Charmes.

Arthur Langley Bell (1874—1956), a civil engineer from England, worked on the design and construction of the outer seawall at Rosyth Dockyard. He developed relationships for lateral pressure and resistance in clay as well as the bearing capacity of shallow foundations in clay. He also used shear-box tests to measure the undrained shear strength of undisturbed clay specimens.

Wolmar Fellenius (1876—1957) developed the stability analysis of saturated clay slopes with the assumption that the critical surface of sliding is the arc of a circle.

Karl Terzaghi (1883—1963) of Austria developed the theory of consolidation for clays as we know it today.

6. Geotechnical Engineering after 1927

The publication of *Erdbaumechanik auf Bodenphysikalisher Grundlage* by Karl Terzaghi in 1925 gave birth to a new era in the development of soil mechanics. Karl Terzaghi is known as the father of modern soil mechanics.

For the next quarter-century, Terzaghi was the guiding spirit in the development of soil mechanics and geotechnical engineering throughout the world. The first conference of the International Society of Soil Mechanics and Foundation Engineering (ISSMFE) was held at Harvard University in 1936 with Karl Terzaghi presiding.

In 1960, Bishop, Alpan, Blight, and Donald provided early guidelines and experimental results for the factors controlling the strength of partially saturated cohesive soils.

1.3 Fundamentals of Rock Mechanics (According to Goodman R. E., 1989)

Rock mechanics is only since about 1960 that it has come to be recognized as a discipline worthy of a special course of lectures in an engineering program.

Awidely accepted definition of rock mechanics is that first offered by the US National Committee on Rock Mechanics in 1964, and subsequently modified in 1974: Rock mechanics is the theoretical and applied science of the mechanical behavior of rock and rock masses; it is that branch of mechanics concerned with the response of rock and rock masses to the force fields of their physical environment. That recognition is an inevitable consequence of new engineering activities in rock, including complex underground installations, deep cuts for spillways, and enormous open pit mines. Rock mechanics deals with the properties of rocks and the special methodology required for design of rock-related components of engineering schemes.

Rock, like soil, is sufficiently distinct from other engineering materials that the process of "design" in rock is really special. In dealing with a reinforced concrete structure, for example, the engineer first calculates the external loads to be applied, prescribes the material on the basis of the strength required (exerting control to ensure that strength is guaranteed), and accordingly determines the structural geometry.

In rock structures, on the other hand, the applied loads are often less significant than the forces deriving from redistribution of initial stresses. Then, since rock structures like underground openings possess many possible failure modes, the determination of material "strength" requires as much judgment as measurement. Finally, the geometry of the structure is at least partly ordained by geological structure and not completely within the designer's freedoms. For these reasons, rock mechanics includes some aspects not considered in other fields of applied mechanics-geological selection of sites rather than control of material properties, measurement of initial stresses, and analysis, through graphics and model studies, of multiple modes of failure. The subject of rock mechanics is therefore closely allied with geology and geological engineering. The rock mass is largely Discontinuous, Anisotropic, Inhomogeneous and Not-Elastic (DIANE) (Harrison and Hudson, 2000).

1. The contents of rock mechanics

As defined by the Committee on Rock Mechanics of the Geological Society of America, "Rock mechanics is the theoretical and applied science of the mechanical behavior of rock; it is that branch of mechanics concerned with the response of rock to the force fields of its physical environment" (Judd W. R., 1964). Rock mechanics mainly discusses the fundamental contents including the material and texture of rocks, stress and strain behavior, deformation and failure of rocks, rock discontinuities and rock surfaces, laboratory testing of rocks, stresses around cavities and excavations, elastic and inelastic behavior of rocks, static and dynamic loading of rock, wave propagation in rocks, hydromechanical behavior of fractures, state of stress underground, and the geological applications.

2. The contents of soil mechanics

Soil mechanics mainly discusses the fundamental contents, including the origin of soil and grain size, weight-volume relationships of soil, plasticity and structure of soil, classification of soil, compaction behavior of soil, permeability characteristics and seepage, stresses in soil masses, shear strength of soil, compressibility of soil, soil slope stability, and soil bearing capacity for shallow foundations.

1.4 Progresses in Geotechnical Engineering

1.4.1 Deep Underground

"After hundreds of years of mining, the more accessible shallow mineral resources are being depleted, and some have now been completely exhausted. This means that the economic exploitation of more of the earth's deeper mineral resources is now required in order to meet society's growing demand for minerals. This demand is not only for traditional metallic ores and energy sources, but also for minerals such as rare earths, which are being used at an increasing rate with the advent of new technologies in the fields of communication, power generation, and power storage, among others", as described by Cai M. F. and Brown E. T. (2017).

Exploitable mineral resources exist at great depth in the form of a number of orebody types in a range of geological and geometrical settings. The current seven deepest mines in the world mine tabular or stratiform gold deposits in the Witwatersrand Basin of South Africa. The deepest of these mines are now around 4 km deep. The next deepest mines in the world are two base metal mines in Canada, which are about 3 km deep. For most metal mines, deep mining always refers to mining at depths of more than 1 km (Cai M. F. and Brown E. T., 2017).

As mentioned by Cai and Brown, the effective development and extraction of deep mineral resources faces a number of engineering challenges arising from factors such as high in situ and induced stresses, and the responses of variable rock masses to these stresses; high in situ temperatures, and the associated ventilation and cooling requirements; the difficulty and cost of exploring deep, and sometimes blind, deposits; the complex and difficult mining conditions that are often encountered; safety concerns leading to the desirability of developing non-entry methods of mining; and methods and costs of handling mined ore at depth and transporting it to the surface. In some extreme cases, new, low-cost, and non-traditional methods of extraction will be required.

Against this background, deep underground, especially deep mining, has been identified as an important topic for research.

1.4.2 Deep Space

With the development of aerospace technology, the exploration on the planets, like Moon, Mars, or some comets, has been conducted. Geotechnical engineering will also face the new challenge of deep space. The novel application of the new principles of soil mechanics or rock mechanics to the next round design of construction, retaining structures in vacuo on the new planets will be the big challenge for human beings. With the development of aerospace science and technology, the requirement for space energy sources has also increased. For deep space exploration or planetary bases far away from the sun, solar energy can no longer satisfy the power demand. New technologies related to the mining of new planets are being emphasized by many scientists.

Considering the special working environment of the space reactor, the energy conversion system must satisfy the following requirements: (1) small mass and compact structure; (2) large output power, high efficiency, and specific power; (3) high reliability; and (4) adaptation to the off-design condition (Zhi L., etc, 2019).

1.4.3 Intelligent Construction

What are intelligent construction technologies? According to the Transtec Group (2020), intelligent construction, also known as smart construction or digital construction, refers to the application of advanced technologies in the construction industry to enhance efficiency, productivity, and safety. It involves the integration of various technologies such as artificial intelligence, robotics, internet of things (IoT), data analytics, and building information modeling (BIM) to optimize the construction process. Intelligent construction technologies (ICT) are a combination of modern science and innovative construction technologies. The applications of ICT to the life cycle of infrastructure, from survey, design, construction, operation, and maintenance/rehabilitation, by adapting to changes in the environments and minimizing risks. The goal of ICT is to improve the quality of construction, save costs, and improve safety. The scope includes intelligent sensing, data analysis, decision-making, and execution. The fields covered include civil engineering, construction machinery, electronic sensor technology, survey and testing technology, information technology and computing, and other related fields.

1. Intelligent Construction

Intelligent construction is an innovative approach to the construction industry that utilizes the latest technology and data analysis to optimize project planning, design, and implementation. By incorporating artificial intelligence, machine learning, and the IoT, intelligent construction can improve project efficiency, reduce costs, and enhance safety.

One of the key benefits of intelligent construction is the ability to gather and analyze large amounts of data in real-time. This data can be used to identify potential issues before they become problems, optimize project scheduling and resource allocation, and improve overall project performance. In addition, intelligent construction can also enhance safety by providing real-time monitoring of worksites and alerting managers to potential hazards. Another important aspect of intelligent construction is the use of building information modeling (BIM), which enables architects, engineers, and contractors to collaborate more effectively throughout the construction process. BIM allows for the creation of detailed 3D models of buildings, which can be used to simulate different scenarios and identify potential issues before construction begins.

It's important to consider challenges like data security, skills training, and integrating new technologies into existing workflows. Nonetheless, the ongoing advancements in intelligent construction hold great potential for transforming the construction industry and driving innovation in the years to come.

Overall, intelligent construction represents a major shift in the construction industry towards a more data-driven, collaborative, and efficient approach. By leveraging the latest technology and data analysis, intelligent construction can help to deliver projects faster, cheaper, and with higher quality than traditional construction methods.

2. What is 3D Printing Concrete?

3D concrete printing is a revolutionary construction technology that allows for the creation of complex and intricate structures using a specially designed 3D printer that extrudes concrete by Xiamen Zhichuangcheng Technology Co. Ltd (Fig. 1-2). This technology has the potential to transform the construction industry by reducing costs, improving efficiency, and enabling the creation of unique designs that were previously impossible to achieve. One of the main advantages of 3D concrete printing is its ability to reduce material waste and construction time. Because the printer only uses the necessary amount of concrete, there is less waste generated than with traditional construction methods. In addition, the printer can work continuously, 24 hours a day, which can significantly reduce construction time. Another benefit of 3D concrete printing is its potential for creating unique and customized designs. With this technology, architects and designers can create complex and intricate shapes that were previously impossible to achieve with traditional construction methods.

This allows for greater creativity and flexibility in design, and can result in more visually striking and interesting buildings.

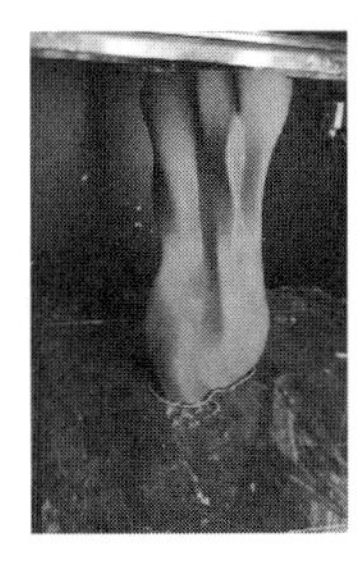

Fig. 1-2 3D printing concrete

With the fast development of 3D printing technology, cement-based 3D printing (3DP) technology has been broadly adopted in civil engineering, and its printability is an important factor for its further application (Jiang Q, etc. 2022). In the past decades, the cement-based 3D printing (3DP) technology has obtained more and more attention of civil engineers. In the construction industry, cement-based 3DP technology can not only improve construction efficiency, but also save costs. The 3DP concrete can be controlled by the computer aided technology. Complex structures could be efficiently established. The 3DP concrete technology has been applied to the construction and reinforcement of bridges and tunnels in traffic engineering. What's more, it is widely used in the manufacture of ceramics and other artworks due to its digital forming characteristics. Using the rock-like cement materials, 3DP has also been employed in physical model tests and other aspects. Updating of printability and mechanical properties of cement-based 3DP can not only improve the quality of printed structural components, but also expand its application to the industry. Based on different purposes, specified additives could be added into cement-based 3DP materials, for example, cementitious supplementary materials and fibers are used to improve their printability. Cementitious supplementary materials include fly ash, silica fume, limestone filler, blast furnace slag, nanosilica, etc., according to Jiang, etc., 2022. Among these additive materials, nanosilica is a very effective cementitious supplementary material that can improve not only the fluidity of mortar, but also the flexibility and durability of concrete. Fibers are also used to enhance the supporting performance of mortar structures, including organic fibers, steel fibers, basalt fibers, carbon fibers and glass fibers. Among these fibers, organic fibers are the most commonly used, including polypropylene (PP) fibers, polyethylene fibers and polyvinyl alcohol fibers. Specifically, PP fibers are often adopted to modifying the toughness of printing structures in 3DP benefitting by their low production cost and high deformation resistance.

Despite the many advantages of 3D concrete printing, there are still some challenges that need to be addressed, such as the limited size of the printers and the need for specialized training and expertise to operate them. However, as the technology continues to develop and improve, it is likely that 3D concrete printing will become an increasingly important tool in the construction industry, with the potential to revolutionize the way we build structures.

1.4.4 Progress in Geotechnical Engineering

One of the key areas of progress in geotechnical engineering is in the use of advanced materials and techniques for soil stabilization and ground improvement. This includes the use of geosynthetics, such as geotextiles and geogrids, to reinforce soil and improve its bearing capacity. In addition, new ground improvement techniques, such as deep soil mixing and jet grouting, are being used to improve the strength and stability of soil and rock formations. Another area of advancement in geotechnical engineering is in the use of advanced sensors and monitoring systems to better understand the behavior of soil and rock structures. This includes the use of fiber optic sensors, acoustic emission sensors, and other advanced technologies to monitor deformation, stress, and other key parameters in real-time. This information can be used to optimize the design and construction of geotechnical structures and to detect potential issues before they become problems. In addition to these technological advancements, there is also growing interest in the use of sustainable and eco-friendly materials in geotechnical engineering. This includes the use of recycled materials, such as crushed concrete and waste tires, for soil stabilization and ground improvement. There is also increased attention being paid to the use of natural materials, such as bamboo and plant-based fibers, for erosion control and slope stabilization.

Overall, the field of geotechnical engineering is rapidly evolving, with new technologies, materials, and techniques being developed to improve the safety, efficiency, and sustainability of infrastructure development. As the demand for more resilient and sustainable infrastructure continues to grow, geotechnical engineering will play an increasingly important role in meeting these needs.

References

[1] Braja M. D.,Sobhan K.. Principles of Geotechnical Engineering [M]. 8 th ed. Stamford,2013.

[2] Cai M. F.,Brown E. T.. Challenges in the Mining and Utilization of Deep Mineral Resources [J]. Engineering, 2017, 3(4): 432-433.

[3] Goodman R. E.. Introduction to Rock Mechanics [M]. 2nd ed. New Jersey: Wiley, 1989.

[4] Harrison J. P., Hudson J. A.. Engineering Rock Mechanics Part 2: Illustrative Worked Examples [M]. Oxford: Pergamon, 2001.

[5] Jiang Q.,Liu Q.,Wu S.,et al. Modification Effect of Nanosilica and Polypropylene Fiber for Extrusion-based 3D Printing Concrete: Printability and Mechanical Anisotropy [J]. Additive Manufacturing, 2022, 56: 102944.

[6] Judd W. R.. State of Stress in the Earth's Crust [C]//Proceedings of the International Conference. Santa Monica, California, 1964: 5-51.

[7] Li Z.,Yang X. Y.,Wang J.,et al. Off-design Performance and Control Characteristics of Space Reactor Closed Brayton Cycle System. Annals of Nuclear Energy, 2019, 128: 318-329.

Chapter 2
The Properties of Rock Materials

2.1 Introduction

Rock mechanics is the theoretical and applied science of the mechanical behavior of rock; it is that branch of mechanics concerned with the response of rock to the force fields of its physical environment (Judd W. R., 1964). The objective of this chapter is: (1) to understand the mechanical behavior of rock materials, rock fractures and rock masses; and (2) to be able to analyze and to determine mechanical properties of rocks for geotechnical engineering applications. First, there is some issues we should know. What is rock mechanics? Rock mechanics is a discipline that uses the principles of mechanics to describe the behavior of the rock of engineering scale. Why is rock mechanics special? Rock at engineering scale is discontinuous, inhomogeneous, anisotropic, and nonlinearly elastic. Rock mechanics deals with the response of rock when the boundary conditions are disturbed by engineering.

2.2 The Mechanism of Rock Formation

2.2.1 Rock Types

Rock is a natural solid substance composed of minerals. Rocks are formed by three origins: igneous rocks from magma, sedimentary rocks from sediments lithification, metamorphic rocks through metamorphism(Fig. 2-1).

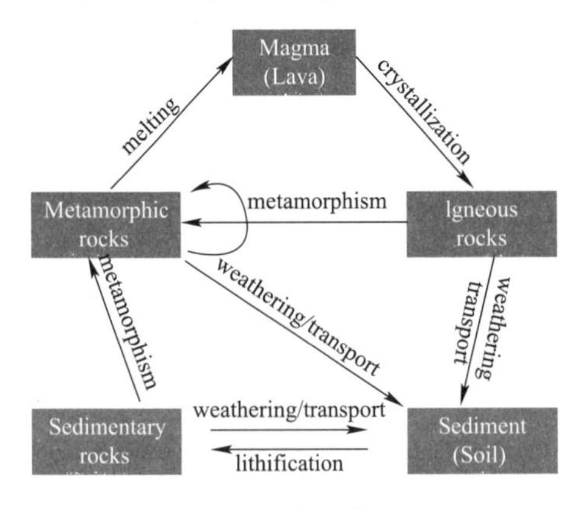

Fig. 2-1 Rock formation cycle

Igneous rocks(Fig. 2-2a) consist of a completely crystalline assemblage of minerals such as quartz, plagioclase, pyroxene, mica, etc. Sedimentary rocks consist of an assemblage of detrital particles and possibly pebbles from other rocks, in a matrix of materials such as clay minerals, calcite, quartz, etc. From their nature, intrusive igneous rocks are generally coarse grained because of longer cooling period allowing crystallisation. Extrusive igneous rocks are fine grained due to fast cooling and none or low crystallisation. Igneous rocks are named primarily according to the mineral compositions and grain sizes.

Sedimentary rocks (Fig. 2-2b) contain voids or empty spaces, some of which may form an interconnected system of pores. Sedimentary rocks are formed in three main ways: (1) deposition of the weathered remains of other rocks (known as 'clastic' sedimentary rocks); (2) deposition of the results of biogenic activity; (3) precipitation from solution. Clastic sedimentary rocks are commonly classified by the particle size of the sediments.

Metamorphic rocks (Fig. 2-2c) are produced by the action of heat, stress, or heated fluids on other rocks, sedimentary or igneous. Metamorphic rock is a new rock transformed from an existing rock, through metamorphism — change due to heat and pressure. Metamorphic rocks can have foliated and non-foliated textures.

Foliation is due to the re-orientation of mica minerals, creating a plane of cleavage or visible mineral alignment feature.

Fig. 2-2 Typical rocks of the three types of rock materials
(a) Igneous rocks; (b) Sedimentary rocks; (c) Metamorphic rocks

2.2.2 Rock Textures

Different rock types have varied rock textures. For the igneous/metamorphic rocks, rock materials are generally with high strength. Sedimentary rocks are with low rock material strength, particularly when cementation is weak. Three key points could be found: different minerals, complex boundings and pores/microcracks.

Anisotropy is defined as properties which are different in different directions. It occurs in both rock materials and rock mass. Rock with obvious anisotropy is slate. Metamorphic phyllite, schist, basalt and sedimentary shale also exhibit anisotropy. Rock mass anisotropy is controlled by (1) joint set, and (2) sedimentary layer.

Rocks surrounding the tunnels are controlled by the distribution of the structure planes. Some cases for the anisotropy of jointed rock mass are shown in Fig. 2-3. All of these minerals are anisotropic in a micro scope. If in a polycrystalline rock there are any preferred orientations of the crystals, this will lead to anisotropy of the rock itself. If the orientations of the crystals are random, the rock itself will be isotropic. The properties of the individual particles in such a specimen may differ widely from one particle to another, and although the individual crystals themselves are often anisotropic, the crystals and the grain boundaries between them interact in a sufficiently random manner so as to imbue the specimen with average homogeneous properties. These average properties are not necessarily isotropic, because the processes of rock formation or alteration often align the structural particles so that their interaction is random with respect to size, composition and distribution, but not with respect to their anisotropy. Nevertheless, specimens of such rocks have gross anisotropic properties that can be regarded as being homogeneous.

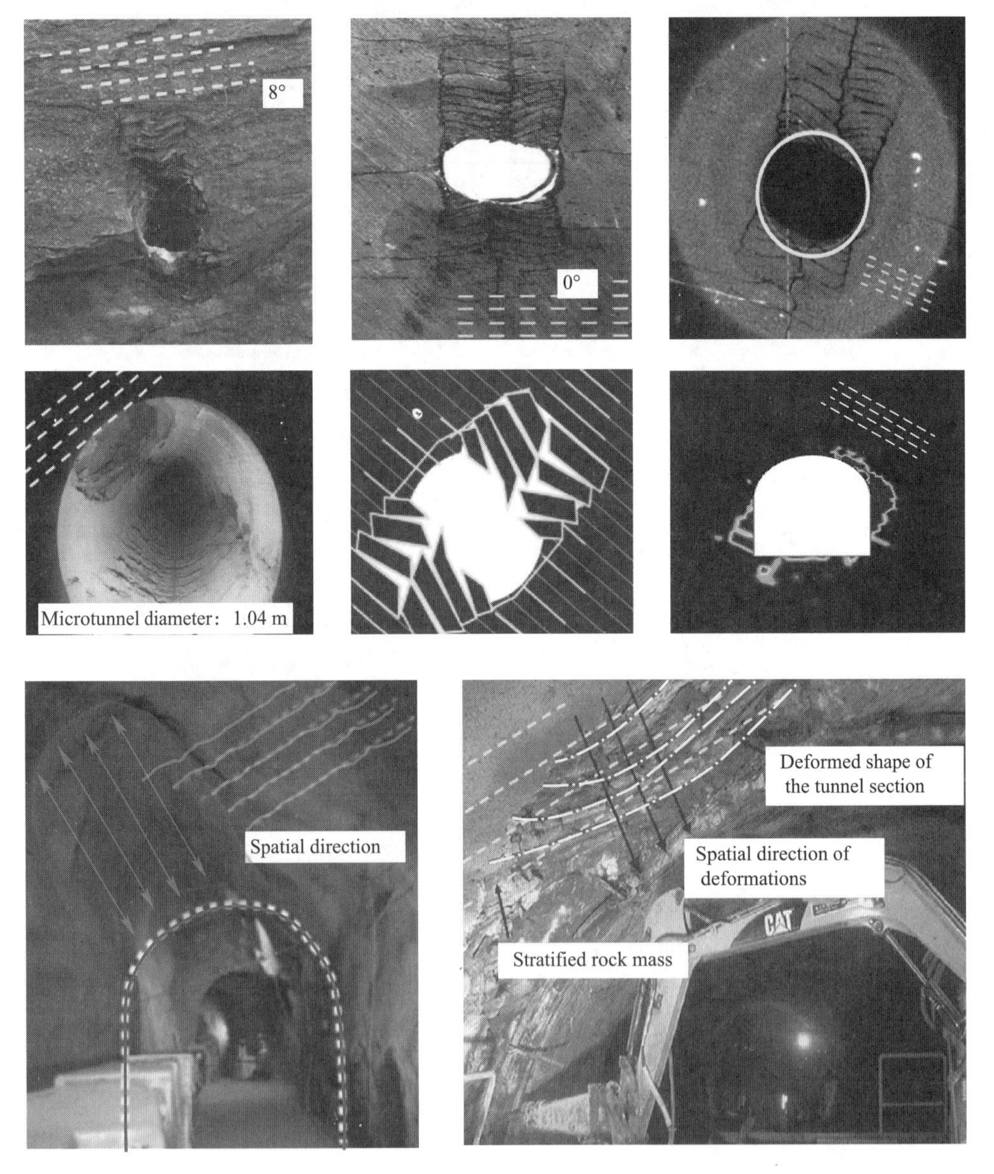

Fig. 2-3 Anisotropic failure patterns of rocks

Inhomogeneity represents property varying with locations. Many construction materials have varied degrees of inhomogeneity. Rock is formed by nature and exhibits great inhomogeneity due to: (1) different minerals in a rock, (2) different bounding between minerals, (3) existence of pores, (4) existence of microcracks (Fig. 2-4). Grain size also has an effect on inhomogeneity properties. For example, there are some correlations in the sedimentary rock between mechanical properties and porosity. A great amount of systematic research has been done on the mechanical properties of single crystals, both with regards to their elastic properties and their plastic deformation. Single crystals show preferred planes for slipping and twinning. Such measurements are an essential preliminary to the understanding of the fabric of deformed rocks, but they have little relevance to the macroscopic behavior of large polycrystalline specimens.

Interlocking structure of a granite

Clastic structure of a sandstone

Granulite

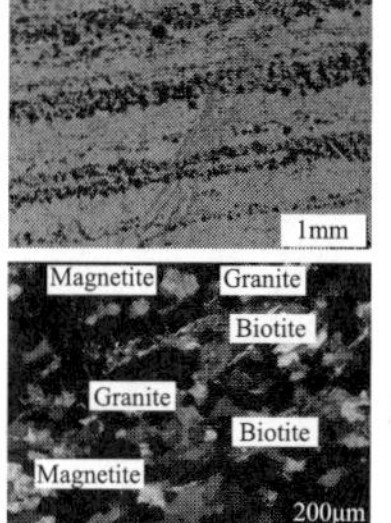

Fig. 2-4 The inhomogeneity of varied rock materials under microscope

Inhomogeneity is thecause of fracture initiation leading to the failure of a rock material. If some elements in the rock material matrix are very weak, they will start to fail early and usually lead to low overall strength of the rock material. Inhomogeneity of a rock mass is primarily due to the existence of the various discontinuities. Rock masses are also inhomogeneous due to the mix of rock types, interbedding and intrusion.

2.2.3 Rock Material and Rock Mass

A rock mass contains (1) rock material, in the form of intact rock blocks of various sizes, and (2) rock discontinuities that cut through the rock, in the forms of fractures, joints, faults, folds, bedding planes, and dykes.

Rock joints: Joints are by far the most common type of geological structure. They are defined as cracks or fractures in rock along which there has been little or no transverse displacement. They are normally in parallel sets. They are generally considered as part of the rock mass. The spacing of joints is usually in the order of a few to a few ten centimeters. For engineering, joints are constant features of the rock mass.

Faults: Faults are planar rock fractures which show evidence of relative movement. Faults have different scales and the largest faults are at tectonic plate boundaries. Faults usually do not consist of a single, clean fracture. They often form fault zones. Large scale fault, fault zone and shear zone, are large and localised feature. They are often dealt separately from the rock mass. There are two types of faults, normal fault and reverse fault.

Folds: Fold is the bended originally flat and planar rock strata, as a result of tectonic force or movement. Folds are usually not considered as part of the rock mass. They are often associated with high degree of fracturing and relatively weak and soft rocks.

Bedding Planes: Bedding plane is the interface between sedimentary rock layers. Bedding planes are isolated geological features due to engineering activities. It mainly creates an interface of two rock materials. However, some bedding planes could also become potential weathering zones and groundwater pockets.

The engineering scale of rock and rock mass are different.

For civil engineering works, e. g., foundations, slopes and tunnels, the scale of projects is usually a few tens to a few hundred meters.

Rock in an engineering scale is generally a mass of rock at the site. This mass of rock, often termed as rock mass, is the whole body of the rock in situ, and consists of intact rock blocks and all types of discontinuities (joints, faults, etc.).

Rock mass behavior is largely governed by joints. Rock joints cut rock into slabs, blocks and wedges, to be free to fall and move. The roles include:

- Acting as a weak plane for sliding and moving.
- Providing water flow channel and creates flow networks.
- Giving large deformation.
- Altering stress distribution and orientation.

On a larger scale, the presence of cracks, joints, bedding and minor faulting raises an important question concerning the continuity of a rock mass. These disturbances may interrupt the continuity of the displacements in a rock mass if they are subjected to tension, fluid pressure, or shear stress that exceeds their frictional resistance to sliding. Where such disturbances are small in relation to the dimensions of a structure in a rock, their effect is to alter the mechanical properties of the rock mass, but this mass may in some cases still be treated as a continuum. Where these disturbances have significant dimensions, they must be treated as part of the structure or as a boundary.

2.3 Laboratory Testing of Rocks

The mechanical properties of a piece of rock depend on its mineral composition, the arrangement of the mineral grains, and any cracks that may have been introduced into it during its long geological history by diagenesis or tectonic forces. Consequently, the mechanical properties of rock vary not only between different rock types but also between different specimens of the nominally same rock. Laboratory testing necessarily plays a large role in rock mechanics. Laboratory testing of rocks is an essential part of geotechnical engineering and geology. These tests provide valuable information about the physical and mechanical properties of rocks, which are crucial for various applications such as construction, mining, petroleum exploration, and environmental studies. In this chapter we describe the basic types of laboratory measurements that are routinely conducted to measure the mechanical properties of rocks. Compressive strength test determines the maximum amount of compressive load a rock sample can withstand before it fails. It helps assess the rock's ability to withstand applied loads and provides insights into its structural integrity. Here, uniaxial compression, triaxial compression and diametral compression will be introduced.

2.3.1 Uniaxial Compression

The uniaxial compression test is a fundamental laboratory test conducted on rock samples to assess their strength and deformation characteristics under compressive loading. It is one of the most commonly performed tests in geotechnical engineering and rock mechanics. During the uniaxial compression test, a cylindrical or cuboidal rock specimen is subjected to a vertically applied compressive force along a single axis. The test measures the stress-strain relationship of the rock sample until it reaches failure. Key parameters determined from this test include:

- Compressive strength: the maximum stress that the rock specimen can withstand before failure occurs. It is typically reported as peak strength or ultimate compressive strength.
- Elastic modulus: also known as Young's modulus or simply modulus of elasticity, it represents the rock's stiffness and resistance to deformation under compressive load. It is calculated as the ratio of stress to strain within the elastic range of the rock's response.
- Poisson's ratio: this parameter indicates the ratio of lateral strain to axial strain when the rock specimen is subjected to compressive loading. It characterizes the material's ability to deform in directions perpendicular to the applied stress.
- Stress-strain curve: the relationship between stress (load per unit area) and strain (deformation) of the rock sample is plotted to obtain the stress-strain curve. This curve provides insights into the rock's behavior, including its deformation properties and failure mechanisms.
- Peak strain: The strain corresponding to the peak stress during the test indicates the deformation capacity of the rock before failure. It helps evaluate the ductility and brittleness of the rock material.

By conducting uniaxial compression tests on rock samples from different locations and lithologies, engineers and geologists can obtain critical data for geotechnical design, slope stability analysis, underground excavations, and understanding the behavior of rock masses in various engineering applications. It's important to note that the behavior of rock under uniaxial compression may vary depending on factors such as mineralogy, porosity, moisture content, temperature, and confining pressure. Therefore, it is common to complement uniaxial compression tests with other laboratory tests and field investigations to obtain a comprehensive understanding of the rock's mechanical properties and behavior. The uniaxial compression test, in which a right circular cylinder or prism of rock is compressed between two parallel rigid plates (as shown in Fig. 2-5), is the simplest mechanical rock test and continues to be widely used. In the simplest version of this test, Fig. 2-5, a cylindrical core is compressed between two parallel metal platens. Hydraulic fluid pressure is typically used to apply the load. The intention of this test is to induce a state of uniaxial stress in the specimen, that is,

$$\tau_{zz}=\sigma, \tau_{xx}=\tau_{yy}=\tau_{xy}=\tau_{yz}=\tau_{xz}=0$$

The axial stress σ is the controlled, independent variable, and the axial strain is the dependent variable. The longitudinal strain can be measured by a strain gauge glued to the lateral surface of the rock. Alternatively, the total shortening of the core in the direction of loading can be measured by an extensiometer that monitors the change in the vertical distance between the platens. In this case, the longitudinal strain is calculated from the relative shortening of the core, that is, $\varepsilon=-\Delta L/L$. If the stress state is indeed uniaxial, then the Young's modulus of the rock could be estimated from $E=\sigma/\varepsilon$. The stress at which the rock fails is known as the unconfined, or uniaxial, compressive strength of the rock.

2.3.2 Triaxial Tests

One of the most widely used and versatile rock mechanics tests is the traditional "triaxial" compression test. Triaxial compression testing is a laboratory technique used to determine the mechanical properties of rock under simultaneous confining pressure and axial compression. This test provides valuable information about the rock's strength, deformation characteristics, and failure behavior under controlled conditions. In triaxial compression testing, a cylindrical rock specimen is placed inside a confining pressure chamber and subjected to three mutually orthogonal stresses: an axial stress σ_1 applied through a loading piston, and two lateral or confining stresses σ_2 and σ_3 applied through flu-id pressure.

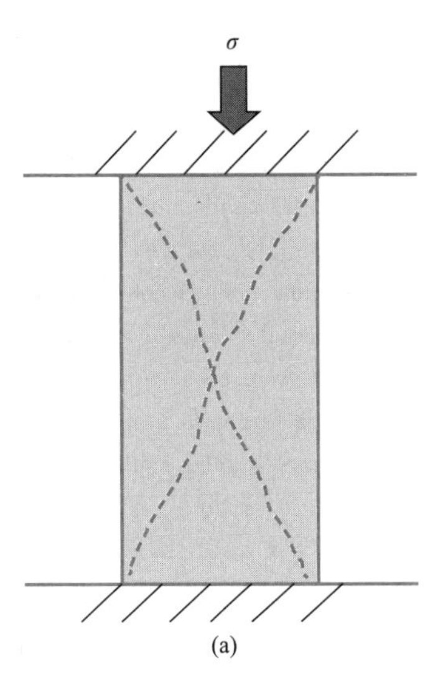

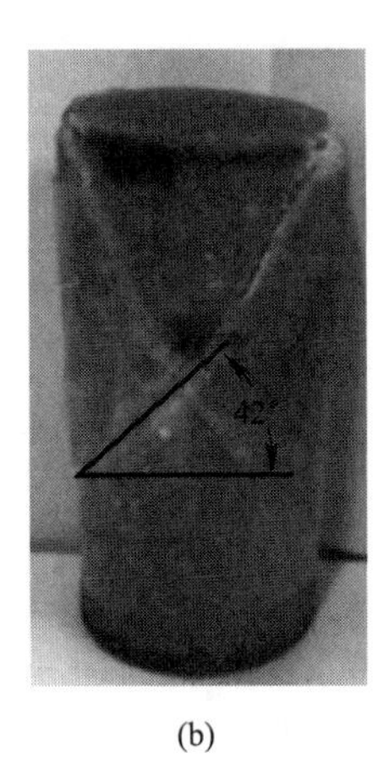

(a) (b)

Fig. 2-5 Unconfined uniaxial compression of a rock for standard configuration

(a) Rock failure initiating at the corners; (b) The failure of sandstone specimen

The confining pressures are typically equal in magnitude, creating a state of hydrostatic stress. Typically, all three stresses are compressive, with the unequal stress more compressive than the two equal stresses, so that $\sigma_1 = \sigma_2 = \sigma_3 > 0$. During the test, measurements are taken as the axial stress is increased incrementally until the rock specimen reaches failure. Key parameters obtained from triaxial compression testing include:

- Mohr-Coulomb strength parameters: the test helps determine the rock's cohesion and friction angle, which are key factors in the Mohr-Coulomb criterion widely used in rock engineering. The cohesion represents the shear strength of the material in the absence of normal stresses, while the friction angle represents the resistance to sliding along potential failure planes.
- Stress-strain behavior: the stress-strain response of the rock sample under varying axial loads and confining pressures provides insights into its elastic, yielding, and plastic deformation characteristics. It helps characterize the rock's strength, stiffness, and deformability.
- Dilatancy: triaxial compression testing allows the measurement of volumetric strain or dilation of the rock specimen. Dilatancy refers to the increase in volume or expansion that occurs when a rock undergoes shearing under elevated stresses.
- Failure mode analysis: by observing the failure patterns and analyzing the developed cracks or fractures within the rock specimen, engineers and geologists can gain insights into the failure mechanisms and identify the type of failure (e. g., tensile, shear, or mixed mode).

Triaxial compression testing is crucial in geotechnical engineering and rock mechanics for applications such as stability analysis of slopes, understanding the behavior of deep-seated rock masses, designing underground excavations, and assessing the performance of rock materials in various engineering projects. It's worth noting that the stress path (loading sequence) during triaxial testing can be modified to simulate specific field conditions and investigate anisotropic or stress-dependent behavior. Different variations of triaxial tests, such as drained and undrained tests, can also be conducted to examine the rock's response under different drainage conditions. Triaxial tests are usually conducted with cylindrical specimens having a length-to-diameter ratio of between 2 : 1 and 3 : 1. It is imperative that the flat surfaces of the specimen should be as nearly parallel as possible, to avoid bending of the specimen under axial stress. The core is jacketed in rubber or thin copper tubing so that the confining fiuid does not penetrate into the pore space (as shown in Fig. 2-6a). A simple triaxial apparatus is the one developed by the US Bureau of Reclamation (Fig. 2-6b).

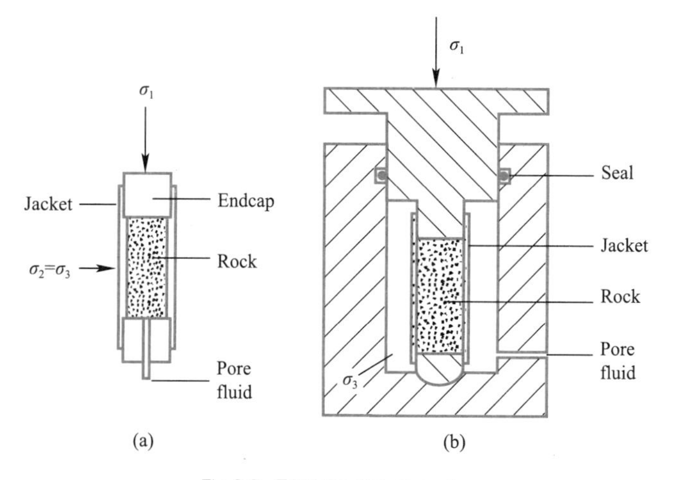

Fig. 2-6 Triaxial testing apparatus

(a) Jacketed cylindrical rock specimen with end-pieces and provision for pore fluid; (b) US Bureau of reclamation cell

2.3.3 True-triaxial Tests

Traditional "triaxial" compression tests involve states of stress in which $\sigma_1 > \sigma_2 = \sigma_3 = 0$. Such tests are incapable of probing the effects of the intermediate principal stress. In order to investigate rock behavior over the full range of stresses that may occur in the subsurface, it would be desirable to conduct tests in which all three principal stresses may have different (compressive) values. Such tests have sometimes been referred to as"polyaxial", although this name has the disadvantage of not being self-explanatory. More recently, the term"true-triaxial", which is inelegant but less open to misinterpretation, has gained acceptance. Several researchers have constructed testing cells that attempt to produce states of homogeneous stress in which the three principal stresses, $\sigma_1 \geqslant \sigma_2 \geqslant \sigma_3 \geqslant 0$, are independently controllable (Fig. 2-7a). Although the designs differ in various ways, in each case a"rectangular" (i.e., parallelepiped-shaped) specimen is used, in contrast to the cylindrical specimens used in traditional triaxial tests.

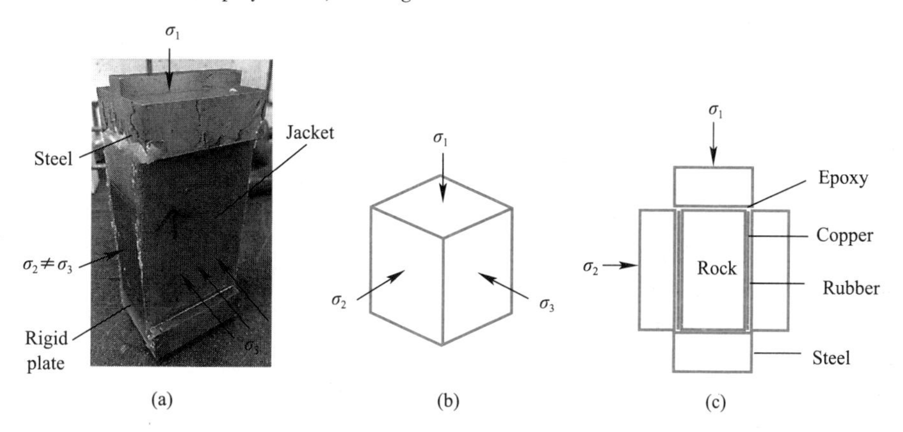

Fig. 2-7 Schematic diagram of triaxial test

(a) Specimen for true-triaxial compression; (b) True-triaxial state of stress applied to a cubical specimen; (c) View along the σ_3 direction of the apparatus

2.3.4 Diametral Compression of Cylinders

The difficulties associated with performing a direct uniaxial tension test on rock have led to the development of a number of "indirect" methods for assessing the tensile strength. Such methods are called indirect because they do not involve the creation of a homogeneous state of tensile stress in rock, but rather involve experimental configurations that lead to inhomogeneous stresses that are tensile in some regions of the specimen. The precise value of the tensile stress at the location where failure initiates must be found by solving the equations of elasticity. The most popular of these tests is the so-called Brazilian test, developed by the Brazilian engineer Fernando Carneiro in 1943 for use in testing concrete. A thin circular disk of rock is compressed between two parallel platens, so that the load is directed along the diameter of the disk (Fig. 2-8a).

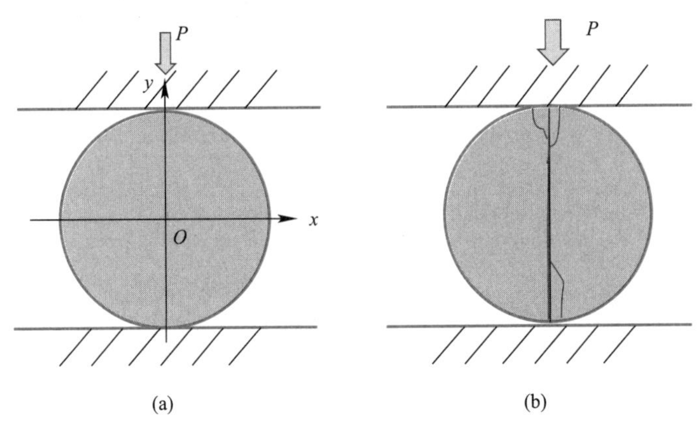

Fig. 2-8 Schematic diagram of Brazilian test

(a) Cylinder compressed between parallel surfaces by a line load P (per unit length into page);

(b) Typical fracture pattern resulting from this loading

2.4 The Mechanical Properties of Rock

The mechanical structure of rock presents several different appearances, depending upon the scale and the detail with which it isstudied. Most rocks comprise an aggregate of crystals and amorphous particles joined by varying amounts of cementing materials. The chemical composition of the crystals may be relatively homogeneous, as in some limestones, or very heterogeneous, as in a granite. Likewise, the size of the crystals may be uniform or variable, but they generally have dimensions of the order of centimeters or small fractions thereof. These crystals generally represent the smallest scale at which the mechanical properties are studied. On the one hand, the boundaries between crystals represent weaknesses in the structure of the rock, which can otherwise be regarded as continuous. On the other hand, the deformation of the crystals themselves provides interesting evidence concerning the deformation to which the rock has been subjected. On a scale with dimensions ranging from a few meters to hundreds of meters, the structure of some rocks is continuous, but more often it is interrupted by cracks, joints, and bedding planes that separate different strata. It is this scale and these continuities which are of most concern in engineering, where structures founded upon or built within rock have similar dimensions. The overall mechanical properties of rock depend upon each of its structural features. However, individual features have varying degrees of importance in different circumstances.

The loads applied to a rock mass are generally due to gravity, and compressive stresses are always encountered. For example, the rock pillars in underground mining are always under compression. For jointed rock masses, the most important factor in connection with the properties and continuity of a rock mass is the friction between surfaces of discontinuities of all sizes in the rock. If conditions are such that sliding is not possible on any surfaces, the system may be treated to a good approximation as a continuum of rock, with the properties of the average test specimen. If sliding is possible on any surface, the system must be treated as a system of discrete elements separated by these surfaces, with frictional boundary conditions over them.

2.4.1 Deformation

1. Experimental Systems

First, rock specimens of various types have been tested at room temperature by the various compression methods, like uniaxial compressive test, biaxial compressive test, triaxial compressive test. The most famous testingapparatus are MTS 810, 815 and 816. The figures of the apparatus are presented in Fig. 2-9. The test specimens are cylinders of 50 mm diameter and 100 mm height, the height/diameter ratio being nearly 2.0 or 2.5.

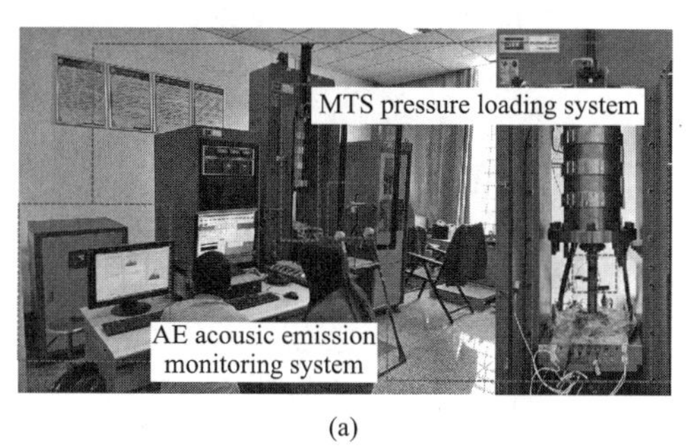

(a)

(b)

(c)

Fig. 2-9 Experimental test systems for rock
(a) MTS 815 system; (b) MTS 816 system; (c) MTS 810 system

For the conventional triaxial compression, the axial stress is applied by a hydraulic testing machine. The confining pressure is supplied independently by another oil compressor. The axial load is measured by a load cell situated between a steel end piece fixed to the test specimen and the bottom of the pressure vessel. The load cell is a closed hollow steel cylinder with an electric resistance strain gage bonded to the inside of the cylinder and measures the axial load without any frictional error. Fig. 2-10 shows a rock testing system for true triaxial test, which is developed by Prof. Joseph Labuz in University of Minnesota. The lateral load σ_2 is applied by a loading cell as shown in Fig. 2-10 (c).

There are several methods of the axial strain measurement (Fig. 2-11). The external method using an extensometer in which the strain is obtained from the displacement of the piston of the press, is depended on the systems. The internal method using the electric resistance strain gage that is bonded directly onto the surface of the rock specimen, is highly sensitive and easy-to-use. Two sets of orthogonal strain gages are usually bonded to measure the strain in both vertical and horizontal directions. The method is unsuitable to measure the mean strain in heterogeneous deformation including microcracks, small faults, etc., and to measure large deformation.

2. Modulus of Elasticity

When the rock specimen is compressed to a point P and then unloaded to zero differential stress and then reloaded, the branches $P2Q$, $Q3P$

Fig. 2-10 The true triaxial compression system in the rock lab of University of Minnesota developed by Joseph Labuz
(a) The true triaxial compression system; (b) The loading cell for lateral compression; (c) The placement of the rock specimen

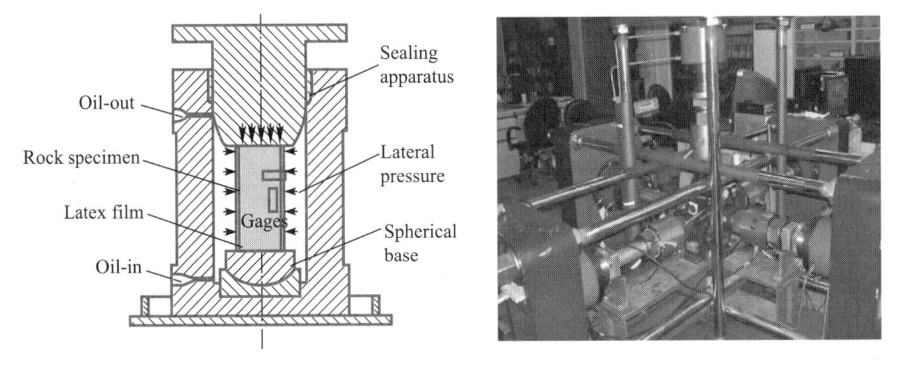

Fig. 2-11 Various methods of the axial strain measurement

are different from the virgin loading curve $O1P$ in many cases, as shown schematically in Fig. 2-12. The unloading and reloading curves usually differ little from each other and form a narrow loop $P2Q3$. The reloading curve $Q34$ passes very near the point P from which the unloading branch drops down and continues to the virgin loading curve $O1P$. This narrow loop circumscribed by the unloading and reloading branches may be approximately substituted by a straight-line PQ. The slope of this line is taken as the mean Young's modulus. The strain of stressed rocks is partly elastic and partly permanent. The elastic strain is obtained as a recovered deformation by unloading to zero differential stress while the permanent strain is obtained as an unrecovered deformation (Fig. 2-12).

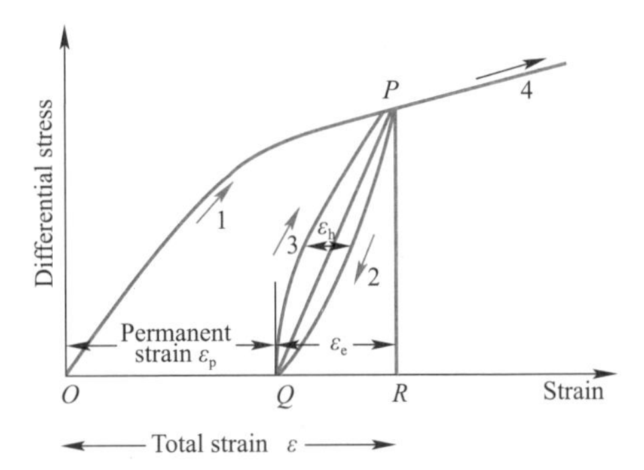

Fig. 2-12 Typical stress-strain curves under cyclic loading

This permanent strain includes various types of unrecoverable deformation due to dislocation, viscous flow, micro-fracturing, etc. It should be noted that the permanent strain is found even at an initial stage of deformation in most rocks. In compact igneous rocks which are very brittle, the permanent strain appears slightly in the nearly linear part of the stress-strain curve and increases abruptly from a proportional limit. In some porous silicate rocks, the appreciable permanent deformation increases linearly at the initial stage and the slope of this curve increases suddenly at some stage. In highly porous silicate rocks, the permanent strain is very large and increases continuously without any such sudden change in the slope of curve, particularly under higher confining pressure.

Generally, the Young's modulus differs considerably at various stages of deformation.

In compact igneous rocks, Young's modulus is nearly constant until their fracture. In other porous rocks, it generally decreases with increasing strain ε.

However, the modulus E of some ductile rocks (tuffs and marble) under high confining pressure begins to increase with strain just after the yielding.

The decrease of Young's modulus E with increasing strain ε is attributed to the micro fracturing in these rock specimens, and the increase of Young's modulus E in a large deformation stage under high confining pressure in soft rocks can be attributed to compaction of fractured rock specimens.

2.4.2 Strength

1. Yield Stress

Strength in the brittle state is the maximum stress achieved during an experiment. This value is definitely determined. However, failure strength in the ductile state, namely yield stress, is usually not so definite (Fig. 2-13). Generally, the yield stress is the stress at which the sudden transition from elastic to plastic deformation state takes place, so the stress at the knee of the stress-strain curves is taken as the yield stress. In some rocks, the knee of the stress-strain curve is clear and the determination of yield stress is easily possible. However, other rocks show a gradual transition from elastic to plastic stage, that is, the slope of the stress-strain curve varies gradually. In this case, the definite determination of yield stress is difficult.

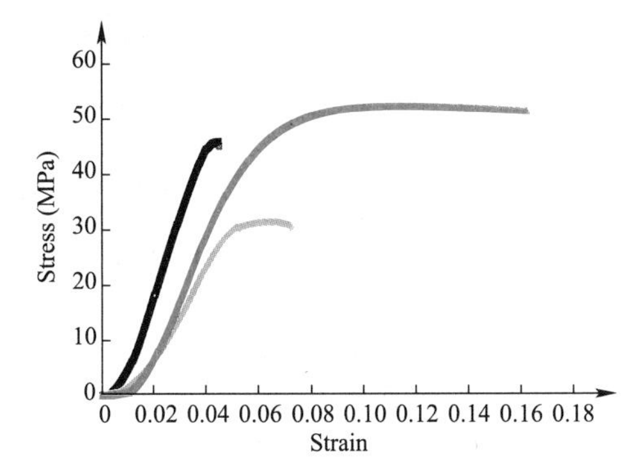

Fig. 2-13 Typical stress-strain curves

2. Uniaxial Compressive Strength

Uniaxial compressive strength (UCS) is the ultimate stress of a cylindrical rock specimen under axial load. It is the most important mechanical properties of rock material, used in design, analysis and modeling. Along with measurements of load, axial and lateral deformations of the specimen are also measured.

3. Tensile Strength

Rock material generally has a low tensile strength, due to the pre-exiting micro cracks in the rock material. The existence of micro cracks may also be the cause of rock failing suddenly in tension with a small strain. Rock material tensile strength can be obtained from several types of tests. The most common tensile test is the Brazilian tests (Fig. 2-14).

4. Shear Strength

Rock resists shear stress by two internal mechanisms, cohesion and internal friction. Cohesion is a measure of internal bonding of the rock material. Internal friction is caused by contact between particles, and is defined by the internal friction angle. Shear strength of rock material can be determined by direct shear tests (Fig. 2-15) and by triaxial compression tests. From a series of triaxial tests, peak stresses σ_1 are obtained at various lateral stresses σ_3. By plotting Mohr circles, the shear envelope is defined and gives the cohesion and internal friction angle. Strength of typical rocks are listed in Table 2-1.

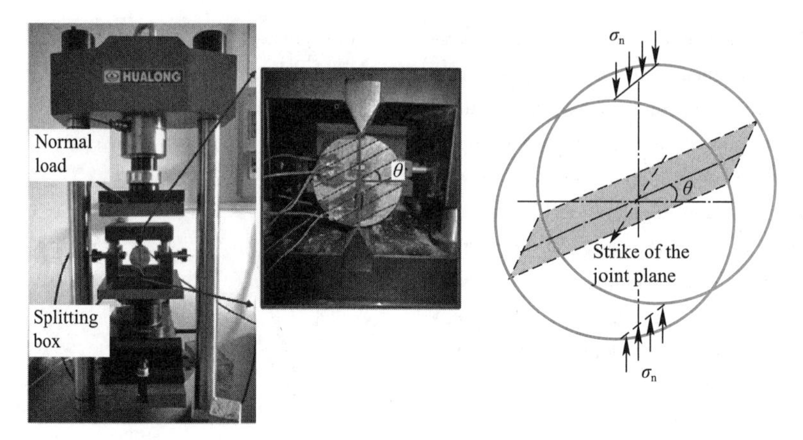

Fig. 2-14 Brazilian tensile test for rocks

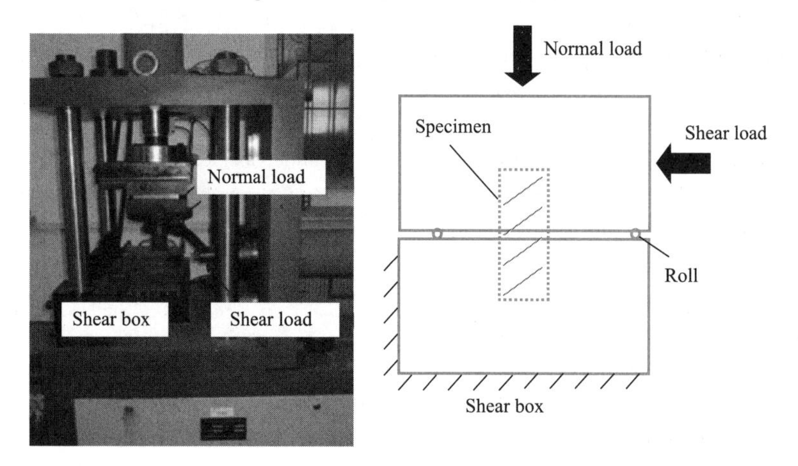

Fig. 2-15 Direct shear test for rock

Strength of typical rocks **Table 2-1**

Rock type	UC strength (MPa)	Tensile strength (MPa)
Granite (花岗岩)	100—300	7—25
Dolerite (辉绿岩)	100—350	7—30
Gabbro (辉长岩)	150—250	7—30
Basalt (玄武岩)	100—350	10—30
Sandstone (砂岩)	20—170	4—25
Shale (页岩)	5—100	2—10
Dolomite (白云石)	20—120	6—15
Gneiss (片麻岩)	100—250	7—20
Slate (板岩)	50—180	7—20
Marble (大理岩)	50—200	7—20
Quartzite (石英岩)	150—300	5—20
Limestone (石灰岩)	30—250	6—25

5. Comparison of Compressive, Tensile and Shear Strengths

Tensile and shear strengths are important as rock fails mostly in tensile and in shear, even the loading may appear to be compression. Rocks generally have high compressive strength, so failure in pure compression is not common. Theoretically, the three strengths are related. This will be discussed in strength criteria.

6. Point Load Index

Point load test is a simple index test for rock material (Fig. 2-16). It gives the standard point load index, $I_{s(50)}$.

Correlation between point load index and strengths, $\sigma_c \approx 22 I_{s(50)}$.

Correction factor can vary between 10 and 30, $\sigma_t \approx 1.25 I_{s(50)}$.

$I_{s(50)}$ should be used as an independent strength index on a cylinder with diameter of 50 mm.

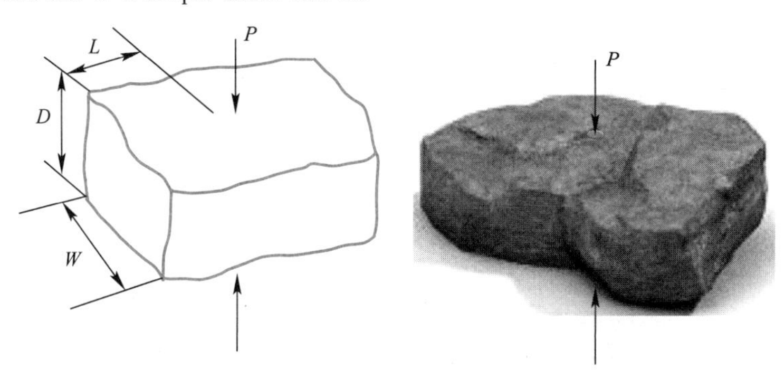

Fig. 2-16 Direct shear test for rock

2.5 Deformation and Failure of Rock

2.5.1 Introduction

Yield strength is defined as the stress at which a material begins to plastically deform. It generally represents an upper limit to the load that can be applied. A yield strength criterion is a hypothesis concerning the limit of stress under any combination of stresses. They are generally described in term of three principal stresses.

Three common strength criteria:

1. Mohr-Coulomb Criterion—Internal Frictional Model Suitable for Brittle Materials

Mohr-Coulomb criterion is a two-parametric criterion. It takes shearing into account. It considers the major and minor principal stresses only (the two principal stresses making the largest difference).

The criterion assumes that a shear failure plane is developed in the rock material. When failure occurs, the stresses developed on the failure plane are on the yield surface (envelope in 2D). Coulomb's shear strength is made up of two parts, a constant cohesion c and a normal stress σ_n dependent frictional component, angle of internal friction φ.

$$\tau = c + \sigma_n \tan \varphi$$

It is a straight line, with an intercept c on the τ-axis and an angle of φ with the σ_n axis.

As shown in Fig. 2-17, rock fails with the formation of a shear failure plane a-b, i. e., the stress condition on the a-b plane satisfies the shear strength condition. In the diagram, when the Mohr circle touches the Mohr-Coulomb strength envelope, the stress condition on the a-b plane meets the strength criterion. From the Mohr circle, the failure plane is defined by θ.

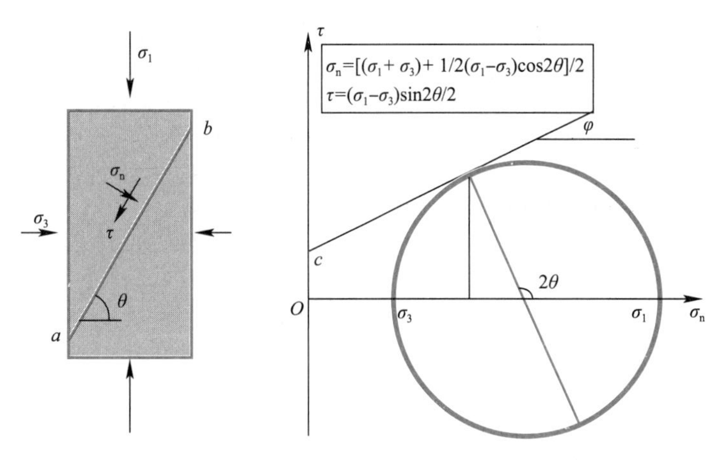

Fig. 2-17 Mohr circle of rock failing with the formation of a shear failure plane

$$\theta=\frac{\pi}{4}+\frac{\varphi}{2}$$

The Mohr-Coulomb criterion is only suitable for the low range of confining stress. At high confining stress, it overestimates the compressive strength and the tensile strength. In most cases, rock engineering deals with shallow problems and low confining stress, so the criterion is widely used, due to its simplicity and popularity.

2. Griffith Criterion — Based on Fracture Initiation at Microscopic Flaws in a Material

Tensile strengths of rock materials are low to be about 10% of the compressive strengths. It is due to the opening and propagation of existing micro-cracks in the rock. Tensile strength of rock materials can be described by the Griffith criterion based on fracture initiation at microscopic flaws in a material. Griffith criterion is based on fracture mechanics of brittle materials, using elastic strain energy concepts. The equation basically states that when a crack is able to propagate enough to fracture a material, that the gain in the surface energy is equal to the loss of strain energy. Under compression, elliptical crack will propagate from the points of maximum tensile stress concentration. When under compression, it gives the following criterion for crack extension.

If $\sigma_1+3\sigma_3>0$

$$(\sigma_1-\sigma_3)^2-8\sigma_t(\sigma_1+\sigma_3)=0$$

When $\sigma_3=0$, $\sigma_1-8\sigma_t=0$ or $\sigma_1=8\sigma_t$.

Uniaxial compressive stress at crack extension is always 8 times the uniaxial tensile strength. It can also be expressed in terms of the shear stress τ and normal stress σ_n acting on the plane containing the major axis of the crack.

$$\tau^2=4\sigma_t(\sigma_n+\sigma_t)$$

When $\sigma_n=0$, $\tau=c=2\sigma_t$.

Griffith criterion suggests when crack start to extend from existing pores, $\sigma_1=8\sigma_t$ and $\tau=2\sigma_t$. The plane compression Griffith theory did not provide a very good model for the peak strength of rock under multiaxial compression. It gives good estimate of tensile strengths, but underestimates compressive strengths, particularly at high lateral stresses. A number of modifications to the Griffith's solution were introduced, but they are not in practical use today.

3. Von Mises Criterion — Based on Distortion Energy Model Suitable for Ductile Materials

At high confining stresses, brittle rocks behave ductile. The compressive strengths at high confining stresses is not as high as estimated by the linear Mohr-Coulomb criterion for brittle failure. Distortion energy based Von Mises yielding criterion is for ductile materials and could be suitable for rock materials at very high confining stresses. Von Mises criterion is also known as the maximum shear strain energy criterion. It suggests that yield begins in a material when the sum of the squares of the principal components of the deviatoric stress reaches a specified critical value, i. e.,

$$(\sigma_1-\sigma_2)^2+(\sigma_1-\sigma_2)^2+(\sigma_1-\sigma_2)^2=c_y=2\sigma_y^2$$

The Von Mises criterion assumes that the materials fail by plastic yielding, depending on the deviator stresses, not on the maximum or hydrostatic stresses. It is not generally appropriate for rocks. However, when rock materials subjected to highconfining pressure, they exhibit ductile behavior, with failure along planes of maximum shear stress at 45° to σ_1, as illustrated by the Von Mises criterion (Table 2-2). The failure of rock corresponding to the three strength criterion is shown in Fig. 2-18.

Comparison of the strength criterion **Table 2-2**

Stress condition	Strength observed	Criterion	Failure mechanism
Tension	Very low tensile strength, 1/20 to 1/8 of UCS	Griffith	Cracking starting at tips of existing microcracks and pores
Compression with low confining stress	Strength increase with confining pressure approximately linearly	Mohr-Coulomb	Brittle shear failure along shear planes at $(45°-\varphi/2)$ to σ_1
Compression with high confining stress	Strength increase with confining pressure at reducing rate	Von Mises	Ductile shear along planes of maximum shear stress at 45° to σ_1

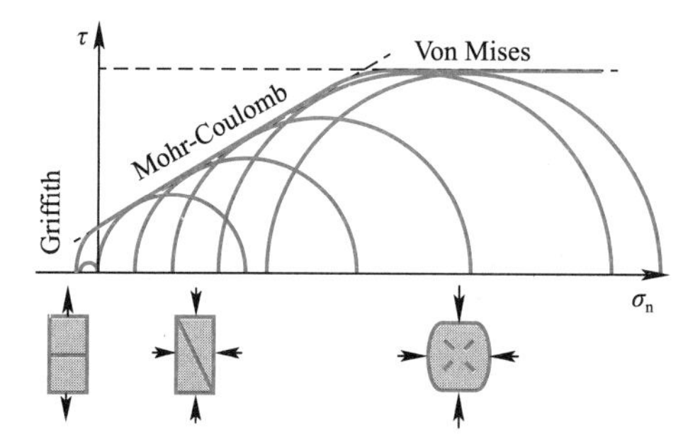

Fig. 2-18 Failure criterion of rock under different stress conditions

2.5.2 The Stress-strain Curve of Rock

The stress-strain curve of rock is a graphical representation that shows the relationship between the applied stress and resulting strain in a rock sample. It provides valuable information about the mechanical behavior and strength of rock under different loading conditions. The stress axis of the curve represents the applied stress on the rock sample, typically measured in megapascals (MPa) or pounds per square inch (psi). The strain axis represents the deformation or elongation of the rock sample in response to the applied stress, usually expressed as a percentage or as a ratio.

The stress-strain curve of rock typically exhibits several distinct phases:

Elastic region: initially, at low levels of stress, the rock deforms elastically, meaning it returns to its original shape once the stress is removed. The relationship between stress and strain in this region is linear, and the rock follows Hooke's Law.

Yielding or plastic deformation: as the applied stress increases, the rock may reach a point called the yield point or plastic limit. At this stage, the rock starts to deform plastically, meaning it undergoes permanent and non-reversible deformation even after the stress is removed. The stress-strain relationship becomes non-linear.

Strain hardening or strain softening: depending on the rock type and composition, it may exhibit strain hardening or strain softening behavior after yielding. Strain hardening refers to an increase in the rock's resistance to further deformation, while strain softening refers to a decrease in resistance.

Ultimate failure: as the stress continues to increase, the rock eventually reaches its maximum strength and undergoes brittle failure. This is accompanied by a sharp drop in stress and significant deformation or fracture.

The shape of the stress-strain curve can vary depending on factors such as rock type, mineral composition, confining pressure, moisture content, and temperature. Different types of rocks, such as igneous, sedimentary, and metamorphic, can exhibit different behaviors on the stress-strain curve. The stress-strain curve is essential in geotechnical engineering and rock mechanics as it helps engineers understand the stability of rock formations, design safe underground excavations, and assess the potential for rockfall or slope failures. It provides valuable insights into the mechanical properties and deformation characteristics of rocks under various loading conditions. Six typical stages are usually observed during the whole loading process, as shown in Fig. 2-19.

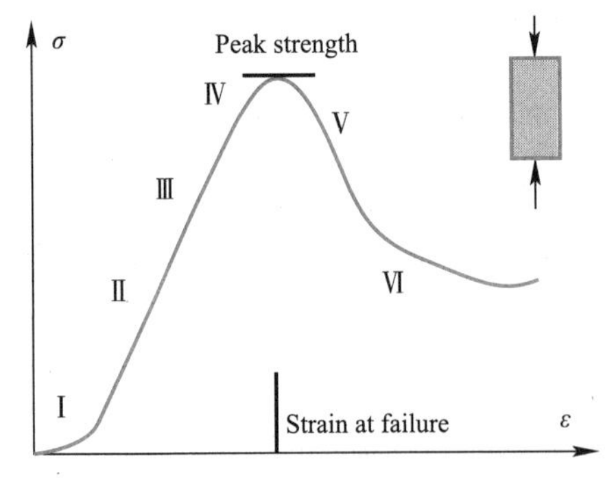

Fig. 2-19 Typical stages which are usually observed during the whole loading process

- Stage Ⅰ— the rock is initially stressed, in addition to deformation. Existing microcracks is closing, causing an initial non-linearity of the curve.
- Stage Ⅱ— the rock basically has a linearly elastic behaviour with linear stress-strain curves, both axially and laterally.
- Stage Ⅲ— the rock behaves near-linear elastic. The axial stress-strain curve is near-linear and is nearly recoverable.
- Stage Ⅳ— the rock is undergone a rapid acceleration of microcracking events and volume increase.
- Stage Ⅴ— the rock has passed peak stress, but is still intact, even though the internal structure is highly disrupt. The specimen is undergone strain softening (failure) deformation.
- Stage Ⅵ— the rock has essentially parted to form a series of blocks rather than an intact structure.

2.5.3 Influence of Testing Machine Stiffness

Whether or not the post-peak portion of the stress-strain curve can be followed and the associated progressive disintegration of the rock studied, depends on the relative stiffnesses of the specimen and the testing machine. The subject is important in assessing the likely stability of rock fracture in mining applications including pillar stability and rockburst potential.

The figures illustrate the interaction between a specimen and a conventional testing machine (Fig. 2-20). The specimen and machine are regarded as springs loaded in parallel. The machine is represented by a linear elastic spring of constant longitudinal stiffness k_m, and the specimen by a non-linear spring of varying stiffness k_s. Compressive forces and displacements of the specimen are taken as positive. Thus, as the specimen is compressed, the machine spring extends.

Fracture is the formation of planes of separation in the rock material. It involves the breaking of bonds to form new surfaces. The onset of fracture is not necessarily synonymous

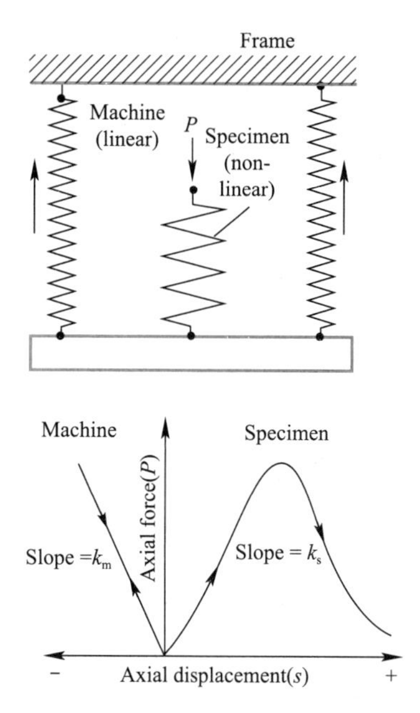

Fig. 2-20 Interaction between a specimen and a conventional testing machine

with failure or with the attainment of peak strength. Strength, or peak strength, is the maximum stress, usually averaged over a plane, that the rock can sustain under a given set of conditions. It corresponds to point B. After its peak strength has been exceeded, the specimen may still have some load-carrying capacity or strength. The minimum or residual strength is reached generally only after considerable post-peak deformation (point C).

The terms brittle and ductile which are used in this book are understood as follows (Fig. 2-21). Brittle behavior is characterized by a sudden change of slope in the stress-strain curve near the yield point followed by a complete loss of cohesion or an appreciable drop in differential stress. Brittle fracture is the process by which sudden loss of strength occurs across a plane following little or no permanent (plastic) deformation. Ductile behavior is characterized by the deformation without any downward slope after the yield point. Ductile deformation occurs when the rock can sustain further permanent deformation without losing load-carrying capacity.

Rocks generally fail at a small strain, typically from 0.2% to 0.4%. Brittle rocks, typically crystalline rocks, have low strain at failure, while soft rock, such as shale and mudstone, tend to have relatively high strain at failure. Most rocks, including all crystalline igneous, metamorphic and sedimentary rocks, behave brittle under uniaxial compression. A few soft rocks, mainly of sedimentary origin, behave ductile.

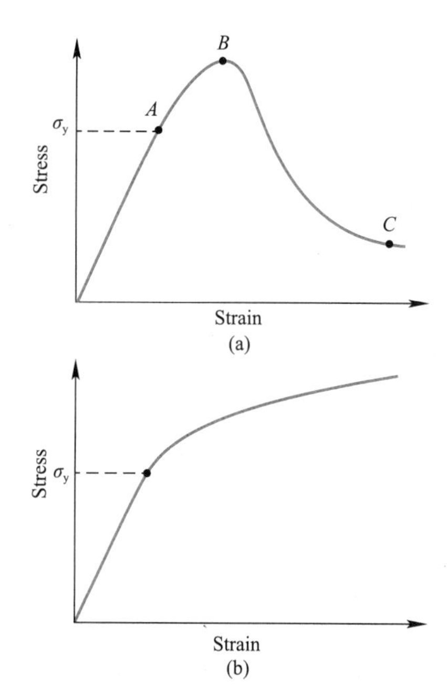

Fig. 2-21 Typical stress-strain curve of brittle and ductile behavior
(a) Brittle behavior; (b) Ductile behavior

2.5.4 Effects of Confining Stress

It has been known since the end of the 19 th century that if the confining stress applied to the sides of a cylindrical specimen during a triaxial compression test is increased, the axial stress required to cause failure will increase, and the rock will show a tendency toward greater ductility. In the classical experiments, oil is used to apply a confining stress $\sigma_2 = \sigma_3$ to the sides of the specimen, while the axial stress σ_1 is slowly increased. The effect that confining pressure has on the axial stress versus axial strain curve is shown in Fig. 2-22. For each value of $\sigma_2 = \sigma_3$, the stress-strain curve initially exhibits a nearly linear elastic portion, with a slope (Young's modulus) that is nearly independent of the confining stress. But both the yield stress and the failure stress increase as the confining stress increases. Finally, there is a small descending portion of the curve, ending in brittle fracture. A different type of behavior is exhibited by other rocks, notably carbonates and some sediments. For sufficiently low confining stresses, exemplified by the curve labeled $\sigma_3=0$, brittle fracture (denoted by the dashed line Fig. 2-5a) occurs as for the quartzite described above. But at higher confining stresses, the rock can undergo a strain as large as 7%, with no substantial loss in its ability to support a load (i. e., no decrease in the axial stress).

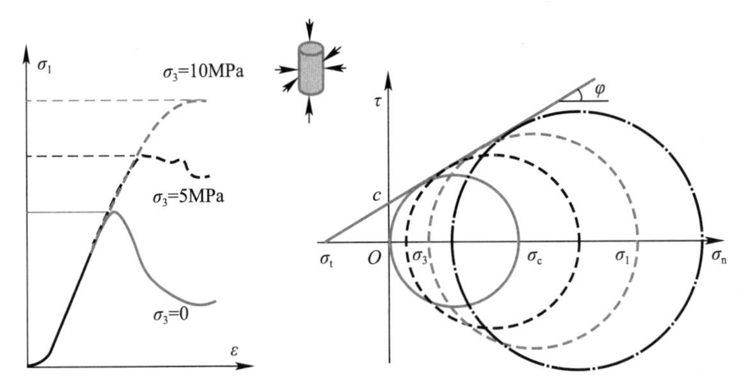

Fig. 2-22 Stress-strain curves for rocks at various confining pressures

2. 5. 5 Types of Fracture

The different types of stress-strain behavior discussed in the previous two sections correspond to different physical processes occurring within the rock. Under unconfined compression, a rock tends to deform elastically, until failure occurs abruptly (Fig. 2-23). This failure is accompanied by somewhat irregular longitudinal splitting (Fig. 2-23a). With a moderate amount of confining pressure, longitudinal fracturing is suppressed, and failure occurs along a clearly defined plane of fracture (Fig. 2-23b). This plane is typically inclined at an angle less than 45° from the direction of σ_1 (the axial direction, in this case). This plane is characterized by shearing displacement along its surface, and is referred to as a shear fracture. Under some circumstances, failure occurs along two conjugate shear planes located symmetrically with respect to the axial direction, but this seems to be an experimental artifact caused by the ends of the specimen being constrained against rotation. If the confining pressure is increased, so that the rock becomes fully ductile,

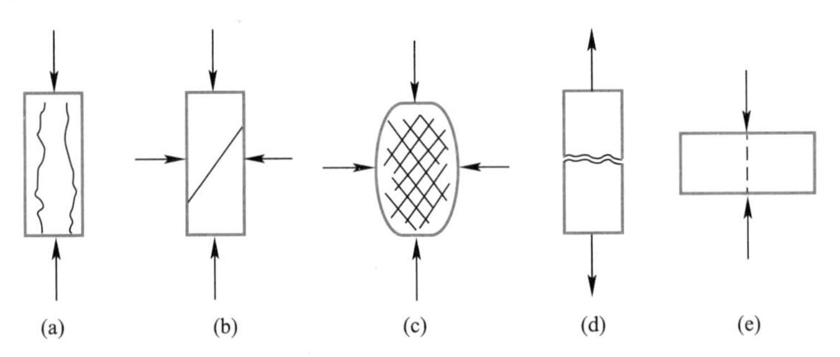

Fig. 2-23 Main features of different fractures

(a) Longitudinal splitting under uniaxial tension; (b) Shear fracture; (c) Multiple shear fractures; (d) Extension fracture; (e) Extension fracture produced by opposing line loads

a network of small shear fractures appears, accompanied by plastic deformation of the individual rock grains (Fig. 2-23c). The second basic type of fracture, an extension fracture, typically appears when a rock fails under uniaxial tension. The main characteristic of this type of fracture is a clean separation of the two halves of the sample, with no tangential offset between the two surfaces (Fig. 2-23d). Under more complicated systems of applied stress, fractures appear which may be regarded as belonging to one or another of these two basic types. If a slab of rock is compressed between two opposing line loads (Fig. 2-23e), an extension fracture appears between the loads. When the fracture surfaces are examined from a specimen that has undergone longitudinal splitting, as in Fig. 2-23 (a), parts of the surfaces will have the appearance of a shear fracture, and other parts will appear to be extension fractures.

2.5.6 Effects of Pore Fluids

Groundwater is part of the hydrological cycle. It also belongs to hydrogeology. Most of the igneous and metamorphic rocks are very dense with interlocked texture. The rocks therefore have extremely low permeability and porosity. Some clastic sedimentary rocks, typically sandstones, can be porous and permeable. Weathered rocks can also be porous and permeable. Fig. 2-24 shows the permeable rock in some permeable rock in mining engineering.

Permeability is a measure of the ability of a material to transmit fluids. It is given by the Darcy's law (Fig. 2-25).

Fig. 2-24 Inflow of water in the permeable rock in mining engineering

$$Q=Ak(h_1-h_2)/L$$

Where,

Q —— flow rate;

K —— coefficient of permeability;

A —— cross section area;

L —— length;

h_1, h_2 —— hydraulic head.

Most rocks have very low permeability. Permeability of rock material is governed by porosity. Porous rocks such as sandstones usually have high permeability while granites have low permeability. Permeability of rock materials, except for porous materials, has limited interests. In the rock mass, flow is concentrated in fractures. Rock masses are fractured. Fractures provide flow paths and flow is governed by the apertures. Flow in a fractured rock mass is also controlled by the connectivity of the fracture system or network, as shown in Fig. 2-26. Although a rock mass can be seen as highly fractured, only a limited percentage of fractures are interconnected and conduct flow. At site, it is common to see only a few fractured with water flowing, while others are dry.

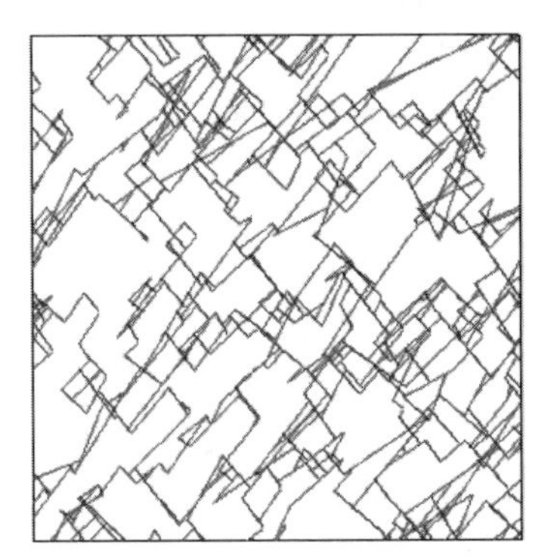

Fig. 2-25 Water flow model by the Darcy's law

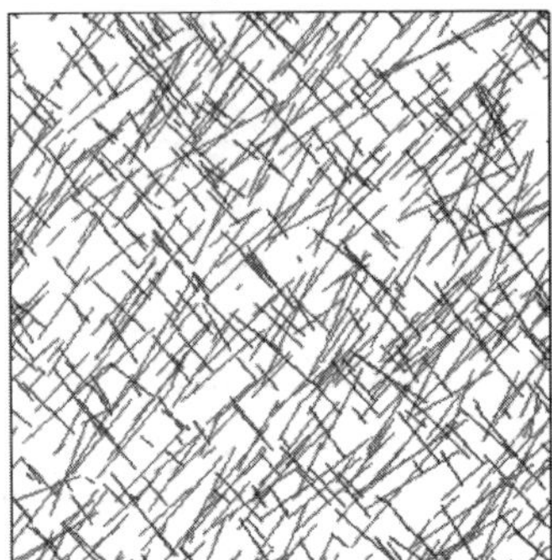

Fig. 2-26 Flow in the fracture network

Groundwater is important to rock mechanics.

(1) Water pressure contributes to the stress field.

(2) Water changes rock parameters, e. g., friction.

(3) When water is present, it increases the complexity of rock engineering, e. g., more difficult to tunnel with water inflow and high water pressure.

2.5.7 Failure under True-triaxial Conditions

At depth, rock is subjected to axial and lateral stresses (triaxial), and compressive strength is higher in triaxial condition. True triaxial compression means three different principal stresses. It is often simplified by making two lateral stresses equal to minor principal stress (axisymmetric triaxial test, as shown in Fig. 2-27). The behaviour of rock in triaxial compression changes with increasing confining pressure: (1) peak strength increases, (2) post peak behavior from brittle gradually changes to ductile. Stress-strain behaviour at elastic region appears the same as uniaxial compression. The effect of intermediate principal stress should be considered. Axisymmetric triaxial test gives strength without considering the effect of intermediate principal stress σ_2, and generally under-estimate the strength. Rock triaxial compressive strength generally increases with σ_2, but σ_3 is fixed. When σ_2 is excessively greater than σ_3, the strength may start to decrease. Young's modulus and Poisson's Ratio can be experimentally determined from the stress-strain curve. They seem to be unaffected by change of confining pressure. High strength rocks also tend to have high Young's modulus, depending on rock type and

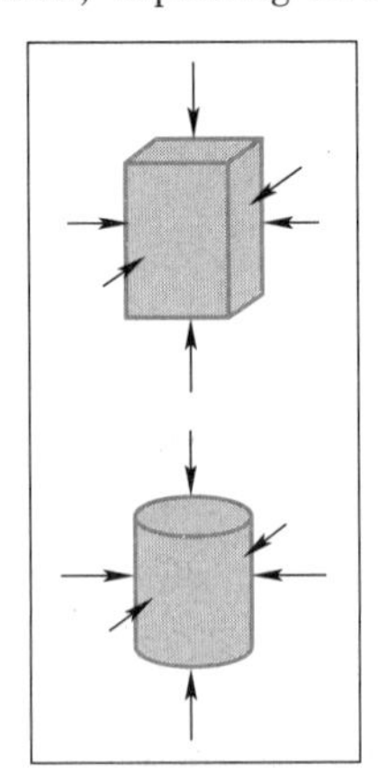

Fig. 2-27 Failure under triaxial conditions

other factors. For most rocks, the Poisson's ratio is between 0. 15 and 0. 4.

2. 5. 8 Special Rocks

1. Weathering

All rocks disintegrate slowly as a result of: (1) mechanical weathering, breakdown of rock into particles without changing the chemical composition of the minerals in the rock; (2) chemical weathering, breakdown of rock by chemical reaction, primarily by water and air. Weathering is progressive, between fresh rock and completed material (soil), rocks can be slightly, moderately and highly weathered. Those weathered rocks are still intact and have the structure and texture as rock. However, due to weathering, their properties have been affected and altered. Weathering causes significant reduction of rock material strength.

2. Soft Rocks and Hard Soils

Soft rocks and hard soils are geological formations that differ in their properties, composition, and strength. Here's a brief overview of each.

Soft rocks: soft rocks refer to sedimentary rocks or loosely consolidated deposits that are relatively easier to excavate or deform compared to harder rock formations. Examples of soft rock include clay, shale, siltstone, and some types of sandstone. These rocks typically have lower strengths, lower resistance to erosion, and higher porosity compared to harder rocks. Soft rocks can have varying levels of stability depending on factors such as moisture content, compaction, and the presence of weak layers or discontinuities. They can be more susceptible to weathering, deformation, and erosion over time. In engineering applications, construction on soft rock formations may require additional considerations for stability and foundation design.

Hard soils: hard soils, also known as stiff or dense soils, are fine-grained soils with high shear strength and resistance to deformation. These soils are often formed through processes like weathering or compaction over extended periods. Examples of hard soils include stiff clay, silty clay, and compacted fill materials. Hard soils exhibit higher cohesion and internal friction, which contribute to their characteristic stiffness and ability to bear loads without significant settlement. Due to their low permeability, these soils typically have good drainage properties. However, hard soils can pose challenges for excavation due to their resistance to penetration and require specialized techniques and equipment.

Engineering projects involving soft rocks and hard soils require careful assessment and consideration of their properties during planning and design stages. Factors such as geotechnical investigations, soil stabilization methods, and foundation design techniques play crucial roles in ensuring the safety and stability of structures built on these formations. It's important to note that site-specific characteristics and local geological conditions can vary significantly, so consulting with a geotechnical engineer or a qualified professional is essential for assessing the specific properties and behavior of soft rock and hard soil formations at a given location.

Sedimentary rocks are formed by sediments such as soil, through long processes of compaction and cementation. The process could be stopped before the sediments have completely solidified. The materials then could be highly consolidated but not fully solidified. Typically, those materials have low strength and high deformability, and when placed in water, they often could be dissolved. When in a dry environment, they behave as weak rock and when in water, it collapses.

3. Swelling Rocks

Some rocks have the characteristics of swelling, that is, when the rock is exposed with water (directly in contact with water or in air), it expands. This is primarily due to the swelling behavior of the minerals of the rock, typically the montmorillonite clay minerals. Rock and soil containing considerable amount of montmorillonite minerals will exhibit swelling and shrinkage characteristics.

Swelling rock is a type of rock or soil that exhibits significant volume changes in response to changes in moisture content. When water is absorbed into the rock or soil, it causes expansion, and when water is lost, the material contracts. This property of swelling and shrinking is known as "swelling potential" or "shrink-swell behavior". The most common type of swelling rock is expansive clay, also referred to as shrink-swell clay. Expansive clay is composed of fine-grained particles that have the ability to absorb and retain water. When expansive clay absorbs water, it undergoes a process called hydration, where the individual clay particles attract and bind water molecules, causing them to expand. The expansion of swelling rock due to moisture absorption can exert significant pressure on structures built above or within it. This pressure can lead to various geotechnical problems, including foundation movement, cracking of walls, and damage to infrastructure. Additionally, the subsequent drying of swelling rock can cause significant shrinkage, leading to ground cracking and subsidence.

It's important to conduct thorough geotechnical investigations before construction to identify the presence and potential extent of swelling rock. Proper site characterization and geotechnical analysis are crucial in designing and implementing appropriate engineering solutions to mitigate the risks associated with swelling rock.

4. Crushed Rocks

Crushed rocks, also known as crushed stones or gravel, are a type of construction aggregate that is made by mechanically crushing and grading large rocks, boulders, or quarry stones. It is commonly used in various construction applications, including road base, concrete production, drainage systems, and landscaping. Here are some key points about crushed rocks.

(1) Size and grades: the crushed rock is produced in different sizes and grades to suit specific construction needs. The size can range from fine particles (such as sand) to larger pieces (such as 2—3 in stones). Common grades include 5 mm, 10 mm, 20 mm, and 40 mm.

(2) Composition: the crushed rock is typically composed of hard and durable materials such as granite, limestone, trap rock, or basalt. These rocks are crushed into smaller fragments, resulting in angular pieces with varied shapes and sizes.

Uses:

(1) Road construction: the crushed rock is often used as a base or sub-base material for road construction. It provides stability, enhances drainage, and serves as a load-bearing layer.

(2) Concrete production: the crushed rock is a key component in the production of concrete. It is mixed with cement, sand, and water to create the solid matrix of the concrete mixture.

(3) Drainage systems: the permeability of crushed rock makes it suitable for use in drainage systems and French drains. Water can flow through the void spaces within the crushed rock, facilitating efficient drainage.

(4) Landscaping: the crushed rock is commonly used in landscaping projects. It can be used as decorative gravel in gardens, walkways, or as a base for patios and retaining walls.

(5) Durability and strength: the crushed rock provides good strength and durability when compacted. Its interlocking angular particles create a stable foundation and resist movement under heavy loads.

(6) Sustainability: using the crushed rock as a construction material can be environmentally sustainable. It is often sourced from local quarries, reducing transportation distances and associated carbon emissions.

The specific type and grade of crushed rock used in a project may vary depending on the engineering requirements and regional availability. It is recommended to consult with local construction professionals or suppliers to determine the most suitable crushed rock for your specific application. Characteristics of highly fractured and crushed rocks are quite different from the massive rock masses. They behave as granular and block materials, depending on the geometry and friction. When such materials are encountered in engineering, they need to be addressed separately.

2.5.9 Resources in Rocks

Resources in rocks refer to valuable substances or materials that are naturally occurring and extracted from rock formations. Rocks can contain a wide variety of resources, including minerals, metals, fossil fuels, and building materials. These resources play a vital role in supporting human activities and industries. Here is an introduction to some notable resources found in rocks.

1. Minerals

Minerals are naturally occurring inorganic substances with specific chemical compositions and crystal structures. Many rocks contain valuable minerals that have important industrial uses. Examples include iron ore in banded iron formations, copper in porphyry deposits, gold in quartz veins, and coal in sedimentary rock layers. Minerals are utilized in various sectors, ranging from construction and manufacturing to electronics and energy production.

2. Metals

Some rocks host concentrations of metallic elements that are economically viable for extraction. These metallic resources include iron, aluminum, copper, zinc, lead, nickel, and many others. Metallic minerals are often concentrated in specific rock types or geological settings, such as sulfide minerals in volcanic deposits or oxide minerals in lateritic weathering profiles. Metals are essential for infrastructure development, transportation, machinery, and consumer goods.

3. Fossil Fuels

Fossil fuels, including coal, oil, and natural gas, are derived from ancient organic matter preserved in rock formations. Coal forms from the accumulation and compaction of plant remains over millions of years, while oil and gas result from the decomposition of marine organisms in sedimentary rocks. Fossil fuels are vital sources of energy for power generation, transportation and heating, although their use has environmental implications. Oil and gas are the naturally occurring hydrocarbons trapped in sedimentary rock formations. They can be found in porousor fractured rocks below the Earth's surface. Shale gas is trapped in shale formations.

4. Building Materials

Rocks provide abundant resources for construction and building materials. Limestone, for example, is a common rock that is used extensively in the construction industry as a building stone, aggregate, and raw material for cement production. Other rocks such as granite, sandstone, marble, and slate are utilized for architectural purposes, flooring, countertops, and decorative elements.

5. Industrial Minerals

Certain rocks are rich in non-metallic industrial minerals, which have various industrial applications. Examples include silica sand used in glass manufacturing, gypsum for plasterboard production, talc for cosmetics and ceramics, kaolin for paper and porcelain, and phosphate rock for fertilizer production. These industrial minerals are essential components of numerous products and industries.

6. Gemstones

Some rocks contain precious and semi-precious gemstones that have aesthetic or ornamental value. Gemstones like diamonds, rubies, sapphires, emeralds, and quartz crystals form in specific geological conditions and are pursued for their beauty and rarity. Gemstone mining and trade form an important part of the jewelry industry and contribute to regional economies.

7. Groundwater

Rocks can act as reservoirs for groundwater storage. Porous and permeable rocks such as sandstone and limestone can hold significant amounts of water within their pore spaces, serving as aquifers. Groundwater is a crucial natural resource for drinking water supplies, agricultural irrigation, and industrial processes. The extraction and utilization of these resources from rocks involves various mining and extraction techniques. Careful management and sustainability practices are necessary to ensure responsible resource extraction, minimize environmental impacts, and preserve these valuable geological assets for future generations.

8. Hot Rock Geothermal Energy

In many locations, granitic rock at great depths (3—5 km) has a temperature up to 250℃. As the heat is continuous supplied from Earth' s interior or by natural radioactive delay, this geothermal energy is renewable. Hot rock geothermal energy, also known as enhanced geothermal systems (EGS) or deep geothermal energy, is a form of renewable energy that harnesses the heat stored within the Earth's crust. Unlike traditional geothermal systems that rely on naturally occurring hot water or steam reservoirs, hot rock geothermal energy extracts heat from subsurface rocks by creating or enhancing existing fractures in the rock formation. Here is an introduction to the concept and process of hot rock geothermal energy.

(1) Geothermal heat source: the Earth's interior generates heat through natural processes like radioactive decay and residual heat from its formation. This heat is continuously conducted towards the surface. In the deeper layers of the Earth's crust, temperatures can reach several hundred degrees Celsius.

(2) Hot rock reservoirs: in hot rock geothermal energy, the objective is to access and extract heat from rocks that are at high temperatures but lack sufficient permeability to allow the free flow of water or steam. These rocks, typically found at depths of several kilometers, act as the heat source for geothermal energy production.

(3) Hydraulic fracturing: to enhance the permeability and enable fluid circulation in the hot rock reservoir, hydraulic fracturing or "fracking" techniques are used. High-pressure fluids are injected into the wellbore, creating fractures within the rock formations to stimulate the flow path for the working fluid.

(4) Injection and extraction wells: hot rock geothermal systems comprise two primary wells—an injection well and an extraction well. The injection well facilitates the circulation of cool water or other working fluids into the reservoir to absorb heat from the rocks. The heated fluid is then pumped up through the extraction well to the surface.

(5) Heat extraction and power generation: at the surface, the extracted fluid's heat is transferred to a secondary fluid in a heat exchanger. The secondary fluid, typically a closed-loop system with a lower boiling point than water, vaporizes and drives a turbine-generator to produce electricity. After transferring its heat, the secondary fluid is cooled down in a condenser and circulated back for reuse.

(6) Advantages and challenges: hot rock geothermal energy has several advantages, including being a renewable and low-emission energy source that can provide baseload power. It has the potential to be highly sustainable and offers long-term reliability. However, challenges exist, such as the need for deep drilling, hydraulic fracturing expertise, and potential seismicity associated with the fracturing process. The economic viability of hot rock geothermal projects also depends on factors like resource availability and upfront investment costs.

Hot rock geothermal energy holds promise as a technology for harnessing the vast amounts of heat stored beneath the Earth's surface. Ongoing research and technological advancements aim to improve its efficiency, reduce costs, and expand its utilization to contribute to a more sustainable and diversified energy mix.

References

[1] Judd W. R.. State of Stress in the Earth's Crust [C]//Proceedings of the International Conference. Santa Monica, California, 1964: 5-51.

Chapter 3
The Properties of Rock Mass

3.1 Introduction

Rock mass is a geologic body composed by the discontinuities which have a critical influence on the deformational behavior of blocky rock systems, as shown in Fig. 3-1.

Fig. 3-1 The fractured rock mass

A rock mass refers to a collection or assemblage of intact rocks and/or discontinuous rock masses that are bound together. Due to the existence of discontinuities, jointed rock mass often exhibits an inherent anisotropy and deformability of rock masses influencing their behavior is an important geomechanical property for the design of rock structures. In jointed rocks, the strength and deformational behavior of rock mass depend principally on the state of intact rock whose obtaining method for the mechanical properties is relatively systematic and existing discontinuities whose strength and distributions are both the key influencers. Therefore, accurate description of joints is an important topic for estimating and evaluating the deformability of rock masses.

It is an essential concept in geotechnical engineering, geology, and rock mechanics. Understanding the behavior and characteristics of rock masses is crucial for various engineering and geological applications, such as slope stability analysis, tunneling, foundation design, and mining.

Jointed rock masses are geological formations consisting of rocks that are characterized by a network of fractures or joints. These fractures can occur naturally due to tectonic forces or can be created by human activities such as mining or drilling. Jointed rock masses are common in many geological settings and are important in a variety of engineering and environmental applications. In engineering, jointed rock masses are of particular interest due to their impact on the stability of structures and the behavior of foundations. The presence of joints can affect the strength and deformation characteristics of the rock mass, which can in turn affect the stability of slopes, tunnels, and other infrastructure. Understanding the orientation, spacing, and persistence of joints is therefore critical in designing and constructing safe and stable structures in jointed rock masses.

3.2 Types and Properties of Discontinuities

3.2.1 Types of Discontinuities

There are many types of rock discontinuities, including:

- Joints—most common, normally in sets.
- Fractures—randomly distributed.
- Faults—singular and large scale.
- Bedding—singular and large scale.
- Interfaces—singular and large scale.

An individual joint is often termed as a fracture. Principal geometrical characteristics of rock joints, as shown in Fig. 3-2.

- Number of joint sets.
- Joint persistence.
- Joint plane orientation.
- Joint spacing, joint frequency, block size, and RQD.
- Joint surface roughness and matching.
- Joint aperture and filling.

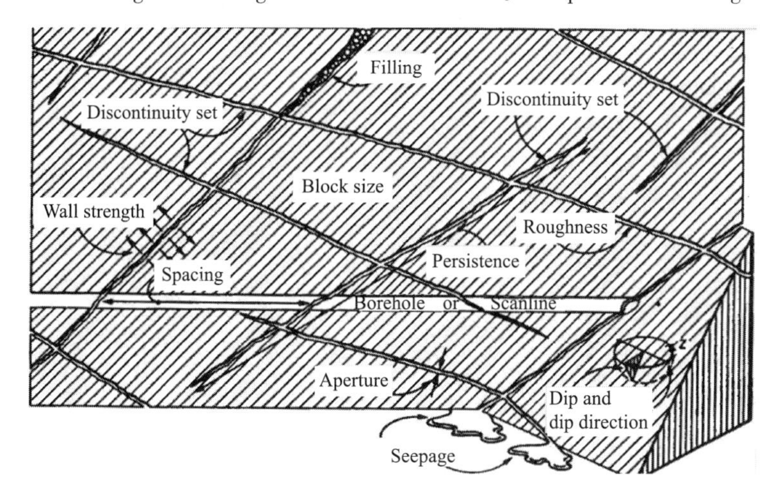

Fig. 3-2 The geometrical parameters of joints

Joints are generally in sets, i. e., parallel joints. The number of joint sets can be up to five(Table 3-1). Typically, one joint set cuts the rock mass into plates, two perpendicular sets cut rock into column and three into blocks, and more sets cut rocks into mixed shapes of blocks and wedges. Mechanical properties of the rock mass is influenced by joint sets. More joint sets provide more possibilities of potential slide planes.

Classification of joints according to joint sets

Table 3-1

Ⅰ	Massive, occasional random fractures
Ⅱ	One joint set
Ⅲ	One joint set plus random fractures
Ⅳ	Two joint sets
Ⅴ	Two joint sets plus random fractures
Ⅵ	Three joint sets
Ⅶ	Three joint sets plus random fractures
Ⅷ	Four or more joint sets
Ⅸ	Crushed rock, earth-like

1. Joint Persistence

Persistence is the length of a discontinuity, and can be quantified by observing the trace lengths of discontinuities on exposed surfaces (Table 3-2). The persistence of joint sets controls large scale sliding or "down-stepping" failure of slope, dam foundation and tunnel excavation.

Classification of joints according to persistence Table 3-2

ISRM suggested description	Surface trace length (m)
Very low persistence	<1
Low persistence	1—3
Medium persistence	3—10
High persistence	10—20
Very high persistence	>20

2. Joint Plane Orientation

Orientation of joint sets controls the possibility of unstable conditions or excessive deformations. The orientation of joints determines the shape of the rock blocks. Orientation is defined by dip angle (inclination) and dip direction (facing) or strike (running).

3. Joint Spacing, Frequency, Block Size, and RQD

Fracturing degree of a rock mass is controlled by the number of joints in the rock mass. More joints mean that less average spacing between joints. Joint spacing controls the size of individual rock blocks (Table 3-3). It controls the mode of failure and flow. For example, a close spacing gives low mass cohesion and circular or even flow failure.

Joint spacing is the perpendicular distance between joints. For the joint set shown in Fig. 3-3, it is usually expressed as the mean spacing of that joint set. Often the apparent spacing is measured. Measurements of joint spacing are different on different measuring faces and directions. For example, in a rock mass with mainly vertical joints, measurements in vertical direction have far greater spacing than those in horizontal direction.

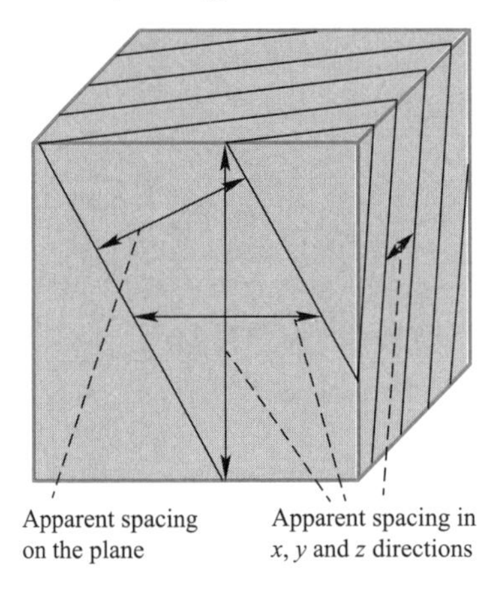

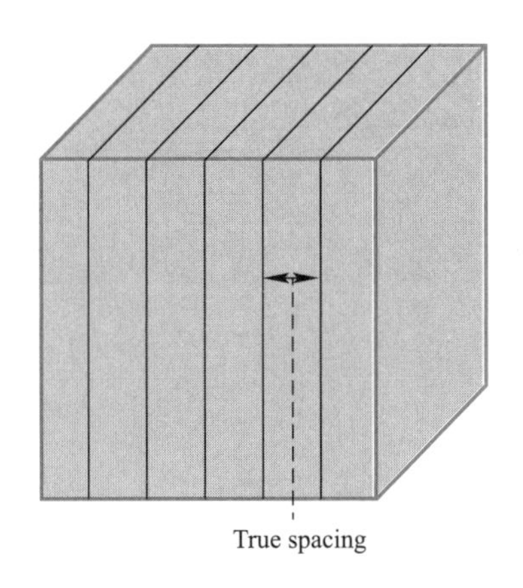

Fig. 3-3 The schematic for joint spacing

Classification of joints according to joint spacing Table 3-3

Description	Joint spacing (m)
Extremely close spacing	<0. 02
Very close spacing	0. 02—0. 06
Close spacing	0. 06—0. 2
Moderate spacing	0. 2—0. 6
Wide spacing	0. 6—2
Very wide spacing	2—6
Extremely wide spacing	>6

Joint frequency λ is defined as the number of joints per meter length. It is therefore simply the inverse of joint spacing s_j, i. e., $\lambda = 1/s_j$, as shown in Fig. 3-4.

Rock quality designation (RQD) is defined as the percentage of rock cores that have length equal or greater than 10 cm over the total drill length (Fig. 3-5).

The length of one core is 100 cm, in which the length sum of rock cores whose length is longer than 10 cm is 61 cm. Then the RQD could be calculated by RQD = (61/100) × 100% = 61%.

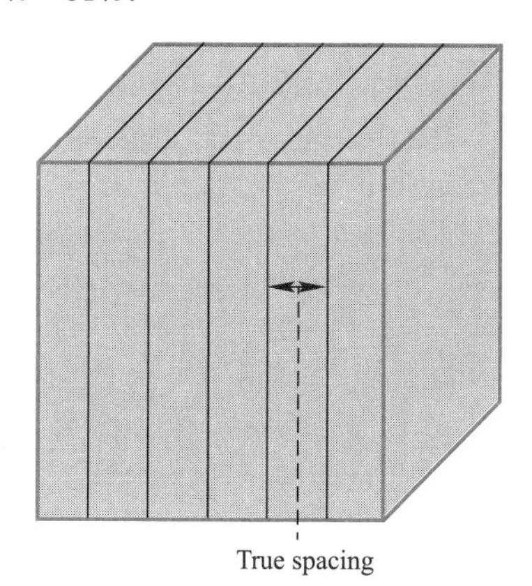

Fig. 3-4 The calculation of joint frequency according to joint spacing

4. Block Size and Volumetric Joint Count

Joint space also defines the size of rock blocks (Table 3-4). When a rock mass contains more joints numbers, the joints have lower average spacing and smaller block size. RQD could be related to volumetric joint count J_v by: RQD = 115−3.3J_v, for J_v between 4.5 and 30. J_v<4.5, RQD = 100%; J_v>30, RQD = 0.

5. Joint Surface Roughness and Matching

A joint is an interface of two contacting surfaces. The surfaces can be smooth or rough; they can be in good contact and matched, or they can be poorly contacted and mismatched (Fig. 3-6). The condition of contact also governs the aperture of the interface. The interface

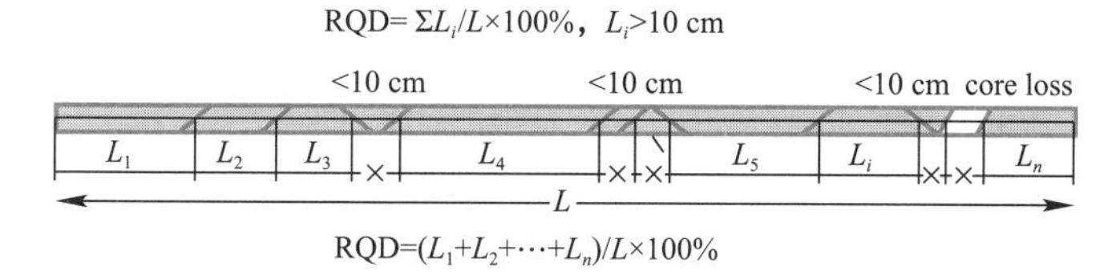

Fig. 3-5 The calculation of RQD

Classification of joints according to block size and volumetric joint count

Table 3-4

Designation	Volumetric joint count, joints(m^3)
Very large blocks	<1
Large blocks	1—3
Medium-sized blocks	3—10
Small blocks	10—30
Very small blocks	>30
Crushed rock	>60

can be filled with intrusive or weathered materials. Joint surface roughness is a measure of surface unevenness and waviness relative to its mean plane. The roughness is characterized by large scale waviness and small scale unevenness (irregularity) of a joint surface. It is the principal governing factor of the direction of shear displacement and shear strength, and in turn, the stability of potentially sliding blocks.

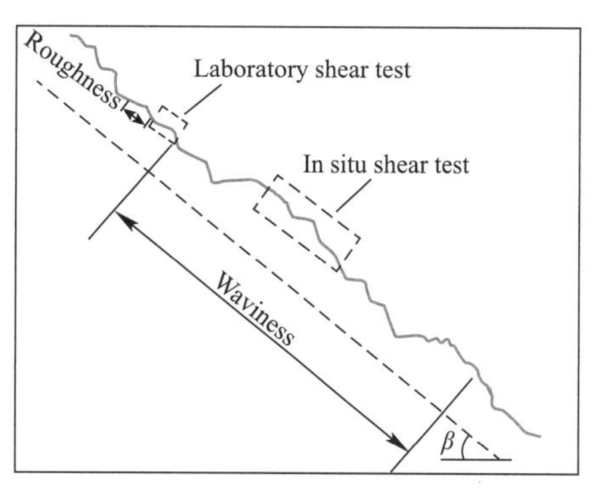

Fig. 3-6 The joint surface waviness and roughness

Roughness should first be described in meter scale (step, undulating, and planar) and then in centimeter scale (rough, smooth), as suggested by ISRM. It is not a quantitative measure. Joint roughness coefficient (JRC) is a quantitative measure of roughness, varying from 0 for the smooth flat surface to 20 for the very rough surface. Joint roughness is affected by geometrical scale. JRC number is obtained by directly comparing the actual joint surface profile with the typical profile in the chart. JRC_{20} is the profile for 20 cm and JRC_{100} for 100 cm. The value of JRC decreases with increasing size (Fig. 3-7).

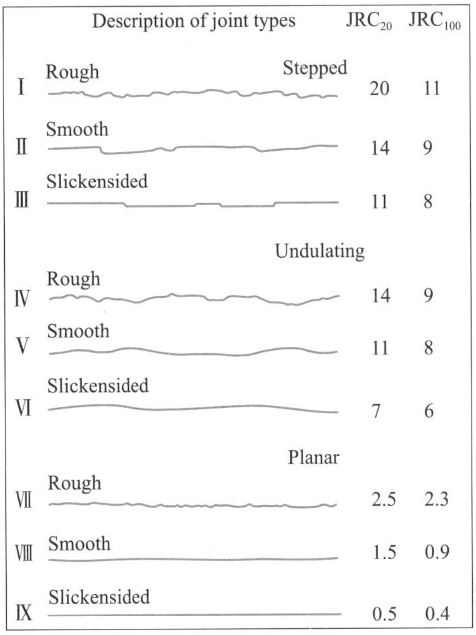

	Description of joint types	JRC_{20}	JRC_{100}
	Stepped		
Ⅰ	Rough	20	11
Ⅱ	Smooth	14	9
Ⅲ	Slickensided	11	8
	Undulating		
Ⅳ	Rough	14	9
Ⅴ	Smooth	11	8
Ⅵ	Slickensided	7	6
	Planar		
Ⅶ	Rough	2.5	2.3
Ⅷ	Smooth	1.5	0.9
Ⅸ	Slickensided	0.5	0.4

Fig. 3-7 The joint surface roughness by JRC

In reality, profiles of joint surfaces are 3D features (Fig. 3-8). ISRM and JRC descriptions are 2D based. It is therefore suggested to take several linear profiles of a surface for the description and JRC indexing. Joint surface is a rough profile that can be described by statistic method and fractal. Fractal method is applicable not only in 2D (linear profile), but also in 3D (surface plane profile). It is a useful tool to quantify the surface profile.

A joint is an interface of two surfaces. Properties of a joint are also controlled by the relative positioning of the two surfaces, in addition to the profiles. For example, joints in fully contacted and interlocked positions have little possibility of movement and is also difficult to shear, as compared to the same rough joints in point contact where movement can easily occur. Often, joints are differentiated as matched and mismatched. A joint matching coefficient (JMC) has been suggested, as shown in Fig. 3-9.

Fig. 3-8 Rough rock surface in 3D

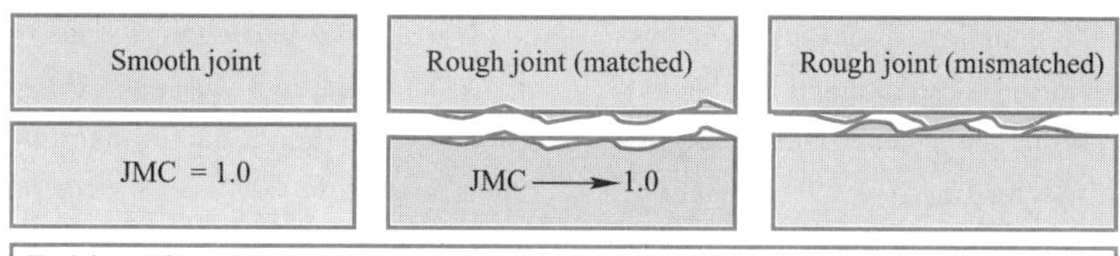

Fig. 3-9 JMC for rock joints

In a natural joint, it is very seldom that the two surfaces are in complete contact. There usually exists an opening or a gap between the two surfaces. The perpendicular distance separating the adjacent rock walls is termed as aperture. Joint opening is either filled with air and water (open joint) or with infill materials (filled joint), as shown in Fig. 3-10. Open or filled joints with large apertures have low shear strength. Aperture also associates with flow and permeability. Aperture can be the real aperture and equivalent hydraulic aperture. The latter is particularly important when permeability is concerned. Filling is material in the rock discontinuities separating the adjacent rock surfaces. In general, properties of the filling material affect shear strength, deformability and permeability of the discontinuities. Change of aperture will lead to significant change of permeability. The description of joint aperture is listed in Table 3-5.

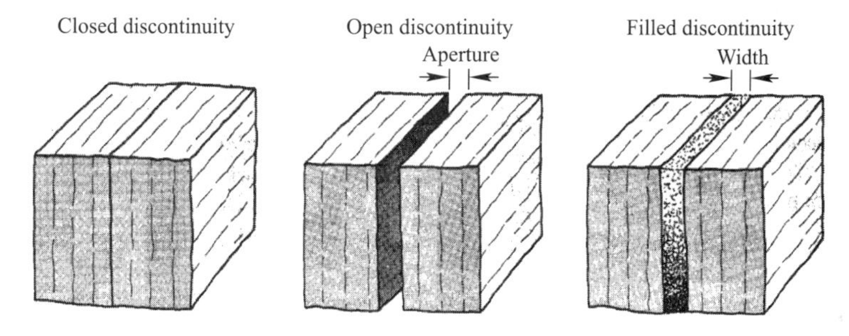

Fig. 3-10 Closed and open joints

Description of different aperture of rock joints **Table 3-5**

Aperture (mm)	Description	Feature
<0. 1	Very tight	"Closed feature"
0. 1—0. 25	Tight	
0. 25—0. 5	Partly open	
0. 5—2. 5	Open	"Gapped feature"
2. 5—10	Widely open	
10—100	Very widely open	"Open feature"
100—1000	Extremely widely open	
>1000	Cavernous	

3. 2. 2 Characterization of Discontinuities

Characterization of discontinuities is important for the estimating of spatial distribution in the rock masses. Rock outcrops or man-made excavations (e. g., borehole, quarry, tunnel and roadcut) could provide vital information of the discontinuities. 3D geometrical representation of complex discontinuities systems could provide a computational model that explicitly represents the geometrical properties of each individual fracture (e. g., orientation, size, position, shape and aperture), and the topological relationships between individual fractures and fracture sets. Discrete fracture network (DFN) model shown in Fig. 3-11 is based on: geological mapping, stochastic generation, geomechanics simulation, the distribution laws (Poisson, lognormal, exponential, power law, etc.).

Fig. 3-11 Discrete fracture network (DFN) model in three-dimension

1. Generation of A Fracture System Model

ShapeMetriX 3D is a powerful tool for the geological and geotechnical data collection and assessment for rock masses. During measuring process, two digital photos taken by a calibrated camera serve for a 3D reconstruction of the rock face geometry which is represented on the computer by a photorealistic spatial characterization. From it, measurements are taken by marking visible rock mass features, such as spatial orientations of joint surfaces and traces, as well as areas, lengths, or positions. Finally, the probability statistical models of discontinuities are established. It is generally applied in the following typical situations such as long rock faces at small height, rock slopes with complex geometries, etc. Almost any rock face can be reconstructed at its optimum resolution by using this equipment and its matching software.

2. Collection of Structure Data

Stereoscopic photogrammetry deals with the measurement of three-dimensional information from two images showing the same object or surface but taken from two different angles, just as shown in Fig. 3-12 (a) and (b). By automatic identification of corresponding points within the image pair, the result of the acquisition is a metric 3D image that covers the geometry of the rock surface together with a real photo. Once the image of a rock wall is ready, geometric measurements can be taken shown in Fig. 3-13. There are three groups of discontinuities in this bench face.

(a)

(b)

Fig. 3-12 Two different angles of stereoscopic photogrammetry
(a) Photo of rock slope looking left; (b) Photo of rock slope looking right

The measured orientation in a hemispherical plot can also be captured as Fig. 3-14. Fig. 3-15 shows the results of the distribution of joints. The determination of the joint density and spacing as well as trance length can be acquired.

Fig. 3-13 Geological mapping and geometric measurements around the generic 3D image

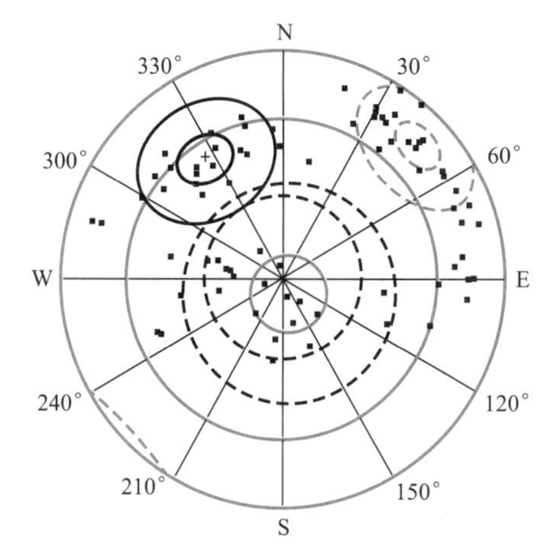

Fig. 3-14 Hemispherical plot of three discontinuity sets

3. Characteristics of the Discontinuities

The probability statistical models of joint traces, as well as dip angle, lengths, or spacing are listed in Table 3-6. Type 1 of the probability statistical model stands for negative exponential distribution, 2 for normal distribution, 3 for logarithmic normal distribution and 4 for uniform distribution. Although ShapeMetriX 3D can give a detailed geological data of the structures, it cannot be directly used by the mechanical simulation model in PFC2D. In the ensuring, the fundamental of stochastic modeling on fracture networks based on the geological data will be discussed in detail.

4. Fundamental Principles of Monte-Carlo

Using the Monte-Carlo Stochastic Simulation (MCSS) method, a random-sampling technique, discontinuities network model is established. Namely through the stochastic simulation, the implement course how to obtain the discontinuities network submitting to the geometry parameter probability model which is established according to the actual statistical is introduced as is shown in Fig. 3-16.

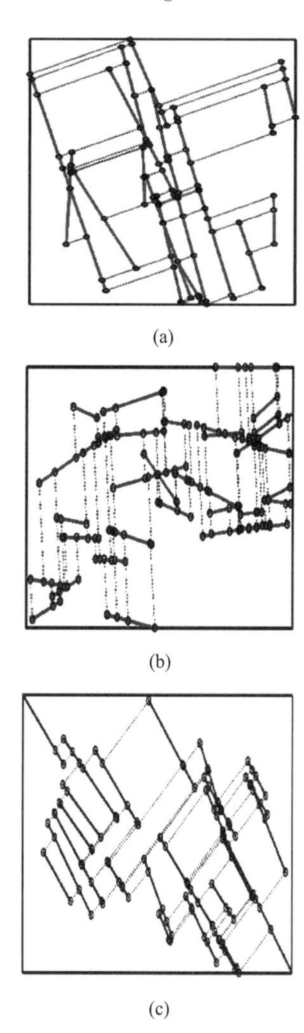

Fig. 3-15 Spacing map of the structural plane
(a) Set 1; (b) Set 2; (c) Set 3

The establishment of the discontinuities network using MCSS is the reverse process of geological date collection. The MCSS could give the basic random data such as dip angles or trace length for the network simulation of rock mass discontinuities based on the acquired probability distribution model of the discontinuities geometric parameters that are measured in-situ.

Characteristics of the discontinuities **Table 3-6**

Set	Density (/m²)	Dip direction (°)	Fracture characteristics											
			Dip (°)			Trace length (°)			Fault throw (m)			Spacing (m)		
			Type	Average	Standard deviation	Type	Average	Standard deviation	Type	Average	Standard deviation	Type	Average	Standard deviation
1	1.9	44.3	2	74.61	2.1	1	1.88	—	4	0.28	0.14	1	0.53	—
2	1.5	147.9	2	17.5	1.9	2	1.46	0.61	4	0.53	0.34	1	0.68	—
3	1.1	326.1	2	54.20	1.3	1	1.5	—	4	0.42	0.27	1	0.90	—

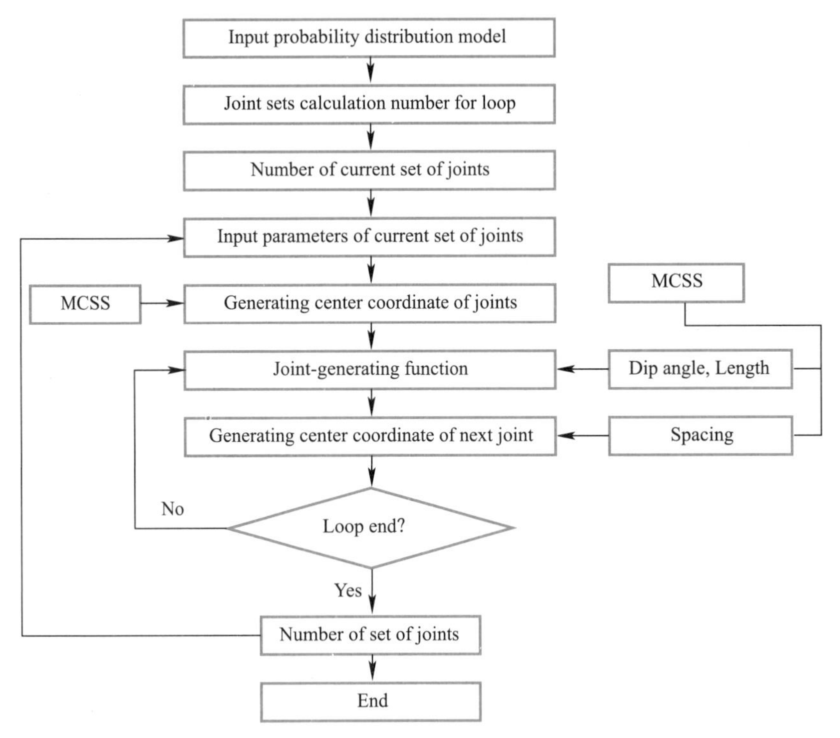

Fig. 3-16 Flow chart of the main routine of the Monte-Carlo Stochastic Simulation (MCSS) method

Network geometric graphics are simulated with the data from MCSS by a second development of PFC2D using the FISH programming language. Randomness of the data such as dip angel is the core of MCSS and it will be generated by using urand, a build-in FISH function uniformly distributed on the interval from 0 to 1 in PFC2D. In the program, area-ratio is defined by trace length and fault throw of one certain set of joints. The process of realizing the basic random data is presented based on the normal distribution function from Eq. (3-1).

$$F(x)=\frac{1}{\sqrt{2\pi}\sigma_x}\int_{-\infty}^{x} e^{-\frac{1}{2}\left(\frac{x-\mu_x}{\sigma_x}\right)^2} dx \quad (3\text{-}1)$$

Where,

μ —— the average value of particular geological parameter such as dip angle, length or spacing of joints;

σ —— the standard deviation of this parameter.

As Eq. (3-1) is a function whose definite integral is hard to obtain, random number n_i is solved based on the centre limit theorem of mathematical statistical theory (Su, etc., 2004;

Jiao, etc., 2010) implied by Eq. (3-2).

$$n_i = \sigma \sum_{i=1}^{12} (r_i - 6) + \mu \qquad (3\text{-}2)$$

In the following section, the impact of joints on the macroscale response will be examined by performing several compression tests. Fig. 3-17 shows the numerical model built in PFC2D, whose size is 5.0 m×10.0 m and made up of approximately 5092 randomly placed particles with the radius of 5.0 cm.

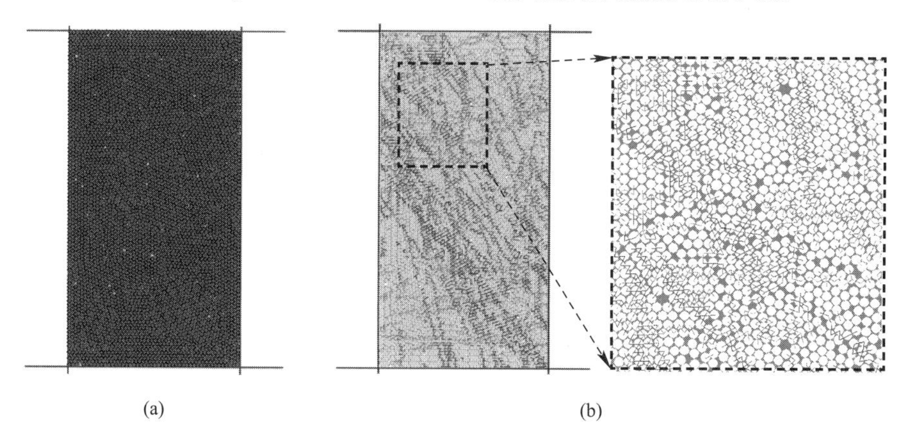

Fig. 3-17 Jointed rock model for the verification
(a) Bonded particle assembly; (b) Jointed network model

3.3 Mechanical Properties of Rock Joints

3.3.1 Stress and Deformation at Discontinuity

Stresses are often disturbed by a discontinuity. For a rock fracture with opening, normal stress on the fracture walls is zero and there are stress concentrations on the contact points. The stress field is no longer the same as in the continuous material. For a closed joint, the stress field may be continuous although strain may not. Displacement at discontinuity is not continuous. For example, at a fracture plane, sliding or shear displacement may occur. There may be much greater normal displacement at fracture than those of the material. Discontinuities can range from a fully welded interface to an opening containing different material. The mechanics vary different.

For a fully welded interface between two different materials, it has the continuities both in stress and displacement. Discontinuity is the change of materials at the interface.

For a fully-contacted smooth interface, the interface represents a weak plane of shearing.

For a locally-contacted fracture with gaps, both stress and displacement are discontinuous.

3.3.2 Normal Stiffness and Displacement

Normal stress and displacement of fully-contact discontinuity is continuous and therefore can be dealt with continuum approach.

For locally-contacted fractures, there are voids between the two sides, and stress-displacement function is discontinuous.

(1) An idealised pillar-contacted fracture (Fig. 3-18).

(2) An idealised prism-contacted fracture (Fig. 3-19).

A natural joint always has opening aperture

of less than 1mm to a few millimetre. With increasing normal stresses, the opening closes, and contact areas of the joint surfaces increase. The normal stress-normal displacement curve is non-linear. The normal stiffness, slope of the curve, is therefore not a constant.

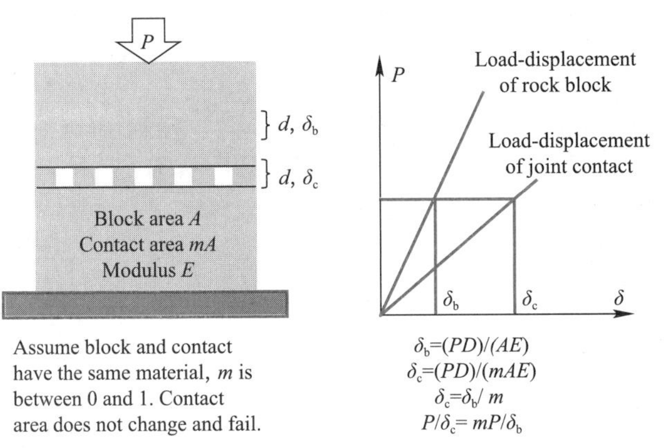

Fig. 3-18 An idealised pillar-contacted fracture

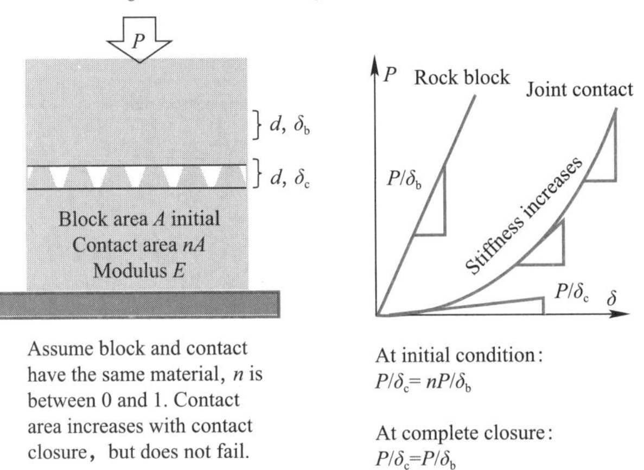

Fig. 3-19 An idealised prism-contacted fracture

When the joint is completely closed, displacement is then only by the deformation of the rock material. Stress-deformation curve of a natural joint in a granite, showing non-linear characteristics of joint stiffness. At high normal stress, joint is closed, the normal stiffness approaches that of rock material. When the joint is completely closed, there is no further closure of the joint, the displacement is then only by the elastic deformation of the rock material.

3.3.3 Shear and Friction between Contact Planes

The most common shear phenomenon of a discontinuity is the sliding between two contact surfaces, shown in Fig. 3-20. The friction theory gives the relationship between the friction angle φ, the normal force N and shear force S, as in Eq. (3-3).

$$S = N\tan\varphi \tag{3-3}$$

When slipping at the surface of contact is about to occur, the maximum static frictional force is proportional to the normal force (Fig. 3-21). When slipping is occurring, the kinetic frictional force is proportional to the normal force.

3.3.4 Shear Strength of Rock Joints

Shear behavior of rock joints is perhaps one of the most important features in geotechnical

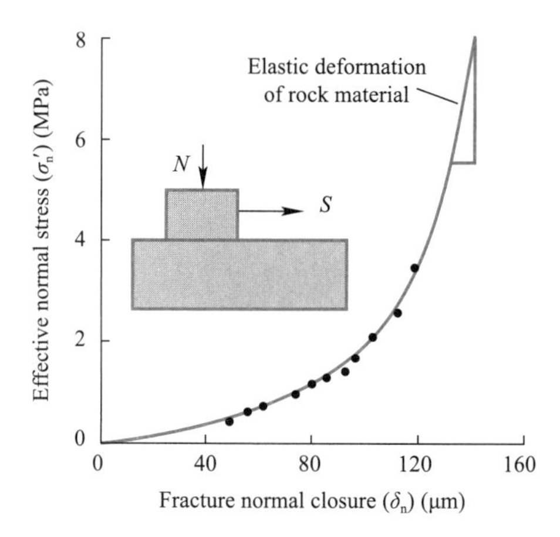

Fig. 3-20 Sliding between two contact surfaces

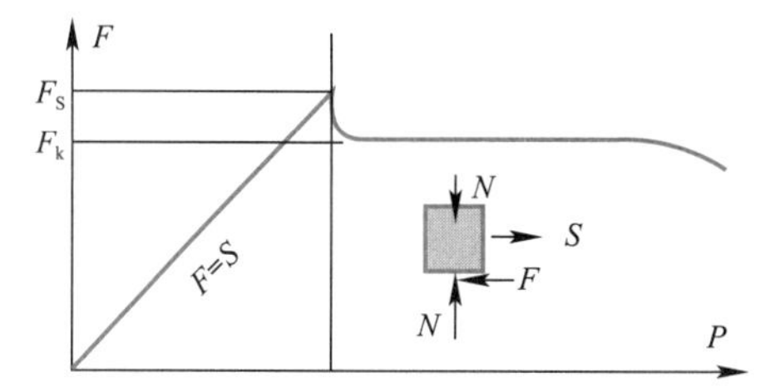

Fig. 3-21 Shearing load with displacement

engineering and rock mechanics. Conditions for sliding of rock blocks along existing joints and faults at slope or excavation opening are governed by the shear strength developed on the sliding rock discontinuities (Fig. 3-22). In slope, shear is subjected to a constant normal load generated by the weight of the blocks. In tunnel, shear is subjected to constant stiffness due to the constraints of lateral displacement.

In tests, shear stress quickly reaches a peak (peak strength). With shearing progressed, shear stress stabilizes to a residual level (residual strength), as shown in Fig. 3-23 (a). For rough joints, peak shears strength is significantly higher than the residual strength.

Peak shear strength does not follow the linear friction law. Gradient of the peak shear strength-normal stress decreases with increasing normal stress. When increasing normal stresses, the shear strength will also increase, as shown in Fig. 3-23 (b). Fig. 3-24 shows the two different direct shear tests with constant normal

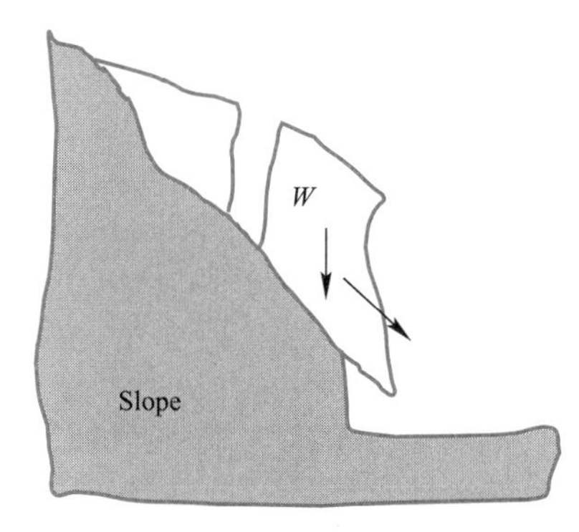

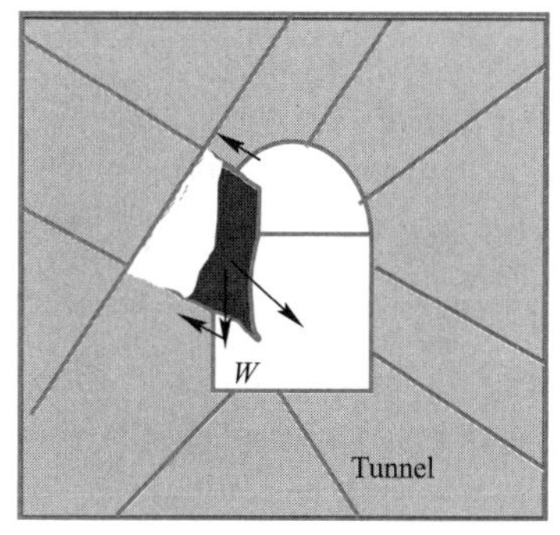

Fig. 3-22 Sliding of rock blocks along existing joints and faults at slope or excavation opening

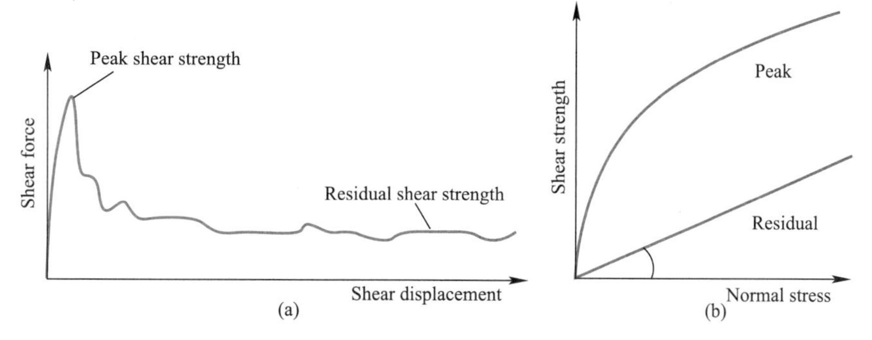

Fig. 3-23 Curve of shear force with shear displacement

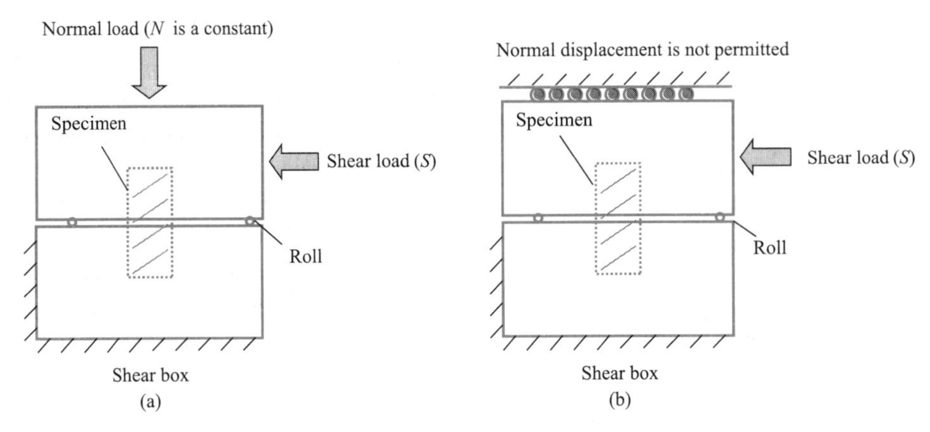

Fig. 3-24 Two different direct shear tests
(a) Constant normal load (CNL); (b) Constant normal displacement (CND)

load and constant normal displacement. At low normal stress, friction angle for rough joints can be as high as 70°, the frictional angle decreases with increasing normal stress; at high normal stress, the frictional angle approaches to φ_b.

3.3.5 Factors Affecting Rock Joint Shear Strength

1. Direction

Joint surface profile is a 3D feature while shearing is a directional activity. Surface profile along a particular direction would be different along another direction and hence gives different shear strength.

2. Matching and Mismatching

Natural joints suffer from weathering and alteration. It changes the degree of matching of joint surfaces. Mismatched joints generally have much lower shear strength than matched joints. It is also evident from cyclic shear tests. Later cycle shear tests give lower strength than the first cycle—due to joint surface damage and reduced matching.

3. Water and Water Pressure

When a joint is wet, it has generally a lower friction angle than a dry joint. Shear strength of a wet joint is calculated using the wet (and drained) friction angle. If a joint is subjected to water pressure, normal stress in the shear strength equation is the effective normal stress, i.e., total stress—water pressure.

4. Scale Effects

Scale effects refer to the phenomenon where the mechanical behavior of a rock mass, particularly jointed rock mass, can change as the scale of observation or testing is altered. In jointed rock masses, the presence of fractures, faults, and joints significantly affects their mechanical properties and behavior (Fig. 3-25). Here are some scale effects observed in jointed rock masses.

Strength and deformation characteristics: the strength and deformation characteristics of jointed rock masses can exhibit scale dependence. As the scale increases, the apparent strength of the rock mass can decrease due to increased discontinuity frequency and reduced effective stress transfer across joints. Similarly, the deformation behavior, such as the dilation or closure of joints, may vary with scale.

Shear behavior: the shear behavior of jointed rock masses can show scale effects. At small scales, the shear behavior may be controlled by individual joint properties, such as friction and cohesion. However, at larger scales, the overall behavior becomes influenced by the interaction between multiple joints, leading to complex and unpredictable responses.

Stress redistribution: in jointed rock masses, stress redistribution occurs within and around the rock mass as load is applied. This redistribution of stress can be influenced by the scale of the rock mass. As the scale increases, local stress concentrations around joints may become more significant, affecting the overall stress distribution and potentially leading to localized failure.

Fig. 3-25 Scale effect of jointed rock mass

Permeability and fluid flow: the permeability and fluid flow characteristics of jointed rock masses can display scale effects. The presence of joints provides preferential pathways for fluid flow, and the connectivity and size of joints can vary with scale. Therefore, the permeability and hydraulic conductivity of the rock mass may change as the scale of observation or testing is altered.

It's important to note that the scale effects in jointed rock mass behavior can make its mechanical response more challenging to predict accurately. Therefore, a thorough understanding of the geological structure, joint characteristics, and appropriate site investigations are essential for reliable engineering design and assessment in jointed rock masses.

3.4 Rock Mass Quality Classification

Rock mass is a matrix consisting of rock material and rock discontinuities. Properties of rock mass therefore are governed by the parameters of rock joints and rock material, as well as boundary conditions (Fig. 3-26). The behaviour of rock changes from continuous elastic for intact rock materials to discontinues running of highly fractured rock masses, depending mainly on the existence of rock joints. The prime parameters governing rock mass properties are listed in Table 3-7.

Fig. 3-26 The environmental conditions of rock engineering

3.4.1 Rock Load Factor

The concept used in this classification system is to estimate the rock load to be carried by the steel arches installed to support a tunnel (Fig. 3-27). It classifies rock mass into 9 classes, listed in Table 3-8. It provides reasonable support pressure estimates for small tunnels with diameter up to 6 m. It gives over-estimates for large tunnels with diameter above 6 m. The estimated support pressure has a wide range for squeezing and swelling rock conditions for a meaningful application.

Prime parameters governing rock mass property **Table 3-7**

Joint parameters	Material parameters	Boundary conditions
Number of joint sets	Compressive strength	Groundwater pressure and flow
Orientation	Modulus of elasticity	In situ stress
Spacing		Temperature
Aperture		
Surface roughness		
Weathering and alteration		

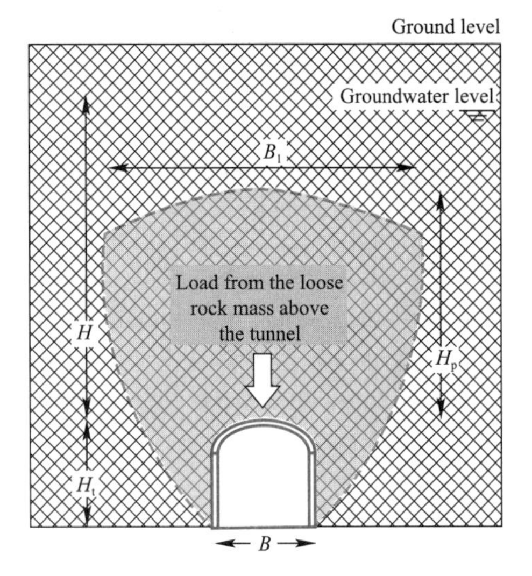

Fig. 3-27 Rock loading surrounding an excavated tunnel

The active span of a tunnel refers to the portion of the tunnel that is actively under construction or in use for vehicle or pedestrian traffic. It represents the length of the tunnel where ongoing construction activities, such as excavation, lining installation, and other necessary operations, are taking place. This section may gradually extend as construction progresses and the tunnel is completed in phases. It's important to note that the active span can vary depending on the specific construction method, tunnel design, and project requirements. In some cases, only a small section of the tunnel may be considered the active span at any given time, while in larger projects, multiple sections may be simultaneously under construction or in use. During

Different rock classes based on rock load factor **Table 3-8**

Rock class	Definition	H_P	Remark
Hard and intact	Hard and intact rock contains no joints and fractures. After excavation the rock may have popping and spalling at excavated face	0	Light lining required only if spalling or popping occurs
Hard stratified and schistose	Hard rock consists of thick strata and layers. Interface between strata is cemented. Popping and spalling at excavated face is common	0—0. 5B	Light support for protection against spalling. Load may change between layers
Massive, moderately jointed	Massive rock contains widely spaced joints and fractures. Block size is large. Joints are interlocked. Vertical walls do not require support. Spalling may occur	0—0. 25B	Light support for protection against spalling

Continued

Rock class	Definition	H_p	Remark
Moderately blocky and seamy	Rock contains moderately spaced joints. Rock is not chemically weathered and altered. Joints are not well interlocked and have small apertures. Vertical walls do not requird support. Spalling may occur	$0.25B$—$0.35(B+H_t)$	No side pressure
Very blocky and seamy	Rock is not chemically weathered, and contains closely spaced joints. Joints have large apertures and appear separated. Vertical walls need support	$(0.35$—$1.1)(B+H_t)$	Little or no side pressure
Completely crushed but chemically intact	Rock is not chemically weathered, and highly fractured with small fragments. The fragments are loose and not interlocked. Excavation face in this material needs considerable support	$1.1(B+H_t)$	Considerable side pressure. Softening effects by water at tunnel base. Use circular ribs or support rib lower end
Squeezing rock at moderate depth	Rock slowly advances into the tunnel without perceptible increase in volume. Moderate depth is considered as 150—1000 m	$(1.1$—$2.1)(B+H_t)$	Heavy side pressure. Invert struts required. Circular ribs recommended
Squeezing rock at great depth	Rock slowly advances into the tunnel without perceptible increase in volume. Great depth is considered as more than 1000 m	$(2.1$—$4.5)(B+H_t)$	Heavy side pressure. Invert struts required. Circular ribs recommended
Swelling rock	Rock volume expands (and advances into the tunnel) due to swelling of clay minerals in the rock at the presence of moisture	Up to 75 m, irrespective of B and H_t	Circular ribs required. In extreme cases use yielding support

the construction phase, safety measures and procedures are implemented to protect workers and maintain traffic flow if applicable. Once the tunnel is fully constructed, the entire length of the tunnel becomes the active span, allowing for regular operation and usage according to its intended purpose. Active span is the longer dimension of an unsupported excavation opening. Active span is governed by the rock mass quality, as shown in Fig. 3-28.

Stand-up time is the length of time that the active span can stand without any support. Stand-up time is related with rock mass quality. Rock mass classes are assigned according to the stand-up time.

3.4.2 Rock Mass Rating (RMR)

Rock mass rating (RMR) is a geotechnical classification system used to assess the engineering properties and stability of rock masses for various construction and excavation projects. It was developed by Bieniawski Z. T. in 1973 and has since been widely applied in rock engineering practice. It is an overall comprehensive index of rock mass quality combining the gover-

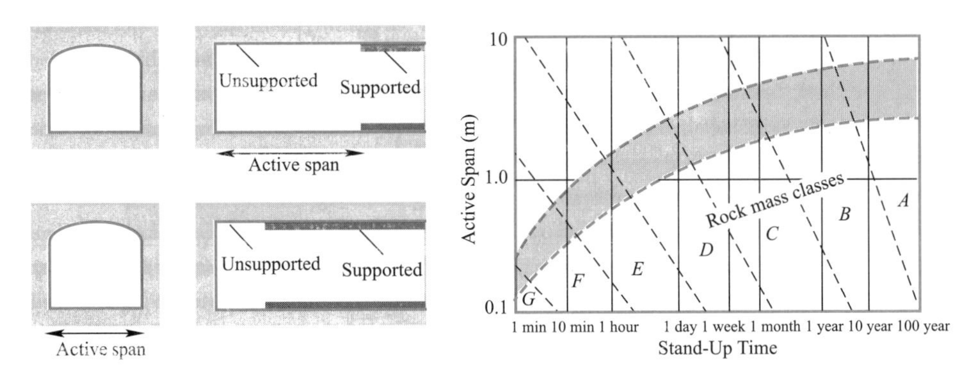

Fig. 3-28 Active span and stand-up time

ning rock parameters. RMR can be used for the analysis and the design of tunnels, mines, slopes, and foundations. By assigning numerical values to these parameters and applying specific weighting factors, RMR provides a quantitative rating ranging from 0 to 100. Higher ratings indicate better rock mass quality and greater stability, while lower ratings suggest poorer rock mass conditions and higher potential for instability.

The RMR values can be used to guide design decisions, support slope stability analysis, determine suitable excavation methods, and facilitate the selection of appropriate ground support measures for rock engineering projects. It's important to note that RMR is just one of many rock mass classification systems used in geotechnical engineering. Other classification methods, such as the geological strength index (GSI) and Q-system, may also be employed depending on the project requirements and geological conditions. Basic RMR is the sum of the rating of five parameters (Table 3-9).

(1) Strength of intact rock material: uniaxial compressive strength or point load index.

(2) RQD.

(3) Spacing of joints: average spacing of all rock discontinuities.

(4) Condition of joints: aperture, surface roughness, weathering and alteration, infilling.

(5) Groundwater conditions: inflow or water pressure.

RMR system Table 3-9

<table>
<tr><td rowspan="3">1</td><td rowspan="2">Strength of intact rock material</td><td>Point load strength index (MPa)</td><td>>10</td><td>4—10</td><td>2—4</td><td>1—2</td><td>—</td><td>—</td><td>—</td></tr>
<tr><td>Uniaxial compressive strength (MPa)</td><td>>250</td><td>100—250</td><td>50—100</td><td>25—50</td><td>5—25</td><td>1—5</td><td><1</td></tr>
<tr><td colspan="2">Rating</td><td>15</td><td>12</td><td>7</td><td>4</td><td>2</td><td>1</td><td>0</td></tr>
<tr><td rowspan="2">2</td><td>RQD (%)</td><td>90—100</td><td colspan="2">75—90</td><td colspan="2">50—75</td><td colspan="2">25—50</td><td><25</td></tr>
<tr><td>Rating</td><td>20</td><td colspan="2">17</td><td colspan="2">13</td><td colspan="2">8</td><td>3</td></tr>
<tr><td rowspan="2">3</td><td>Joint spacing (m)</td><td>>2</td><td colspan="2">0.6—2</td><td colspan="2">0.2—0.6</td><td colspan="2">0.06—0.2</td><td><0.06</td></tr>
<tr><td>Rating</td><td>20</td><td colspan="2">15</td><td colspan="2">10</td><td colspan="2">8</td><td>5</td></tr>
<tr><td rowspan="2">4</td><td>Condition of joints</td><td>Not continuous, very rough surfaces, unweathered, no separation</td><td colspan="2">Slightly rough surfaces, slightly weathered, separation <1 mm</td><td colspan="2">Slightly rough surfaces, highly weathered, separation <1 mm</td><td colspan="2">Continuous, slickensided surfaces, or gouge < 5 mm thick, or separation 1—5 mm</td><td>Continuous joints, soft gouge >5 mm thick, or separation >5 mm</td></tr>
<tr><td>Rating</td><td>30</td><td colspan="2">25</td><td colspan="2">20</td><td colspan="2">10</td><td>0</td></tr>
</table>

Continued

5	Ground-water	Inflow per 10 m tunnel length (l/min), or	none	<10	10—25	25—125	>125
		joint water pressure/ major in situ stress, or	0	0—0. 1	0. 1—0. 2	0. 2—0. 5	>0. 5
		general conditions at excavation surface	completely dry	damp	wet	dripping	flowing
	Rating		15	10	7	4	0

1. Obtain Parameter Ratings

For R1, R2, and R3, use the parameter values to get the corresponding ratings; for R5, in addition to values, flow observations are also used; R4 may require some judgments. Use the rating that makes the most appropriate matching to the descriptions.

2. Adjustment for Joint Orientation

RMR also includes adjustment for joint orientation with respect to the engineering project (Table 3-10).

Adjustment rating for joint orientation **Table 3-10**

	Very favourable	Favourable	Fair	Unfavourable	Very unfavourable
Tunnel	0	-2	-5	-10	-12
Foundation	0	-2	-7	-15	-25
Slope	0	-5	-25	-50	-60

Adjusted RMR = Basic RMR + Adjustment Rating

The relationship between unsupported span and stand-up time is shown in Fig. 3-29.

RMR ratings	81—100	61—80	41—60	21—40	<20
Rock mass class	A	B	C	D	E
Description	Very good rock	Good rock	Fair rock	Poor rock	Very poor rock
Average stand-up time	10 year for 15 m span	6 months for 8 m span	1 week for 5 m span	10 hours for 2. 5 m span	30 minutes for 0. 5 m span
Rock mass cohesion (kPa)	>400	300—400	200—300	100—200	<100
Rock mass friction angle	>45°	35°—45°	25°—35°	15°—25°	<15°

Fig. 3-29 Vnsupported span and stand-up time (one)

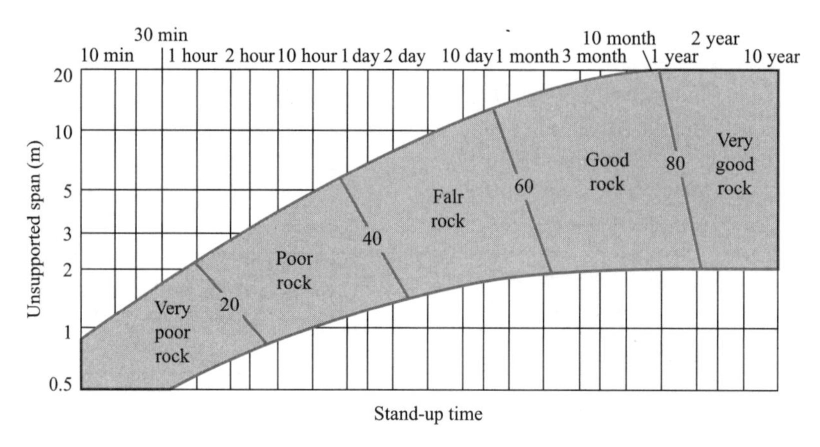

Fig. 3-29 Vnsupported span and stand-up time (two)

3. 4. 3 Q-system

The Q-system is a geotechnical classification system used for assessing the stability and support requirements of rock masses in underground excavations. It was developed by Barton, etc., in 1974 and has since been widely used in tunneling, mining, and other underground engineering projects. The Q-system takes into account various geological and geotechnical parameters to classify rock masses and estimate their behavior under different conditions(Eq. 3-4).

- Uniaxial compressive strength (UCS): this parameter represents the strength of intact rock and is an important factor in assessing the stability of a rock mass.
- Rock quality designation (RQD): RQD provides information about the continuity and quality of the rock core samples and reflects the degree of jointing and fracturing within the rock mass.
- Spacing of discontinuities: the spacing between joints or fractures within the rock mass indicates the frequency and width of potential failure planes. Closer discontinuity spacing generally implies more fractured and weaker rock mass conditions.
- Condition of discontinuities: the condition of joints or fractures, including roughness, alteration, weathering, and infilling, isconsidered in evaluating the shear strength and stability of the rock mass.
- Groundwater conditions: the presence of water and its effect on rock mass stability are taken into account. Factors such as percolation, pore pressure, and groundwater inflow are considered.
- Orientation of discontinuities: the orientation and inclination of joints or discontinuities affect the structural behavior and stability of the rock mass.

Based on these parameters, the Q-system assigns values and grades to different parameters and calculates a Q-value, which ranges from 0 to over 100. A higher Q-value indicates a more competent and stable rock mass, while a lower value suggests poorer rock mass conditions requiring additional support measures. The Q-system provides guidance on the required rock support and excavation methods based on the Q-value (Table 3-11). It helps engineers make informed decisions regarding the design of support systems, such as bolts, shotcrete, steel sets, and rock anchors, to ensure safe underground excavations. It's important to note that the Q-system is one of several rock mass classification systems used in geotechnical engineering. The choice of a classification system depends on project-specific requirements, geological conditions, and local experience. Q-system is developed for rock tunnel support design, based on extensive engineering projects. It also recognizes that the rock mass quality is influenced by a number of parameters. The Q-system is given as:

$$Q = \underbrace{(RQD/J_n)}_{\text{Block size}} \underbrace{(J_r/J_a)}_{\text{Inter-block strength}} \underbrace{(J_w/SRF)}_{\text{Active stress}} \tag{3-4}$$

Where,

RQD —— rock quality designation;

J_n —— joint set number;

J_r —— joint roughness number;

J_a —— joint alteration number indicating the degree of weathering, alteration and filling;

J_w —— joint water reduction factor;

SRF —— stress reduction factor.

Q-value and rock mass classes **Table 3-11**

Q-value	Class	Rock mass quality
400—1000	A	Exceptionally good
100—400	A	Exeremely good
40—100	A	Very good
10—40	B	Good
4—10	C	Fair
1—4	D	Poor
0. 1—1	E	Very poor
0. 01—0. 1	F	Extremely poor
0. 001—0. 01	G	Exceptionally poor

Parameters for RQD and joint set are based on values. For other parameters, judgment is required, by making the most appropriate matching to the descriptions in the Q-table, and then assign the parameters (Table 3-12). Each parameter may have different conditions.

The values of different parameter **Table 3-12**

1. Rock quality designation		RQD
A	Very poor	0—25
B	Poor	25—50
C	Fair	50—75
D	Good	75—90
E	Excellent	90—100

Note: (1) Where RQD is reported or measured as less than or equal to 10 (including 0), a nominal value of 10 is used to evaluate Q

(2) RQD interval of 5, i. e., 100, 95, 90, etc., are sufficientil accurate

2. Joint set number		J_n
A	Massive, no or few joints	0. 5—1
B	One joint set	2
C	One joint set plus random joints	3
D	Two joint set	4
E	Two joint set plus random joints	6
F	Three joint set	9

Continued

2. Joint set number			J_n
G	Three joint set plus random joints		12
H	Four or more joint sets, heavily jointed		15
J	Crushed rock, earthlike		20
Note: (1) For intersections, use $(3.0J_n)$ (2) For portals, use $(2.0J_n)$			
3. Joint roughness number			J_r
Rock-wall contact, and Rock wall contact before 10 cm shear			
A	Discontinuous joints		4
B	Rough or irregular, undulating		3
C	Smooth, undulating		2
D	Slickensided, undulating		1.5
E	Rough or irregular, planar		1.5
F	Smooth, planar		1.0
G	Slickensided, planar		0.5
Note: Descriptions refer to small and intermediate scale features, in that order			
No rock-wall contact when sheared			
H	Zone containing clay minerals thick enough to prevent rock-wall contact		1.0
J	Sandy, gravelly or crushed zone thick enough to prevent rock-wall contact		1.0
Note: (1) Add 1.0 if the mean spacing of the relevant joint set≥3m (2) $J_r=0.5$ can be used for planar slickensided joints having lineations, provided the lineations are oriented for minimum strength (3) J_r and J_a classification is applied to the joint set or discontinuity that is least favourable for stability both from the point of view of orientation and shear resistance			
4. Joint alteration number		ϕ_r (approximately)	J_a
Rock-wall contact (no mineral filings, only coatings)			
A	Tight healed, hard, non-softening, impermeable filling, i. e., quartz or epidote	—	0.75
B	Unaltered joint walls, surface staining only	25—35°	1.0
C	Slightly altered joint walls. Non-softening mineral coating, sandy particles, clay-free disintegrated rock, etc.	25—30°	2.0
D	Sitty-or sandy-clay coatings, small clay fraction (non-softening)	20—25°	3.0
E	Softening or low friction mineral coatings, i. e., kaolinite or mica. Also chlorite, talc, gypsum, graphite, etc., and small quantities of swelling clays	8—16°	4.0
Rock wall contact before 10 cm shear (thin mineral fillings)			
F	Sandy particles, clay-free disintegrated rock, etc.	25—30°	4.0

Continued

4. Joint alteration number		ϕ_r(approximately)	J_a
G	Strongly over-consolidated non-softening clay mineral fillings (continuous, but<5 mm thickness)	16—24°	6.0
H	Medium or low over-consolidated softening clay mineral fillings (continuous, but<5 mm thickness)	12—16°	8.0
J	Swelling-clay fillings, i.e., montmorillonite (continuous, but<5 mm thickness). Value of Ja depends on percent of swelling clay size particles, and access to water, etc.	6—12°	8—12
No rock-wall contact when sheared (thick mineral filings)			
K, L, M	Zones or bands of disintegrated or crushed rock and clay (see G, H, J for description of clay condition)	6—24°	6,8,8—12
N	Zones or bands of silty-or sandy-clay, small clay fraction (non-softening)	—	5
O, P, R	Thick, continuous zones or bands of clay (see G, H, J for clay condition description)	6—24°	10,13, 13—20
5. Joint water reduction factor		Water pressure	J_w
A	Dry excavation or minor inflow, i.e., <5 l/min locally	<1 kg/cm^2	1.0
B	Medium inflow or pressure, occasional outwash of joint fillings	1—2.5	0.66
C	Large inflow or high pressure in competent rock with unfilled joints	2.5—10	0.5
D	Large inflow or high pressure, considerable outwash of joint fillings	2.5—10	0.33
E	Exceptionally high inflow or water pressure at blasting, decaying with time	>10	0.2—0.1
F	Exceptionally high inflow or water pressure continuing without noticeable decay	>10 kg/cm^2	0.1—0.05
Note: (1) Factors C to F are crude estimates. Increase J_w if drainage measures are installed (2) Special problems caused by ice formation are not considered			

6. Stress reduction factor		SRF
Weakness zones intersecting excavation, which may cause loosening of rock mass when tunnel is excavated		
A	Multiple occurrences of weakness zones containing clay or chemically disintegrated rock, very loose surrounding rock (any depth)	10

Continued

6. Stress reduction factor		SRF
B	Single weakness zone containing clay or chemically disintegrated rock (depth of excavation≤50 m)	5
C	Single weakness zone containing clay or chemically disintegrated rock (depth of excavation>50 m)	2.5
D	Multiple shear zones in competent rock (clay-free) (depth of excavation≤50 m)	7.5
E	Single shear zone in competent rock (clay-free) (depth of excavation≤50 m)	5
F	Single shear zone in competent rock (clay-free) (depth of excavation>50 m)	2.5
G	Loose, open joint, heavily jointed (any depth)	5

Note: Reduce SRF value by 25%—50% if the relevant shear zones only influence but not intersect the excavation

Competent rock, rock stress problems		σ_c/σ_1	σ_θ/σ_c	SRF
H	Low stress, near surface, open joints	>200	<0.01	2.5
J	Medium stress, favourable stress condition	200—10	0.01—0.03	1
K	High stress, very tight structure. Usually favourable to stability, may be unfavourable to wall stability	10—5	0.3—0.4	0.5—2
L	Moderate slabbing after>1 hour in massive rock	5—3	0.5—0.65	5—50
M	Slabbing and rock burst after a few minutes in massive rock	3—2	0.65—1	50—200
N	Heavy rock burst (strain-burst) and immediate dynamic deformation in massive rock	<2	>1	200—400

Note: (1) For strongly anisotropic virgin stress field (if measured): when $5\leq\sigma_1/\sigma_3\leq10$, reduce σ_c to 0.75 σ_c; when $\sigma_1/\sigma_3>10$, reduce σ_c to 0.5 σ_c; where σ_c is unconfined compressive strength, σ_1 and σ_3 are major and minor principal stresses, and σ_θ is maximum tangential stress (estimated from elastic theory)

(2) Few cases records available wher depth of crown below surface is less than span width. Suggest SRF increase from 2.5 to 5 for such cases (see H)

$$RMR = 9\ln Q + (44\pm18)$$

Q and RMR should alwaysbe assessed directly. It is not recommended to use one to derive the other.

3.4.4 Geological Strength Index (GSI) System

The Hoek-Brown failure criterion for rock masses is widely accepted and has been applied in a large number of projects around the world. While, in general, it has been found to be satisfactory, there are some uncertainties and inaccuracies that have made the criterion inconvenient to apply and to incorporate into numerical models and limit equilibrium programs Relevant studies are shown in the Table 3-14.

The famous rock mass classification scheme known as the geological strength index (GSI) was introduced by Hoek E. (2002), as shown in Table 3-13. The GSI is developed to overcome some of the deficiencies that have been

Schematic of geological strength index (GSI) system Table 3-13

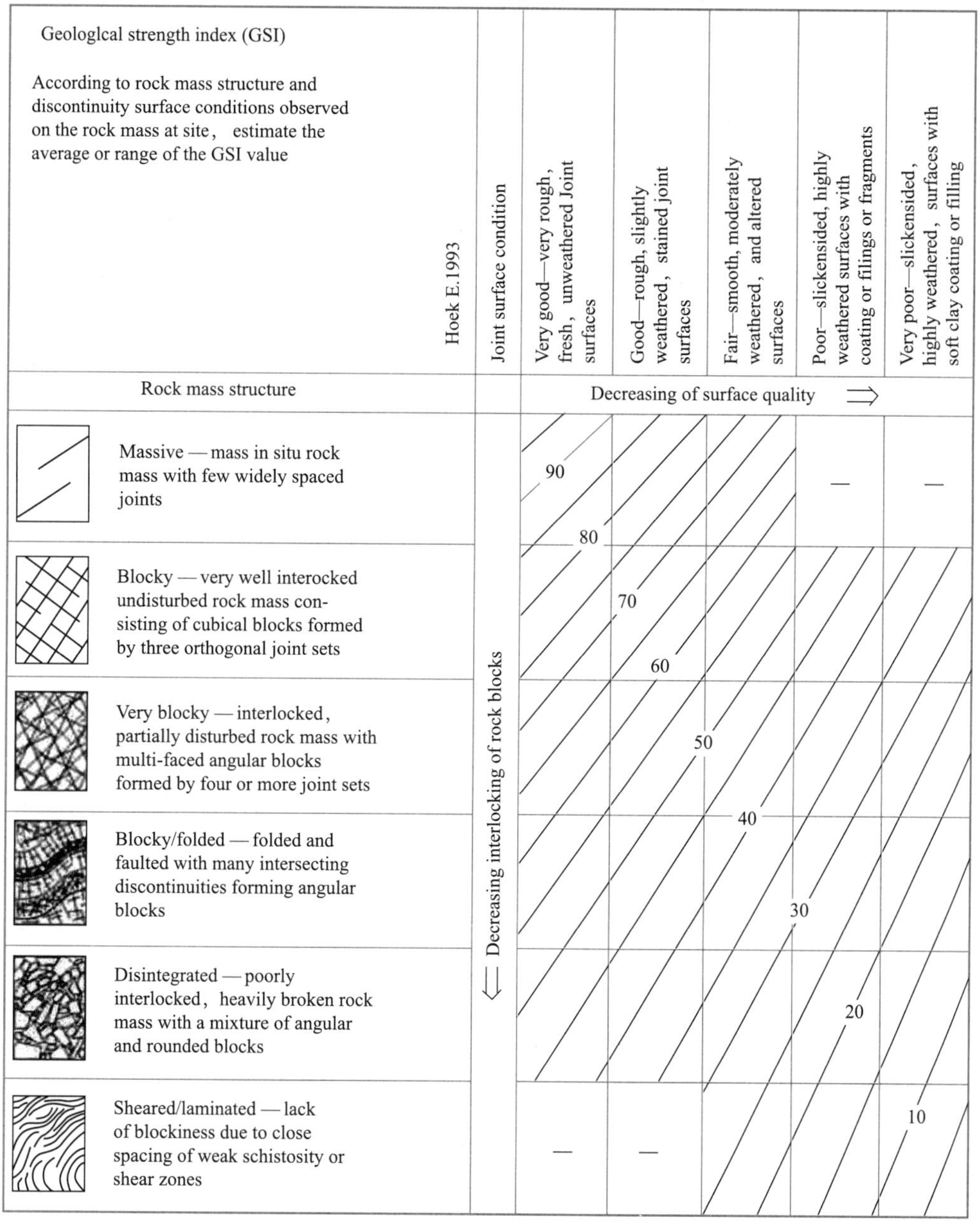

Geologlcal strength index (GSI) According to rock mass structure and discontinuity surface conditions observed on the rock mass at site, estimate the average or range of the GSI value	Hoek E.1993	Joint surface condition	Very good—very rough, fresh, unweathered Joint surfaces	Good—rough, slightly weathered, stained joint surfaces	Fair—smooth, moderately weathered, and altered surfaces	Poor—slickensided, highly weathered surfaces with coating or filings or fragments	Very poor—slickensided, highly weathered, surfaces with soft clay coating or filling
Rock mass structure			Decreasing of surface quality ⟹				
Massive — mass in situ rock mass with few widely spaced joints		⟸ Decreasing interlocking of rock blocks	90 80			—	—
Blocky — very well interocked undisturbed rock mass con-sisting of cubical blocks formed by three orthogonal joint sets				70 60			
Very blocky — interlocked, partially disturbed rock mass with multi-faced angular blocks formed by four or more joint sets					50 40		
Blocky/folded — folded and faulted with many intersecting discontinuities forming angular blocks						30	
Disintegrated — poorly interlocked, heavily broken rock mass with a mixture of angular and rounded blocks						20	
Sheared/laminated — lack of blockiness due to close spacing of weak schistosity or shear zones			—	—			10

identified in using the RMR scheme with the rock mass strength criterion. The GSI is developed specifically as a method of accounting for those properties of a discontinuous or jointed rock mass which influence its strength and deformability. GSI is aimed to estimate the reduction in rock mass strength for different geological conditions. The system gives a GSI value estimated from rock mass structure and rock discontinuity surface condition. The direct application of GSI value is to estimate the parameters in the Hoek-Brown strength criterion for rock masses.

$$RMR = 9\ln Q + (44 \pm 18) \tag{3-5}$$

$$RMR = 13.5\log Q + 43 \tag{3-6}$$

$$GSI = RMR - 5 \quad (\text{for } GSI > 25) \tag{3-7}$$

Relevant Studies **Table 3-14**

Publication	Coverage	Equations
Hoek & Brown 1980	Original criterion for heavily jointed rock masses with no fines. Mohr envelope was obtained by statistical curve fitting to a number of (σ'_n, τ) pairs calculated by the method published by Balmer σ'_1, σ'_3 are major and minor effective principal stresses at failure, respectively σ_t is the tensile strength of the rock mass m, s are material constants σ'_n, τ are effective normal and shear stresses, respeetively	$\sigma'_1=\sigma'_3+\sigma_{ci}\sqrt{m\sigma'_3/\sigma_{ci}+s}$ $\sigma_t=\frac{\sigma_{ci}}{2}(m-\sqrt{m^2+4s})$ $\tau=A\sigma_{ci}[(\sigma'_n-\sigma_t)/\sigma_{ci}]^B$ $\sigma'_n=\sigma'_3+[(\sigma'_1-\sigma'_3)/(1+\partial\sigma'_1/\partial\sigma'_3)]$ $\tau=(\sigma'_1-\sigma'_3)\sqrt{\partial\sigma'_1/\partial\sigma'_3}$ $\partial\sigma'_1/\partial\sigma'_3=m\sigma_{ci}/2(\sigma'_1-\sigma'_3)$
Hoek 1983	Original criterion for heavily jointed rock masses with no fines with a discussion on anisotropic failure and an exact solution for the Mohr envelope by Dr Bray J. W.	$\sigma'_1=\sigma'_3+\sigma_{ci}\sqrt{m\sigma'_3/\sigma_{ci}+s}$ $\tau=(\cot\varphi'_i-\cos\varphi'_i)m\sigma_{ci}/8$ $\varphi'_i=\arctan(1/\sqrt{4h\cos^2\theta-1})$ $\theta=[90+\arctan(1/\sqrt{h^3-1})]/3$ $h=1+[16(m\sigma'_n+s\sigma_{ci})/(3m^2\sigma_{ci})]$
Hoek & Brown 1988	As for Hoek E. (1983) but with the addition of relationships between constants m and s and a modified form of RMR (Bieniawski) in which the groundwater rating was assigned a fixed value of 10 and the adjustment for joint orientation was set at 0. Also a distinction between disturbed and undisturbed rock masses was introduced together with means of estimating deformation modulus E (after Serafim and Pereira)	Disturbed rock masses: $m_b/m_i=e^{(RMR-100)/14}$ $s=e^{(RMR-100)/6}$ Undisturbed or interlocking rock masses: $m_b/m_i=e^{(RMR-100)/28}$ $s=e^{(RMR-100)/9}$ $E=10^{(RMR-10)/40}$ m_b, m_i are for broken and intact rock, respectively

Continued

Publication	Coverage	Equations
Hoek, Wood & Shah 1992	Modified criterion to account for the fact the heavily jointed rock masses have zero tensile strength. Balmer's technique for calculating shear and normal stress pairs was utilised	$\sigma'_1=\sigma'_3+\sigma_{ci}(m_b\sigma'_3/\sigma_{ci})^{\alpha}$ $\sigma'_n=\sigma'_3+[(\sigma'_1-\sigma'_3)/(1+\partial\sigma'_1/\partial\sigma'_3)]$ $\tau=(\sigma'_1-\sigma'_3)\sqrt{\partial\sigma'_1/\partial\sigma'_3}$ $\partial\sigma'_1/\partial\sigma'_3=1+\alpha m_b^{\alpha}(\sigma'_3/\sigma_{ci})^{-1}$
Hoek 1994 Hoek, Kaiser & Bawden 1995	Introduction of the Generalised Hoek-Brown criterion, incorporating both the original criterion for fair to very poor quality rock masses and the modified criterion for very poor quality rock masses with increasing fines content. The Geological Strength Index GSI was introduced to overcome the deficiencies in Bieniawski's RMR for very poor quality rock masses. The distinction between disturbed and undisturbed rock masses was dropped on the basis that disturbance is generally induced by engineering activities and should be allowed for by downgrading the value of GSI	$\sigma'_1=\sigma'_3+\sigma_c(m\sigma'_3/\sigma_{ci}+s)^{\alpha}$ for GSI > 25 $m_b/m_i=e^{(GSI-100)/28}$ $s=e^{(GSI-100)/9}$ $a=0.5$ for GSI < 25 $s=0$ $a=0.65-GSI/200$

Continued

Publication	Coverage	Equations
Hoek, Carranza-Torres, & Corkum 2002	An "exact" method for calculating the cohesive strength and angle of friction is presented and appropriate stress ranges for tunnels and slopes are given. A rock mass damage criterion is introduced to account for strength reduction due to stress relaxation and blast damage in slope stability and foundation problems. The "switch" at GSI = 25 for the coefficients s and a, is eliminated, which gives smooth continuous transitions for the entire range of GSI values	$\sigma'_1=\sigma'_3+\sigma_{ci}\left(m_b\frac{\sigma'_3}{\sigma_{ci}}+s\right)^a$ $m_b=m_i e^{\frac{GSI-100}{28-14D}}$ $s=e^{\frac{GSI-100}{9-3D}}$ $a=\frac{1}{2}+\frac{1}{6}(e^{-GIS/15}-e^{-20/3})$ $\sigma_c=\sigma_{ci}-s^a$ $\sigma_t=-\frac{s\sigma_{ci}}{m_b}$ $\sigma'_n=\frac{\sigma'_1+\sigma'_3}{2}-\frac{\sigma'_1-\sigma'_3}{2}\cdot\frac{d\sigma'_1/d\sigma'_3-1}{d\sigma'_1/d\sigma'_3+1}$ $\tau=(\sigma'_1-\sigma'_3)\frac{\sqrt{d\sigma'_1/d\sigma'_3}}{d\sigma'_1/d\sigma'_3+1}$ $d\sigma'_1/d\sigma'_3=1+am_b(m_b\sigma'_3/\sigma_{ci}+s)^{a-1}$ $E_m=\left(1-\frac{D}{2}\right)\sqrt{\frac{\sigma_{ci}}{100}}\cdot 10^{(GSI-10)/40}(\text{sigci}\leqslant 100)$ $E_m=\left(1-\frac{D}{2}\right)\cdot 10^{(GSI-10)/40}(\text{sigci}>100)$ $\phi'=\sin^{-1}\left[\frac{6am_b(s+m_b\sigma'_{3n})^{a-1}}{2(1+a)(2+a)+6am_b(s+m_b\sigma'_{3n})^{a-1}}\right]$ $c'=\frac{\sigma_{ci}[(1+2a)s+(1-a)m_b\sigma'_{3n}](s+m_b\sigma'_{3n})^{a-1}}{(1+a)(2+a)\sqrt{1+[6am_b(s+m_b\sigma'_{3n})^{a-1}]/[(1+a)(2+a)]}}$ $\sigma_{3n}=\sigma'_{3max}/\sigma_{ci}$ $\sigma'_{cm}=\frac{2c'\cos\phi'}{1-\sin\phi'}$ $\sigma'_{cm}=\sigma_{ci}\cdot\frac{[m_b+4s-a(m_b-8s)](m_b/4+s)^{a-1}}{2(1+a)(2+a)}$ $\frac{\sigma'_{3max}}{\sigma'_{cm}}=0.47\left(\frac{\sigma'_{cm}}{\gamma H}\right)^{-0.94}$ (Tunnels) $\frac{\sigma'_{3max}}{\sigma'_{cm}}=0.72\left(\frac{\sigma'_{cm}}{\gamma H}\right)^{-0.91}$ (Slopes)

References

[1] Hoek E., Carranza-Torres C., Corkum B.. Hoek-Brown Failure Criterion—2002 Edition [C]//Proceedings NARMS-TAC Conference Toronto, 2002: 267-273.

[2] Brady B. H. G., Brown E. T.. Rock Mechanics for Underground Mining [M]. 3rd ed. Boston: Kluwer Academic Publishers, 2004.

Chapter 4
The Physical Properties of Soil

4.1 Introduction

Soil is a crucial component of the Earth's ecosystem and plays a vital role in supporting life. It is a complex mixture of minerals, organic matter, water, air, and living organisms that exists on the Earth's surface. Soil comprises mineral particles such as sand, silt, and clay, which vary in size and provide different properties to the soil. Organic matter, derived from decomposed plant and animal material, is another integral component. Soil formation occurs through weathering processes that break down rocks and minerals over long periods. Factors such as climate, topography, parent material, organisms, and time influence soil formation.

Soil has distinct layers or horizons that develop vertically. These include the topsoil (rich in organic matter and nutrients), subsoil (accumulation of clay and other minerals), and various underlying layers. Soil has physical characteristics that impact its behavior and functionality. This includes texture (proportions of sand, silt, and clay), structure (arrangement of soil particles), porosity (pore spaces for air and water movement), and permeability (ability to allow water to pass through). Soil chemistry affects nutrient availability and pH levels. It involves elements like nitrogen, phosphorus, potassium, and others essential for plant growth. The soil's pH indicates its acidity or alkalinity, which influences nutrient availability to plants. Soil is teeming with diverse organisms, including bacteria, fungi, insects, worms, and small mammals. These organisms contribute to decomposition, nutrient cycling, soil aeration, and the overall health of the ecosystem. Soils fulfill critical ecological functions. They support plant growth by providing anchorage, nutrients, and water storage. Soils also regulate the flow and purification of water, store carbon, influence climate patterns, and provide habitats for countless organisms.

Understanding soil is important in various fields such as agriculture, environmental science, geology, and ecology. Proper soil management is essential for sustainable land use practices, conservation, and ensuring food security.

4.2 Origin of Soil and Grain Size

4.2.1 Introduction

The origin of soil can be attributed to a combination of factors, including geological processes, weathering, and the activities of living organisms. Here is an introduction to the origin of soil and the concept of grain size.

(1) Geological processes: soil originates from the breakdown and weathering of rocks and minerals over long periods due to physical, chemical, and biological processes. These processes include the action of wind, water, ice, and changes in temperature, which contribute to the physical disintegration of rock materials.

(2) Weathering: weathering refers to the gradual breakdown of rocks into smaller particles through processes like mechanical weathering and chemical weathering. Mechanical weathering involves physical forces such as frost action, wind abrasion, and the growth of plant roots that cause rocks to fragment into smaller pieces. Chemical weathering occurs when rocks react

with water, gases, or other substances, leading to chemical changes and the formation of new minerals.

(3) Organic matter: the presence of organic matter derived from the decomposition of dead plant and animal material contributes to soil formation. Organic matter enriches the soil by providing nutrients, improving soil structure, enhancing moisture retention, and supporting the growth of beneficial soil microorganisms.

(4) Erosion and deposition: erosion plays a role in soil formation by removing weathered materials from their original location and depositing them elsewhere. Erosion can occur through water (rivers, streams), wind, glaciers, or gravity. When these eroded materials settle in a new location, they contribute to the formation of soil layers.

Grain size refers to the particle size distribution within a soil sample and plays a significant role in determining various soil properties and characteristics. Grain size is generally classified into four categories (Fig. 4-1).

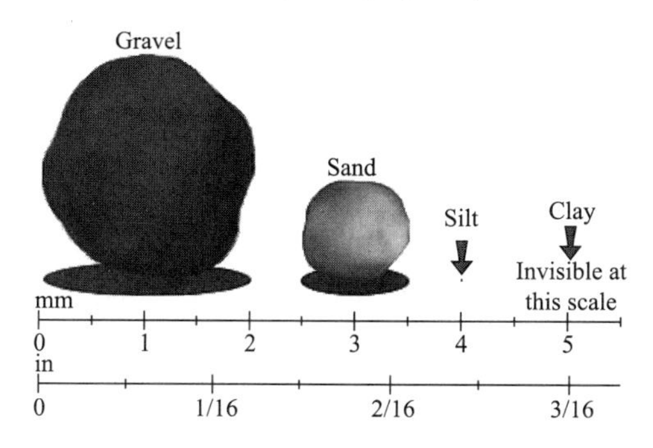

Fig. 4-1 The size of gravel, sand, silt and clay (Suryakanta, 2014)

(1) Gravel: gravel is a loose aggregation of small rock fragments that range in size from 2 to 75 mm in diameter. It is commonly used as a construction aggregate and can be found in various sizes, shapes, and colors. When using gravel, it's important to consider local regulations, as some areas may have specific requirements regarding the use of gravel for driveways or other purposes. Consulting with local professionals or suppliers will ensure you select the appropriate type and size of gravel for your specific project needs.

(2) Sand: sand particles are the largest among the three categories, varying in size from 0.05 to 2 mm in diameter. Sandy soils have larger pore spaces, allowing for good drainage but lower water-holding capacity. They tend to be more coarse-textured and have lower fertility.

(3) Silt: silt particles are smaller than sand, ranging from 0.002 to 0.05 mm in diameter. Silty soils have moderate water-holding capacity and are often fertile due to their ability to retain nutrients. They have a smooth, floury texture and can feel slippery when wet.

(4) Clay: clay particles are the smallest, less than 0.002 mm in diameter. Clayey soils have high water-holding capacity but poor drainage. They tend to be sticky when wet and hard when dry. Clay soils can be nutrient-rich but challenging to work with due to their cohesive nature.

Soil with a balanced mixture of sand, silt, and clay particles is known as loam soil, which is generally regarded as ideal for plant growth due to its good drainage, water retention, and nutrient availability. Understanding the origin of soil and grain size helps in assessing soil suitability for various purposes such as agriculture, construction, and landscaping.

4.2.2 Rock Cycle and the Origin of Soil

The rock cycle is a continuous process that describes the transformation of rocks from one form to another over time. Itinvolves various geological processes and plays a crucial role in the origin of soil. Here's an overview of the rock cycle and its relationship to the formation of soil.

(1) Igneous rocks: the rock cycle begins with the formation of igneous rocks through the solidification of molten magma or lava. This process occurs deep within the Earth's crust or on the surface during volcanic activity. Examples of igneous rocks include granite, basalt, and obsidian.

(2) Weathering: over time, igneous rocks

are exposed to weathering forces such as temperature changes, wind, water, and ice. These processes gradually break down the rocks into smaller particles through physical and chemical weathering. Physical weathering involves the mechanical disintegration of rocks into fragments, while chemical weathering involves the alteration of the composition of minerals within the rocks.

(3) Sedimentation: the weathered rock particles, called sediment, are transported by wind, water, or ice and eventually settle in bodies of water or on land surfaces. This process is known as sedimentation. Sediments may accumulate in lakes, rivers, oceans, and other environments.

(4) Sedimentary rocks: through compaction and cementation, the accumulated sediment undergoes lithification, transforming it into sedimentary rocks. Compaction occurs when the weight of overlying sediments compresses the lower layers. Cementation happens when minerals dissolved in groundwater bind the sediment particles together. Common examples of sedimentary rocks are sandstone, limestone, and shale.

(5) Uplift and erosion: geological processes, such as tectonic movements and erosion, expose buried sedimentary rocks to the Earth's surface. Uplift, caused by tectonic forces, brings deeper rock layers closer to the surface. Erosion, driven by wind, water, or ice, erodes the sedimentary rocks, breaking them into smaller pieces.

(6) Transport and deposition: eroded rock fragments are transported by rivers, streams, glaciers, or wind to new locations. The particles settle and accumulate in different environments, such as river deltas, ocean basins, or on land surfaces. Sediments deposited in these locations form new layers.

(7) Lithification and metamorphism: as new layers of sediments accumulate, the weight and pressure from overlying sediments can cause compaction and lithification, forming new sedimentary rocks. Through further geological processes like heat and pressure, sedimentary rocks can undergo metamorphism, transforming into new types of rocks, such as metamorphic rocks like marble, slate, or gneiss.

(8) Weathering and soil formation: once exposed at the Earth's surface, the newly formed rocks undergo weathering again, leading to the production of smaller particles. Organic matter accumulates from decaying plants and animals.

The transportation of weathering products is an important process in the formation and distribution of soil. When rocks undergo weathering, they break down into smaller particles that can be transported by various agents. Here are some common mechanisms of transportation for weathering products:

(1) Water erosion: streams, rivers, and rainfall can carry away weathered rock fragments and sediments. Water erosion can occur through sheet erosion (even removal of a thin layer of soil), rill erosion (small channels formed on slopes), gully erosion (deeper and wider channels), or via river systems.

(2) Wind erosion: when strong winds blow over exposed soil surfaces, they can pick up loose particles and transport them over long distances. This process is often seen in arid and semi-arid regions with sparse vegetation cover.

(3) Glacial transport: glaciers, massive bodies of ice, have the ability to transport and rework rock debris as they move. As glaciers advance and retreat, they can carry large amounts of sediment, including boulders, sand, and clay, depositing them when the ice melts.

(4) Gravity-driven processes: gravity plays a significant role in transporting rock fragments downslope. Mass wasting events, such as landslides, rockslides, and soil creep, involve the movement of weathered materials under the influence of gravity.

(5) Biological agents: living organisms, such as animals and plants, can contribute to the transportation of weathering products. For example, burrowing animals can relocate soil particles, while plants with deep roots can help bring nutrients and moisture from deeper layers to the

surface.

These transportation processes redistribute weathered rock fragments and sediments to new locations, where they may accumulate and contribute to the formation of new soil layers. Over time, the deposited sediment undergoes further weathering and mixing with organic matter, leading to soil development. It is important to note that transportation is not a standalone process, but rather an integral part of the rock cycle and soil formation. The transportation of weathering products connects various stages of the cycle, including weathering, erosion, deposition, and ultimately the formation of new rocks or soil.

4.2.3 Soil Particle Size

Soil particle size refers to the classification and distribution of mineral particles in the soil. It is an important characteristic that influences various soil properties, including drainage, water-holding capacity, aeration, and nutrient availability. Soil particles are classified into three main categories based on their size. As mentioned above, sand particles have the largest particle size among the three categories. They range in size from 0.05 to 2 mm. Sandy soils have a gritty texture and tend to drain quickly due to the large spaces between particles. Although sand soils provide good aeration, they have lower water-holding capacity and often require more frequent irrigation and nutrient management. Silt particles are smaller than sand particles and range in size from 0.002 to 0.05 mm. Silt soils have a smooth or floury texture and possess intermediate properties between sand and clay. They hold more water and nutrients than sandy soils and exhibit moderate drainage and fertility. However, silt soils can become easily compacted when wet, affecting root growth and water infiltration. Clay particles are the smallest among the three categories, with a size less than 0.002 mm. Clay soils have a sticky and plastic texture when wet and become hard and compacted when dry. Due to their small particle size, clay soils have high water and nutrient-holding capacities, but can be poorly drained and prone to waterlogging. Clay soils also tend to have higher nutrient retention and cation exchange capacity, making them potentially fertile.

The relative proportions of sand, silt, and clay in a soil determine its texture and influence its overall characteristics. Soils with a balanced mixture of these particle sizes, known as loam soils, tend to have excellent water-holding capacity, drainage, aeration, and fertility. Soil scientists use various methods to determine the particle size distribution of a soil sample, such as sedimentation, sieving, and laser diffraction. This information helps in understanding soil behavior, determining soil classification, and making appropriate management decisions for agriculture, construction, and other applications.

1. Grain Size

Grain size refers to the average diameter of individual grains or particles in a material, such as sediment, soil, or rock. It is an important physical characteristic that can provide insights into the origin, transport, and depositional environment of the material. The grain size of a substance can be determined through visual examination or by using various measurement techniques. In sedimentology and geology, grain size is often classified using several standard scales.

(1) Wentworth scale: the Wentworth scale, also known as the Udden-Wentworth scale, classifies sediment particle sizes based on their diameter range. It categorizes grain sizes from largest to smallest as follows.

- Boulder (>256 mm)
- Cobble (64—256 mm)
- Pebble (4—64 mm)
- Granule (2—4 mm)
- Sand (0.0625—2 mm)
- Silt (0.0039—0.0625 mm)
- Clay (<0.0039 mm)

(2) φ scale: The φ scale represents grain sizes on a logarithmic scale, which allows for easier mathematical calculations. The φ scale is

defined as:

$$\varphi=-\log^2 d$$

Where,

d —— the grain diameter (mm).

On the φ scale, each whole number increase corresponds to a doubling of the grain size. For example, a grain with a φ value of 3 is twice as large as a grain with a φ value of 2. The analysis of grain size provides valuable information about sedimentary processes, such as erosion, transportation, and deposition. In addition to grain size, other characteristics like sorting (distribution of grain sizes), shape, and roundness of grains can be analyzed to gain further insights into sedimentary environments and processes. It's worth noting that grain size analysis is applicable to various materials beyond sediments, including granular soil, crushed rock, and mineral samples. The determination of grain size has practical implications in fields such as geotechnical engineering, construction, geology, and sedimentology, aiding in understanding material behavior and making informed decisions related to soil stability, filtration, permeability, and more.

2. Grain Size Diagram (Arnold, 2018)

A grain size diagram, also known as a grain size distribution curve or cumulative frequency curve, is a graphical representation of the relative abundance of different grain sizes in a sediment or rock sample. It illustrates the percentage or cumulative percentage of grains within specific size ranges. Typically, the x axis of the diagram represents the logarithmic scale of grain size, often measured in φ units. The y axis represents the percentage or cumulative percentage of grains. The curve on the graph shows the distribution of grain sizes, with larger grains on the left side and smaller grains on the right side. Each point on the graph indicates the proportion of grains falling within a specific size range. For example, if the curve reaches 80% at $\varphi=2$, it means that 80% of the grains in the sample have a size equal to or smaller than $\varphi=2$. The shape of the curve provides insights into sediment characteristics. A relatively steep curve suggests a narrow range of grain sizes, indicating well-sorted sediments, while a flatter curve suggests a wider range of grain sizes and poor sorting. Additionally, the position of the curve along the x axis indicates the dominant grain size present, such as sand-dominated, silt-dominated, or clay-dominated sediments. Grain size diagrams are valuable tools for interpreting sedimentary environments, distinguishing between different depositional processes, and understanding the geological history of an area. They help scientists and engineers analyze sediment properties, evaluate soil behavior, assess reservoir quality, and make informed decisions in various applications related to earth sciences and engineering disciplines.

The size of the particles in a certain soil can be represented graphically in a grain size diagram, see Fig. 4-2. Such a diagram indicates the percentage of the particles smaller than a certain diameter, measured as a percentage of the weight. A steep slope of the curve in the diagram indicates a uniform soil, a shallow slope of the diagram indicates that the soil contains particles of strongly different grain sizes. For rather coarse particles, say larger than 0.05mm, the grain size distribution can be determined by sieving. The usual procedure is to use a system of sieves having different mesh sizes, stacked on top of each other, with the coarsest mesh on top and the finest mesh at the bottom. After shaking the assembly of sieves, by hand or by a shaking machine, each sieve will contain the particles larger than its mesh size, and smaller than the mesh size of all the sieves above it. In this way, the grain size diagram can be determined. Special standardized sets of sieves are available, as well as convenient shaking machines.

The example shown in Fig. 4-2 illustrates normal sand. In this case, there appear to be no grains larger than 5 mm.

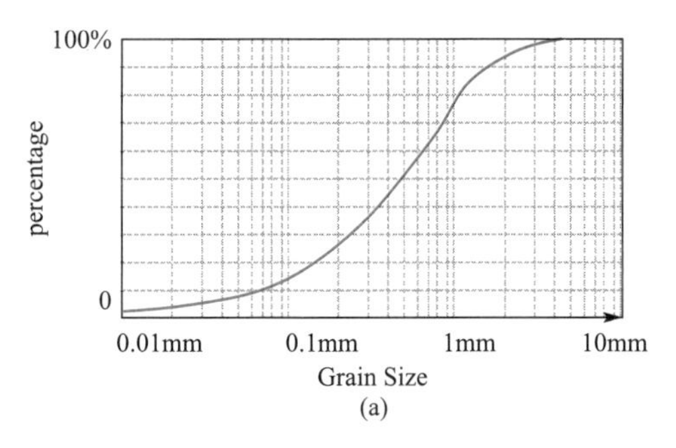

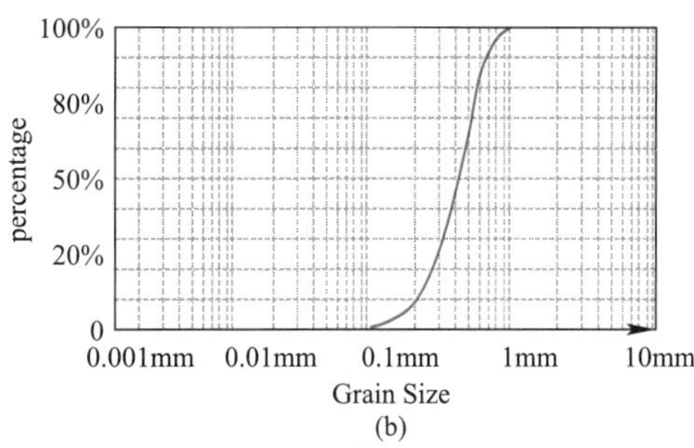

Fig. 4-2 Grain size diagram

4.3 Weight-volume Relationships

4.3.1 Definition

Weight-volume relationships, also known as density relationships, describe the relationship between the weight (mass) and volume of a material. These relationships are commonly used to characterize and quantify the properties of various substances, including solids, liquids, and gases. Here are a few key concepts related to weight-volume relationships. Density is a fundamental property that quantifies the mass of a substance per unit volume. It is calculated by dividing the weight (mass) of an object by its volume. The formula for density is:

Density = Mass/Volume

Density is often expressed in units such as grams per cubic centimeter (g/cm^3) or kilograms per cubic meter (kg/m^3). Specific gravity is a dimensionless quantity that compares the density of a substance to the density of a reference substance, usually water at a specified temperature and pressure. It represents the ratio of the density of the material to the density of water. Specific gravity provides a way to compare the relative densities of different materials without using specific units. Buoyancy is the upward force exerted by a fluid (liquid or gas) on an object immersed in it. The buoyant force is equal to the weight of the displaced fluid and depends on the density of the fluid and the volume of the object. Archimedes' principle states that an object submerged in a fluid experiences an upward buoyant force equal to the weight of the fluid it displaces. This principle is used to determine the density and volume of irregularly-shaped objects by measuring the change in fluid level or weight when the object is immersed.

The understanding of weight-volume relationships is essential in numerous scientific and engineering applications. For example, in materials science and geotechnical engineering, knowledge of density helps assess the suitability of materials for construction or determine their behavior under load. In fluid mechanics, understanding buoyancy and density is crucial in ship design, aerodynamics, and fluid flow analysis. These concepts are also relevant in chemistry, physics, and other fields that involve the measurement and characterization of materials.

4.3.2 Weight-volume Relationships

Fig. 4-3 (a) shows an element of soil of volume V and weight W as it would exist in a natural state. To develop the weight-volume relationships, we must separate the three phases (that is, solid, water, and air) as shown in Fig. 4-3 (b). Thus, the total volume of a given soil sample can be expressed as:

$$V = V_s + V_v = V_s + V_w + V_a \tag{4-1}$$

Where,

V_s—— volume of soil solids;

V_v—— volume of voids;

V_w—— volume of water in the voids;

V_a—— volume of air in the voids.

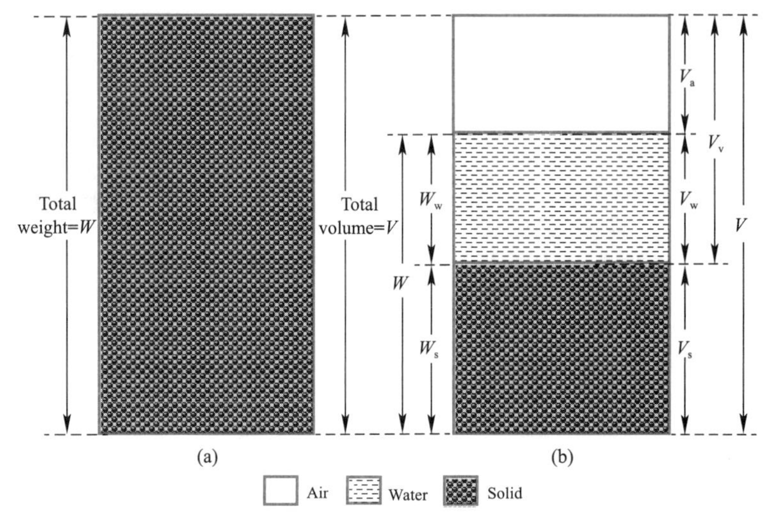

Fig. 4-3 Schematic diagram of weight-volume relationship

(a) Soil element in natural state; (b) Three phases of the soil element

Assuming that the weight of the air is negligible, we can express the total weight of the sample as:

$$W = W_s + W_w \tag{4-2}$$

Where,

W_s—— weight of soil solids;

W_w—— weight of water.

The volume relationships commonly used for the three phases in a soil element are void ratio, porosity, and degree of saturation. Void ratio (e) is defined as the ratio of the volume of voids to the volume of solids. Thus:

$$e = \frac{V_v}{V_s} \tag{4-3}$$

Porosity (n) is defined as the ratio of the volume of voids to the total volume, or:

$$n = \frac{V_v}{V} \tag{4-4}$$

The degree of saturation (S) is defined as the ratio of the volume of water to the volume of voids, or:

$$S = \frac{V_w}{V_v} \tag{4-5}$$

It is commonly expressed as a percentage.

The relationship between void ratio and porosity can be derived from Eq. (4-1), Eq. (4-3), and Eq. (4-4) as follows:

$$e = \frac{V_v}{V_s} = \frac{V_v}{V - V_v} = \frac{\frac{V_v}{V}}{1 - \frac{V_v}{V}} = \frac{n}{1-n} \tag{4-6}$$

Also, from Eq. (4-6):

$$n = \frac{e}{1+e} \tag{4-7}$$

The common terms used for weight relationships are moisture content and unit weight. Moisture content (w) is also referred to as water content and is defined as the ratio of the weight of water to the weight of solids in a given volume of soil.

$$w = \frac{W_u}{W_s} \tag{4-8}$$

Unit weight (γ) is the weight of soil per unit volume. Thus:

$$\gamma = \frac{W}{V} \tag{4-9}$$

The unit weight can also be expressed in terms of the weight of soil solids, the moisture content, and the total volume. From Eq. (4-2), Eq. (4-8), and Eq. (4-9):

$$\gamma = \frac{W}{V} = \frac{W_s + W_w}{V}$$

$$=\frac{W_s\left[1+\left(\frac{W_w}{W_s}\right)\right]}{V}=\frac{W_s(1+w)}{V} \quad (4\text{-}10)$$

Soils engineers sometimes refer to the unit weight defined by Eq. (4-9) as the moist unit weight.

Often, to solve earthwork problems, one must know the weight per unit volume of soil, excluding water. This weight is referred to as the dry unit weight (γ_d). Thus:

$$\gamma_d=\frac{W_s}{V} \quad (4\text{-}11)$$

From Eq. (4-10) and Eq. (4-11), the relationship of unit weight, dry unit weight, and moisture content can be given as:

$$\gamma_d=\frac{\gamma}{1+w} \quad (4\text{-}12)$$

In SI (System International), the unit used is kiloNewtons per cubic meter (kN/m^3). Because the Newton is a derived unit, working with mass densities (ρ) of soil may sometimes be convenient. The SI unit of mass density is kilograms per cubic meter (kg/m^3). We can write the density equations (similar to Eq. 4-9 and Eq. 4-11) as:

$$\rho=\frac{M}{V} \quad (4\text{-}13)$$

And,

$$\rho_d=\frac{M_s}{V} \quad (4\text{-}14)$$

Where,

ρ —— density of soil (kg/m^3);

ρ_d —— dry density of soil (kg/m^3);

M —— total mass of the soil sample (kg);

M_s —— mass of soil solids in the sample (kg);

The unit of total volume, V, is m^3.

The unit weight in kN/m^3 can be obtained from densities in kg/m^3 as Eq. (4-15).

$$\gamma=\frac{g\rho}{1000} \quad (4\text{-}15)$$

And,

$$\gamma_d=\frac{g\rho_d}{1000} \quad (4\text{-}16)$$

Where,

g —— acceleration due to gravity, g = 9.81 m/s^2.

Note that unit weight of water (γ_w) is equal to 9.81 kN/m^3 or 1000 kgf/m^3.

1. Porosity

Soils usually consist of particles, water and air. Porosity of soil refers to the amount of pore space present within the soil material. Pores are the empty spaces or voids between soil particles, and porosity quantifies the volume fraction of these pores relative to the total volume of the soil. The porosity of soil is influenced by various factors, including the arrangement and size of particles, compaction, organic matter content, and the presence of cementing agents. Generally, soils with larger particles (such as sandy soils) have higher porosity compared to soils with smaller particles (such as clay soils).

Porosity can be expressed as a percentage or a decimal fraction. Here's the formula for calculating porosity.

$$\text{Porosity}=(\text{Volume of voids}/\text{Total volume of soil})\times 100\% \quad (4\text{-}17)$$

Porosity is an important property of soil as it affects crucial aspects like water-holding capacity, drainage characteristics, soil fertility, and air circulation within the soil. Soils with higher porosity typically have better water infiltration rates and drainage, allowing plants' roots to access oxygen more easily. On the other hand, excessively high porosity can lead to rapid drainage and low water-holding capacity, which may pose challenges for plant growth.

Scientists and engineers analyze soil porosity to assess its suitability for various applications such as agriculture, geotechnical engineering, groundwater management, and environmental studies. Measurements of porosity help understand soil behavior, predict soil-water relationships, and guide decision-making processes in fields related to land use, construction, and natural resource management.

2. Degree of Saturation

The degree of saturation is a measure of how much of the pore space in a soil or rock is filled with water, expressed as a percentage. It indicates the extent to which the voids or pores are occupied by water relative to their total capacity.

The degree of saturation can becalculated using the following formula.

Degree of saturation = (Volume of water/ Total volume of voids) × 100% (4-18)

In this equation, the volume of water refers to the volume of water occupying the pore spaces, and the total volume of voids represents the entire pore space available within the soil or rock. A degree of saturation value of 100% indicates that all the voids are filled with water, achieving maximum saturation. A value less than 100% indicates that some voids are filled with air or other fluids besides water, implying partial saturation.

Understanding the degree of saturation is essential for various engineering and geotechnical applications, including assessing soil stability, analyzing groundwater flow, and predicting soil behavior under different loading and environmental conditions. It helps determine the water content and drainage characteristics of soils, which influence factors like slope stability, foundation design, and seepage analysis. Additionally, the degree of saturation is also relevant in areas such as agriculture, where proper water management is crucial for crop growth and irrigation practices.

3. Density

The density of soil refers to the mass per unit volume of soil material. It quantifies how much mass is contained within a given volume of soil. There are two common types of density that can be measured in soil.

(1) Bulk density: this represents the total mass of soil (including both solids and pore spaces) per unit volume. Bulk density is calculated by dividing the dry mass of soil by its total volume, including voids.

Bulk density = Dry mass of soil/ Total volume of soil (4-19)

Bulk density is usually expressed in units of grams per cubic centimeter (g/cm^3) or kilograms per cubic meter (kg/m^3).

(2) Particle density: this refers to the mass of solid particles alone per unit volume, excluding pore spaces. Particle density is determined by measuring the mass of the solid particles and dividing it by the total volume of soil occupied by solid particles.

Particle density = Mass of solid particles/ Volume of soil occupied by solids (4-20)

Particle density is also commonly reported in units of grams per cubic centimeter (g/cm^3) or kilograms per cubic meter (kg/m^3). It's important to note that the bulk density of soil is generally lower than the particle density due to the presence of pore spaces. The difference between these densities is related to the porosity of the soil, which represents the volume fraction of pores or voids within the soil. Soil density plays a significant role in various applications, such as agriculture, construction, geotechnical engineering, and environmental studies. It influences factors like soil compaction, water-holding capacity, nutrient availability, permeability, and load-bearing capacity. Understanding soil density helps determine soil health, assess its suitability for different uses, and make informed decisions related to land management, foundation design, irrigation practices, and soil conservation.

For thedescription of the density and the volumetric weight of a soil, the densities of the various components are needed. The density of a substance is the mass per unit volume of that substance. For water, this is denoted by ρ_w, and its value is about 1000 kg/m^3. Small deviations from this value may occur due to temperature differences or variations in salt content.

In soilmechanics, these are often of minor importance, and it is often considered accurate enough to assume that:

$$\rho_w = 1000\ kg/m^3$$

For the analysis of soil mechanics problems, the density of air can usually be disregarded.

The density of the solid particles depends upon the actual composition of the solid material. In many cases, especially for quartz sands, its value is about:

$$\rho_p = 2650 \text{ kg/m}^3$$

This valuecan be determined by carefully dropping a certain mass of particles (W_p) in a container partially filled with water. The precise volume of the particles can be measured by observing the rise of the water table in the glass. This is particularly easy when using a graduated measuring glass. The rising of the water table indicates the volume of the particles (V_p). Their mass (W_p) can be measured most easily by measuring the weight of the glass before and after dropping the particles into it. The density of the particle material then follows immediately from its definition.

$$\rho_p = W_p / V_p \tag{4-21}$$

For sand the value of ρ_p usually is about 2650 kg/m^3.

4. Water Content

The water content is another useful parameter, especially for clays. It has been used in the previous chapter. By definition, the water content (w) is the ratio of the weight (or mass) of the water and the solids.

$$w = W_w / W_p \tag{4-22}$$

It may be noted that this is not a new independent parameter, because:

$$w = S \frac{n}{1-n} \frac{\rho_w}{\rho_p} = Se \frac{\rho_w}{\rho_p} \tag{4-23}$$

For a completely saturated soil ($S=1$) and assuming that $\rho_p/\rho_w = 2.65$, it follows that void ratio e is about 2.65 times the water content.

References

[1] Verruijt A.. An Introduction to Soil Mechanics [M]. Cham: Springer International Publishing, 2018.

[2] Braja M. D., Sobhan K.. Principles of Geotechnical Engineering [M]. 8th ed. Stamford: Cengage Learning, 2013.

Chapter 5
Constitutive Relationships of Rocks

5.1 Stress and Infinitesimal Strain

5.1.1 Problem Definition

This part in Sections 5.1, 5.2 and 5.3 is according to Brady B. H. G. and Brown E. T., 2006. We keep the original content because the description of the professional terms of rock mechanics. Some of the figures are redrawn to make them much clearer. The following are the main content about the constitutive relationships of rock.

The engineering mechanics problem posed by underground mining is the prediction of the displacement field generated in the ore body and surrounding rock by any excavation and ore extraction processes. The stress and strain of rock refer to the physical response of rock materials under external forces or loads. Here are the problem definitions for stress and strain in the context of rocks.

Stress is a measure of the internal forces acting within a rock material per unit area. It represents the intensity of the forces applied to the rock that cause deformation or changes in shape. The problem of stress in rocks involves determining the distribution and magnitude of these internal forces. Stress in rock can be categorized into three main types, compressive stress, tensile stress and shear stress. Compressive stress occurs when forces act to squeeze or compress the rock material, causing it to shorten in one or more directions. Tensile stress arises when forces act to stretch or pull the rock material apart, leading to elongation. Shear stress occurs when forces act parallel to each other but in opposite directions, causing one part of the rock to slide past another. Understanding the distribution and magnitude of stress in rock is crucial for various applications such as rock mechanics, geotechnical engineering, tunneling, mining, and slope stability analysis. It helps assess the potential for rock failure, determine load-bearing capacity, design support systems, and mitigate hazards associated with rock masses.

Strain refers to the deformation or change in shape experienced by a rock material in response to applied stresses. It quantifies the relative displacement or elongation between different parts of the rock. The problem of strain in rock involves measuring and characterizing these deformations. There are different types of strain, including linear strain, volumetric strain and shear strain.

Linear strain: measures the change in length of a rock sample along a specific direction relative to its original length.

Volumetric strain: describes the change in volume of a rock sample due to an applied stress field.

Shear strain: represents the change in the shape or distortion of a rock material caused by shear forces.

The study of strain in rocks helps engineers and geologists understand the behavior of rock masses, predict deformation patterns, evaluate the performance of engineering structures, and monitor stability over time. It is essential for applications such as rock slope stability analysis, rock burst prediction in mining operations, and the design of rock anchors and reinforcement systems. Understanding the stress and strain characteristics of rock materials allows for proper evaluation and management of geological hazards, optimization of engineering designs, and the development of strategies to ensure safe and sustainable use of rock resources.

The following part is cited from Brady B. H. G. and Brown E. T., 2006.

5.1.2 Force and Stress

Stress is a property at a point. It is a tensor. There are normal stresses and shear stresses. There are nine stress components on a small cube, including three normal stresses σ_{xx}, σ_{yy}, σ_{zz} and six shear stresses τ_{xy}, τ_{yx}, τ_{xz}, τ_{zx}, τ_{yz}, τ_{zy}. These stress components can be listed out in matrix form. Corresponding shear stresses are equal. Hence matrix can be reduced to symmetrical. Six shear stresses $\tau_{xy} = \tau_{yx}$, $\tau_{xz} = \tau_{zx}$, $\tau_{yz} = \tau_{zy}$.

There is an inclination of the axes at which all shear stresses disappear (stress transformation). The remaining stresses are the principal stresses. Strains are deformations per lengths caused by stresses. In elastic region, they can be related by Young's modulus.

Strain in stress direction always causes strains in other directions. The ratio of strains is Poisson's ratio. Stresses and strains are related by constitutive laws. Plane stresses and strains can be represented by Mohr circles.

The concept of stress is used to describe the intensity of internal forces set up in a body under the influence of a set of applied surface forces. The idea is quantified by defining the state of stress at a point in a body in terms of the area intensity of forces acting on the orthogonally oriented surfaces of an elementary free body centered on the point. If a Cartesian set of reference axes is used, the elementary free body is a cube whose surfaces are oriented with their outward normal parallel with the co-ordinate axes.

Fig. 5-1(a) illustrates a finite body in equilibrium under a set of applied surface forces P_j. To assess the state of loading over any interior surface, one could proceed by determining the load distribution over the interior surface required to maintain equilibrium of part of the body. Suppose, over an element of surface ΔA surrounding a point O, the required resultant force to maintain equilibrium is ΔR, as shown in Fig. 2-1 (b). The magnitude of the resultant stress σ_r at O, or the stress vector, is then defined by:

$$\sigma_r = \lim_{\Delta A \to 0} \frac{\Delta R}{\Delta A}$$

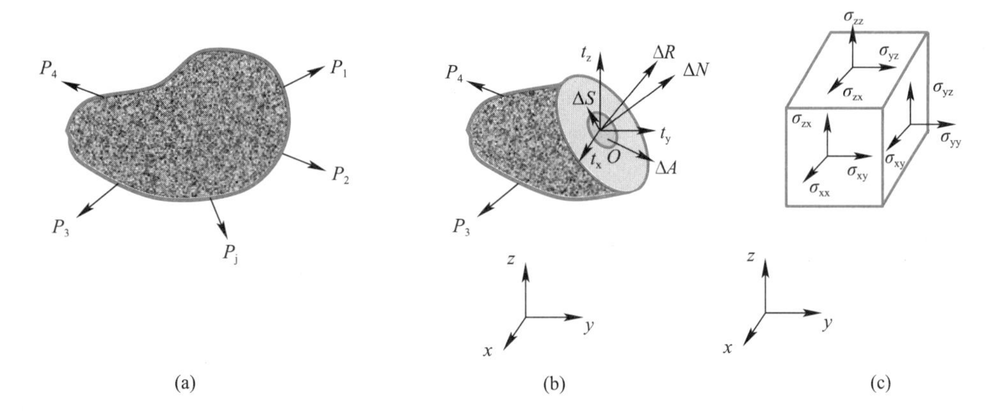

Fig. 5-1 Force state of a finite body subjected to a surface load

(a) A finite body subject to surface loading; (b) Determination of the forces, and related quantities, operating on an internal surface; (c) Specification of the state of stress at a point in terms of the traction components on the face of a cubic free body

If the vector components of ΔR acting normally and tangentially to ΔA are ΔN, ΔS, the normal stress component, σ_n, and the resultant shear stress component, τ, at O are defined by:

$$\sigma_n = \lim_{\Delta A \to 0} \frac{\Delta N}{\Delta A}, \qquad \tau = \lim_{\Delta A \to 0} \frac{\Delta S}{\Delta A}$$

The stress vector, σ_v, may be resolved into components t_x, t_y, t_z directed parallel to a set of reference axes x, y, z. The quantities t_x, t_y,

t_z, shown in Fig. 5-1 (b) are called traction components acting on the surface at the point O. As with the stress vector, the normal stress σ_n, and the resultant shear stress τ, the traction components are expressed in units of force per unit area. A case of particular interest occurs when the outward normal to the elementary surface ΔA is oriented parallel to a co-ordinate axis, e. g., the x axis. The traction components acting on the surface whose normal is the x axis are then used to define three components of the state of stress at the point of interest.

$$\sigma_{xx}=t_x, \quad \sigma_{xy}=t_y, \quad \sigma_{xz}=t_z \quad (5\text{-}1)$$

In thedoubly-subscripted notation for stress components, the first subscript indicates the direction of the outward normal to the surface, the second the sense of action of the stress component. Thus σ_{xz} denotes a stress component acting on a surface whose outward normal is the x axis, and which is directed parallel to the z axis. Similarly, for the other cases where the normal to elements of surfaces are oriented parallel to the y and z axes respectively, stress components on these surfaces are defined in terms of the respective traction components on the surfaces, i. e.:

$$\sigma_{yx}=t_x, \quad \sigma_{yy}=t_y, \quad \sigma_{yz}=t_z \quad (5\text{-}2)$$

$$\sigma_{zx}=t_x, \quad \sigma_{zy}=t_y, \quad \sigma_{zz}=t_z \quad (5\text{-}3)$$

The senses of action of the stress components defined by these expressions are shown in Fig. 5-1(c), acting on the visible faces of the cubic free body.

It is convenient to write the nine stress components, defined by Eq. (5-1), Eq. (5-2), Eq. (5-3), in the form of a stress matrix $[\sigma]$, defined by:

$$[\sigma]=\begin{bmatrix} \sigma_{xx} & \sigma_{xy} & \sigma_{xz} \\ \sigma_{yx} & \sigma_{yy} & \sigma_{yz} \\ \sigma_{zx} & \sigma_{zy} & \sigma_{zz} \end{bmatrix} \quad (5\text{-}4)$$

The form of the stress matrix defined in Eq. (5-4) suggests that the state of stress at a point is defined by nine independent stress components. However, by consideration of moment equilibrium of the free body illustrated in Fig. 5-1(c), it is readily demonstrated that:

$$\sigma_{xy}=\sigma_{yx}, \quad \sigma_{yz}=\sigma_{zy}, \quad \sigma_{zx}=\sigma_{xz}$$

Thus, only six independent stress components are required to define completely the state of stress at a point. The stress matrix may then be written as:

$$[\sigma]=\begin{bmatrix} \sigma_{xx} & \sigma_{xy} & \sigma_{xz} \\ \sigma_{xy} & \sigma_{yy} & \sigma_{yz} \\ \sigma_{xz} & \sigma_{yz} & \sigma_{zz} \end{bmatrix} \quad (5\text{-}5)$$

5.1.3 Stress Transformation

The choice of orientation of the reference axes in specifying a state of stress is arbitrary, and situations will arise in which a differently oriented set of reference axes may prove more convenient for the problem at hand. Fig. 5-2 illustrates a set of old (x, y, z) axes and new (l, m, n) axes. The orientation of a particular axis, e. g., the l axis, relative to the original x, y, z axes may be defined by a row vector (l_x, l_y, l_z) of direction cosines. In this vector, l_x represents the projection on the x axis of a unit vector oriented parallel to the l axis, with similar definitions for l_y and l_z. Similarly, the orientations of the m and n axes relative to the original axes are defined by row vectors of direction cosines, (m_x, m_y, m_z) and (n_x, n_y, n_z) respectively. Also, the state of stress at a point may be expressed, relative to the l, m, n axes, by the stress matrix $[\sigma^*]$, defined by:

$$[\sigma^*]=\begin{bmatrix} \sigma_{ll} & \sigma_{lm} & \sigma_{nl} \\ \sigma_{lm} & \sigma_{mm} & \sigma_{mn} \\ \sigma_{nl} & \sigma_{mn} & \sigma_{nn} \end{bmatrix}$$

The analytical requirement is to express the components of $[\sigma^*]$ in terms of the components of $[\sigma]$ and the direction cosines of the l, m, n axes relative to the x, y, z axes.

Fig. 5-2 shows a tetrahedral free body, O-abc, generated from the elementary cubic free body used to define the components of the stress matrix. The material removed by the cut abc has been replaced by the equilibrating force, of magnitude t per unit area, acting over abc. Suppose the outward normal OP to the surface abc is defined by a row vector of direction cosines ($\lambda_x, \lambda_y, \lambda_z$). If the area of abc is A, the projections of abc on the planes whose normals are the x, y, z axes are given, respectively, by:

$$\text{Area } Oac = A_x = A\lambda_x$$

$$\text{Area } Oab = A_y = A\lambda_y$$

$$\text{Area } Obc = A_z = A\lambda_z$$

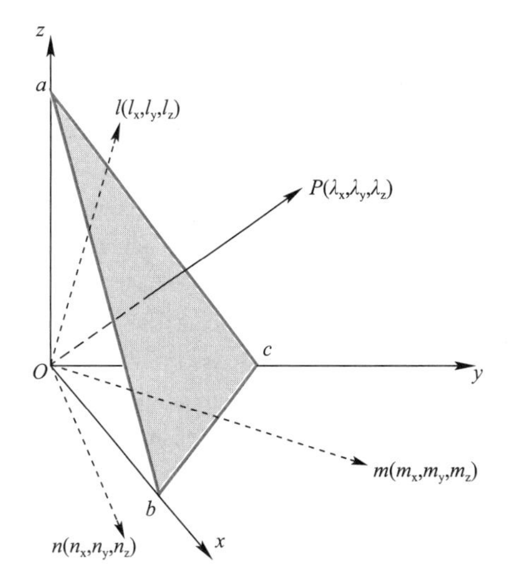

Fig. 5-2 Free-body diagram for establishing the stress transformation equations, principal stresses and their orientations

Suppose the traction vector t has components t_x, t_y, t_z. Application of the equilibrium requirement for the x direction, for example, yields:

$$t_x A - \sigma_{xx} A\lambda_x - \sigma_{xy} A\lambda_y - \sigma_{zx} A\lambda_z = 0 \quad (5\text{-}6)$$

Or,

$$t_x = \sigma_{xx}\lambda_x + \sigma_{xy}\lambda_y + \sigma_{zx}\lambda_z$$

Eq. (5-6) represents an important relation between the traction component, the state of stress, and the orientation of a surface through the point. Developing the equilibrium equations, similar to Eq. (5-6), for the y and z directions, produces analogous expressions for t_y and t_z. The three equilibrium equations may then be written as:

$$\begin{bmatrix} t_x \\ t_y \\ t_z \end{bmatrix} = \begin{bmatrix} \sigma_{xx} & \sigma_{xy} & \sigma_{zx} \\ \sigma_{xy} & \sigma_{yy} & \sigma_{yz} \\ \sigma_{zx} & \sigma_{yz} & \sigma_{zz} \end{bmatrix} \begin{bmatrix} \lambda_x \\ \lambda_y \\ \lambda_z \end{bmatrix} \quad (5\text{-}7)$$

Or,

$$[t] = [\sigma][\lambda] \quad (5\text{-}8)$$

Proceeding in the same way for another set of co-ordinate axes l, m, n maintaining the same global orientation of the cutting surface to generate the tetrahedral free body, but expressing all traction and stress components relative to the l, m, n axes, yields the relations:

$$\begin{bmatrix} t_l \\ t_m \\ t_n \end{bmatrix} = \begin{bmatrix} \sigma_{ll} & \sigma_{lm} & \sigma_{nl} \\ \sigma_{lm} & \sigma_{mm} & \sigma_{mn} \\ \sigma_{nl} & \sigma_{mn} & \sigma_{nn} \end{bmatrix} \begin{bmatrix} \lambda_l \\ \lambda_m \\ \lambda_n \end{bmatrix} \quad (5\text{-}9)$$

Or,

$$[t^*] = [\sigma^*][\lambda^*] \quad (5\text{-}10)$$

In Eq. (5-8) and Eq. (5-10), $[t]$, $[t^*]$, $[\lambda]$, $[\lambda^*]$ are vectors, expressed relative to the x, y, z and l, m, n co-ordinate systems. They represent traction components acting on, and direction cosines of the outward normal to, a surface with fixed spatial orientation. From elementary vector analysis, a vector $[v]$ is transformed from one set of orthogonal reference axes x, y, z to another set, l, m, n, by the transformation equation:

$$\begin{bmatrix} v_l \\ v_m \\ v_n \end{bmatrix} = \begin{bmatrix} l_x & l_y & l_z \\ m_x & m_y & m_z \\ n_x & n_y & n_z \end{bmatrix} \begin{bmatrix} v_x \\ v_y \\ v_z \end{bmatrix}$$

Or:

$$[v^*] = [R][v] \quad (5\text{-}11)$$

Inthis expression, $[R]$ is the rotation matrix, whose rows are seen to be formed from the row vectors of direction cosines of the new axes relative to the old axes.

As discussed by Jennings (1977), a unique property of the rotation matrix is that its inverse is equal to its transpose, i. e. :

$$[R]^{-1} = [R]^{\mathrm{T}} \quad (5\text{-}12)$$

Returning now to the relations between $[t]$ and $[t^*]$, and $[\lambda]$ and $[\lambda^*]$, the results expressed in Eq. (5-11) and Eq. (5-12) indicate that:

$$[t^*] = [R][t]$$

Or,

$$[t] = [R]^{\mathrm{T}}[t^*]$$

And,

$$[\lambda^*] = [R][\lambda]$$

Or,

$$[\lambda] = [R]^{\mathrm{T}}[\lambda^*]$$

Then,

$$\begin{aligned}[t^*] &= [R][t] \\ &= [R][\sigma][\lambda] \\ &= [R][\sigma][R]^{\mathrm{T}}[\lambda^*]\end{aligned}$$

But since,

$$[t^*]=[\sigma^*][\lambda^*]$$

Then,

$$[\sigma^*]=[R][\sigma][R]^{T} \tag{5-13}$$

Eq. (5-13) is the required stress transformation equation. It indicates that the state of stress at a point is transformed, under a rotation of axes, as a second-order tensor.

Eq. (5-13) when written in expanded notation becomes:

$$\begin{bmatrix} \sigma_{ll} & \sigma_{lm} & \sigma_{nl} \\ \sigma_{lm} & \sigma_{mm} & \sigma_{mn} \\ \sigma_{nl} & \sigma_{mn} & \sigma_{nn} \end{bmatrix} = \begin{bmatrix} l_x & l_y & l_z \\ m_x & m_y & m_z \\ n_x & n_y & n_z \end{bmatrix} \begin{bmatrix} \sigma_{xx} & \sigma_{xy} & \sigma_{zx} \\ \sigma_{xy} & \sigma_{yy} & \sigma_{yz} \\ \sigma_{zx} & \sigma_{yz} & \sigma_{zz} \end{bmatrix} \begin{bmatrix} l_x & m_x & n_x \\ l_y & m_y & n_y \\ l_z & m_z & n_z \end{bmatrix}$$

Proceeding with the matrix multiplication on the right-hand side of this expression, in the usual way, produces explicit formula for determining the stress components under a rotation of axes, given by:

$$\sigma_{ll}=l_x^2\sigma_{xx}+l_y^2\sigma_{yy}+l_z^2\sigma_{zz}+2(l_xl_y\sigma_{xy}+l_yl_z\sigma_{yz}+l_zl_x\sigma_{zx}) \tag{5-14}$$

$$\sigma_{lm}=l_xm_x\sigma_{xx}+l_ym_y\sigma_{yy}+l_zm_z\sigma_{zz}+(l_xm_y+l_ym_x)\sigma_{xy}+(l_ym_z+l_zm_y)\sigma_{yz}+(l_zm_x+l_xm_z)\sigma_{zx} \tag{5-15}$$

Expressions for the other four components of the stress matrix are readily obtained from these equations by cyclic permutation of the subscripts.

5.1.4 Principal Stresses and Stress Invariants

The discussion above has shown that the state of stress at a point in a medium may be specified in terms of six components, whose magnitudes are related to arbitrarily selected orientations of the reference axes. In some rock masses, the existence of a particular fabric element, such as a foliation or a schistosity, might define a suitable direction for a reference axis. Such a feature might also determine a mode of deformation of the rock mass under load. However, in an isotropic rock mass, any choice of a set of reference axes is obviously arbitrary, and a non-arbitrary way is required for defining the state of stress at any point in the medium. This is achieved by determining principal stresses and related quantities which are invariant under any rotations of reference axes. It is shown that the resultant stress on any plane in a body can be expressed in terms of a normal component of stress, and two mutually orthogonal shear stress components. A principal plane is defined as one on which the shear stress components vanish, i. e. it is possible to select a particular orientation for a plane such that it is subject only to normal stress. The magnitude of the principal stress is that of the normal stress, while the normal to the principal plane defines the direction of the principal stress axis. Since there are, in any specification of a state of stress, three reference directions to be considered, there are three principal stress axes. There are thus three principal stresses and their orientations to be determined to define the state of stress at a point.

Suppose that in Fig. 5-2, the cutting plane *abc* is oriented such that the resultant stress on the plane acts normal to it, and has a magnitude σ_p. If the vector $(\lambda_x, \lambda_y, \lambda_z)$ defines the outward normal to the plane, the traction components on abc are defined by:

$$\begin{bmatrix} t_x \\ t_y \\ t_z \end{bmatrix} = \sigma_p \begin{bmatrix} \lambda_x \\ \lambda_y \\ \lambda_z \end{bmatrix} \tag{5-16}$$

The traction components on the plane *abc* are also related, through Eq. (5-7), to the state of stress and the orientation of the plane. Subtracting Eq. (5-16) from Eq. (5-7) yields the equation:

$$\begin{bmatrix} \sigma_{xx}-\sigma_p & \sigma_{xy} & \sigma_{zx} \\ \sigma_{xy} & \sigma_{yy}-\sigma_p & \sigma_{yz} \\ \sigma_{zx} & \sigma_{yz} & \sigma_{zz}-\sigma_p \end{bmatrix} \begin{bmatrix} \lambda_x \\ \lambda_y \\ \lambda_z \end{bmatrix} = [0] \tag{5-17}$$

The matrix Eq. (5-17) represents a set of three simultaneous, homogeneous, linear equations in λ_x, λ_y, λ_z. The requirement for a non-trivial solution is that the determinant of the coefficient matrix in Eq. (5-17) must vanish. Expansion of the determinant yields a cubic equation in σ_p, given by:

$$\sigma_p^3 - I_1\sigma_p^2 + I_2\sigma_p - I_3 = 0 \qquad (5\text{-}18)$$

In this equation, the quantities I_1, I_2 and I_3, are called the first, second and third stress invariants. They are defined by the expressions:

$$I_1 = \sigma_{xx} + \sigma_{yy} + \sigma_{zz}$$

$$I_2 = \sigma_{xx}\sigma_{yy} + \sigma_{yy}\sigma_{zz} + \sigma_{zz}\sigma_{xx} - (\sigma_{xy}^2 + \sigma_{yz}^2 + \sigma_{zx}^2)$$

$$I_3 = \sigma_{xx}\sigma_{yy}\sigma_{zz} + 2\sigma_{xy}\sigma_{yz}\sigma_{zx} - (\sigma_{xx}\sigma_{yx}^2 + \sigma_{yy}\sigma_{zx}^2 + \sigma_{zz}\sigma_{xy}^2)$$

It is to be noted that since the quantities I_1, I_2, I_3 are invariant under a change of axes, any quantities derived from them are also invariants.

Solution of the characteristic Eq. (5-18) by some general methods, such as a complex variable method, produces three real solutions for the principal stresses.

These are denoted σ_1, σ_2, σ_3, in order of decreasing magnitude, and are identified respectively as the major, intermediate and minor principal stresses.

Each principal stress value is related to a principal stress axis, whose direction cosines can be obtained directly from Eq. (5-17) and a basic property of direction cosines. The dot product theorem of vector analysis yields, for any unit vector of direction cosines (λ_x, λ_y, λ_z), the relation:

$$\lambda_x^2 + \lambda_y^2 + \lambda_z^2 = 1 \qquad (5\text{-}19)$$

Introduction of a particular principal stress value, e. g., σ_1, into Eq. (5-17), yields a set of simultaneous, homogeneous equations in λ_{x1}, λ_{y1}, λ_{z1}. These are the required direction cosines for the major principal stress axis. Solution of the set of equations for these quantities is possible only in terms of some arbitrary constant K, defined by:

$$\frac{\lambda_{x1}}{A} = \frac{\lambda_{y1}}{B} = \frac{\lambda_{z1}}{C} = K$$

Where,

$$A = \begin{vmatrix} \sigma_{yy} - \sigma_1 & \sigma_{yz} \\ \sigma_{yz} & \sigma_{zz} - \sigma_1 \end{vmatrix}$$

$$B = -\begin{vmatrix} \sigma_{xy} & \sigma_{yz} \\ \sigma_{zx} & \sigma_{zz} - \sigma_1 \end{vmatrix} \qquad (5\text{-}20)$$

$$C = \begin{vmatrix} \sigma_{xy} & \sigma_{yy} - \sigma_1 \\ \sigma_{zx} & \sigma_{yz} \end{vmatrix}$$

Substituting for λ_{x1}, λ_{y1}, λ_{z1} in Eq. (5-19), gives:

$$\lambda_{x1} = A/(A^2 + B^2 + C^2)^{1/2}$$

$$\lambda_{y1} = B/(A^2 + B^2 + C^2)^{1/2}$$

$$\lambda_{z1} = C/(A^2 + B^2 + C^2)^{1/2}$$

Proceeding in a similar way, the vectors of direction cosines for the intermediate and minor principal stress axes, i. e., (λ_{x2}, λ_{y2}, λ_{z2}) and (λ_{x3}, λ_{y3}, λ_{z3}), are obtained from Eq. (5-20) by introducing the respective values of σ_2 and σ_3.

The procedure for calculating the principal stresses and the orientations of the principal stress axes is simply the determination of the eigenvalues of the stress matrix, and the eigenvector for each eigenvalue. Some simple checks can be performed to assess the correctness of solutions for principal stresses and their respective vectors of direction cosines. The condition of orthogonality of the principal stress axes requires that each of the three dot products of the vectors of direction cosines must vanish, i. e. :

$$\lambda_{x1}\lambda_{x2} + \lambda_{y1}\lambda_{y2} + \lambda_{z1}\lambda_{z2} = 0$$

This is similar to the result for the (2, 3) and (3, 1) dot products. Invariance of the sum of the normal stresses requires that:

$$\sigma_1 + \sigma_2 + \sigma_3 = \sigma_{xx} + \sigma_{yy} + \sigma_{zz}$$

In the analysis of some types of behaviour in rock, it is usual to split the stress matrix into two components—a spherical or hydrostatic component $[\sigma_m]$, and a deviatoric component $[\sigma_d]$. The spherical stress matrix is defined by:

$$[\sigma_m] = \sigma_m[I] = \begin{bmatrix} \sigma_m & 0 & 0 \\ 0 & \sigma_m & 0 \\ 0 & 0 & \sigma_m \end{bmatrix}$$

Where,

$$\sigma_m = I_1/3$$

The deviator stress matrix is obtained from the stress matrix $[\sigma]$ and the spherical stress matrix, and is given by:

$$[\sigma_d] = \begin{bmatrix} \sigma_{xx} - \sigma_m & \sigma_{xy} & \sigma_{zx} \\ \sigma_{xy} & \sigma_{yy} - \sigma_m & \sigma_{yz} \\ \sigma_{zx} & \sigma_{yz} & \sigma_{zz} - \sigma_m \end{bmatrix}$$

Principal deviator stresses S_1, S_2, S_3 can be established either from the deviator stress matrix, in the way described previously, or from the principal stresses and the hydrostatic stress, i. e. :

$$S_1 = \sigma_1 - \sigma_m$$

Where,

S_1—— the major principal deviator stress.

The principal directions of the deviator stress matrix $[\sigma_d]$ are the same as those of the stress matrix $[\sigma]$.

5.1.5 Differential Equations of Static Equilibrium

Problems in solid mechanics frequently involve description of the stress distribution in a body in static equilibrium under the combined action of surface and body forces. Determination of the stress distribution must take account of the requirement that the stress field maintains static equilibrium throughout the body. This condition requires satisfaction of the equations of static equilibrium for all differential elements of the body.

Fig. 5-3 shows a small element of a body, in which operate body force components with magnitudes X, Y, Z per unit volume, directed in the positive x, y, z co-ordinate directions. The stress distribution in the body is described in terms of a set of stress gradients, defined by $\partial\sigma_{xx}/\partial x$, $\partial\sigma_{xy}/\partial y$, etc. Considering the condition for force equilibrium of the element in the x direction yields the equation:

$$\frac{\partial\sigma_{xx}}{\partial x}\cdot dx\cdot dy\cdot dz+\frac{\partial\sigma_{xy}}{\partial y}\cdot dx\cdot dy\cdot dz+\frac{\partial\sigma_{zx}}{\partial z}\cdot dx\cdot dy\cdot dz+X\cdot dx\cdot dy\cdot dz=0$$

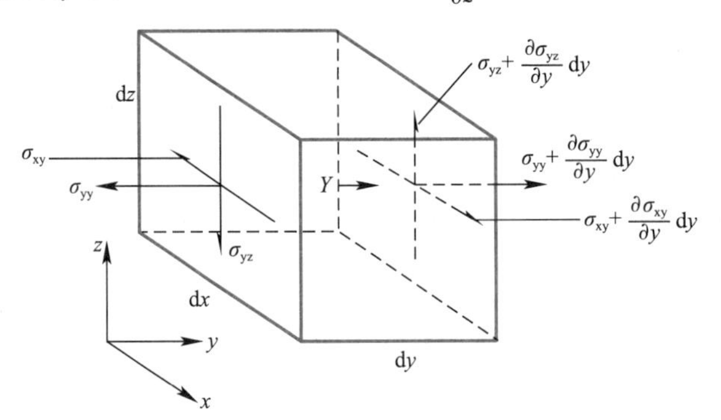

Fig. 5-3 Free-body diagram for development of the differential equations of equilibrium

Applying the same static equilibrium requirement to the y and z directions, and eliminating the term dx, dy, dz, yields the differential equations of equilibrium:

$$\frac{\partial\sigma_{xx}}{\partial x}+\frac{\partial\sigma_{xy}}{\partial y}+\frac{\partial\sigma_{zx}}{\partial z}+X=0$$

$$\frac{\partial\sigma_{xy}}{\partial x}+\frac{\partial\sigma_{yy}}{\partial y}+\frac{\partial\sigma_{yz}}{\partial z}+Y=0$$

$$\frac{\partial\sigma_{zx}}{\partial x}+\frac{\partial\sigma_{yz}}{\partial y}+\frac{\partial\sigma_{zz}}{\partial z}+Z=0 \qquad (5\text{-}21)$$

These expressions indicate that the variations of stress components in a body under load are not mutually independent. They are always involved, in one form or another, in determining the state of stress in a body. A purely practical application of these equations is in checking the admissibility of any closed-form solution for the stress distribution in a body subject to particular applied loads. It is a straightforward matter to determine if the derivatives of expressions describing a particular stress distribution satisfy the equalities of Eq. (5-21).

5.1.6 Plane Problems and Biaxial Stress

Many underground excavation design analysis involving openings where the length to cross section dimension ratio is high, are facilitated considerably by the relative simplicity of the excavation geometry. For example, an excavation such as a tunnel of uniform cross section along its length might be analyzed by assuming that the stress distribution is identical in all planes perpendicular to the long axis of the excavation.

Suppose a set of reference axes, x, y, z, is established for such a problem, with the long axis of the excavation parallel to the z axis, as shown in Fig. 5-4. As shown above, the state of stress at any point in the medium is described by six stress components.

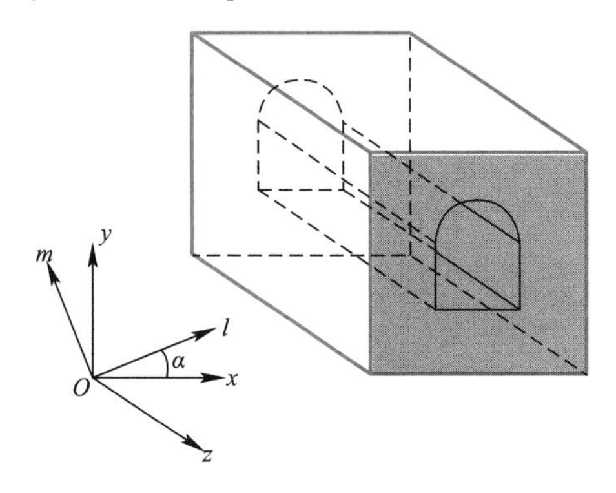

Fig. 5-4 A long excavation, of uniform cross section, for which a contracted form of the stress transformation equations is appropriate

For plane problems in the x, y plane, the six stress components are functions of (x, y) only. In some cases, it may be more convenient to express the state of stress relative to a different set of reference axes, such as the l, m, z axes shown in Fig. 2-4. If the angle lOx is α, the direction cosines of the new reference axes relative to the old set are given by:

$$l_x=\cos\alpha,\quad l_y=\sin\alpha,\quad l_z=0$$

$$m_x=-\sin\alpha,\quad m_y=\cos\alpha,\quad m_x=0$$

Introducing these values into the general transformation equations, i. e., Eq. (5-14) and Eq. (5-15), yields:

$$\begin{aligned}
\sigma_{ll}&=\sigma_{xx}\cos^2\alpha+\sigma_{yy}\sin^2\alpha+2\sigma_{xy}\sin\alpha\cos\alpha\\
\sigma_{mm}&=\sigma_{xx}\sin^2\alpha+\sigma_{yy}\cos^2\alpha-2\sigma_{xy}\sin\alpha\cos\alpha\\
\sigma_{lm}&=\sigma_{xy}(\cos^2\alpha-\sin^2\alpha)-(\sigma_{xx}-\sigma_{yy})\sin\alpha\cos\alpha\\
\sigma_{mz}&=\sigma_{yz}\cos\alpha-\sigma_{zx}\sin\alpha\\
\sigma_{zl}&=\sigma_{yz}\sin\alpha-\sigma_{zx}\cos\alpha
\end{aligned}\tag{5-22}$$

And the σ_{zz} component is clearly invariant under the transformation of axes. The set of Eq. (5-22) is observed to contain two distinct types of transformation: those defining σ_{ll}, σ_{mm}, σ_{lm}, which conform to second-order tensor transformation behaviour, and σ_{mz} and σ_{zl}, which are obtained by an apparent vector transformation.

The latter behaviour in the transformation is due to the constancy of the orientation of the element of surface with its normal to the z axis. The rotation of the axes merely involves a transformation of the traction components on this surface.

For problems which can be analyzed in terms of plane geometry, Eq. (5-22) indicates that the state of stress at any point can be defined in terms of the plane components of stress σ_{xx}, σ_{yy}, σ_{xy} and the antiplane components σ_{zz}, σ_{yz}, σ_{zx}. In the particular case where the z direction is a principal axis, the antiplanet shear stress components vanish. The plane geometric problem can then be analyzed in terms of the plane components of stress, since the σ_{zz}, component is frequently neglected. A state of biaxial (or two-dimensional) stress at any point in the medium is defined by three components, in this case σ_{xx}, σ_{yy}, σ_{xy}.

The stress transformation equations related to σ_{ll}, σ_{mm}, σ_{lm} in Eq. (5-22), for the biaxial state of stress, may be recast in the form:

$$\begin{aligned}
\sigma_{ll}&=\frac{1}{2}(\sigma_{xx}+\sigma_{yy})+\frac{1}{2}(\sigma_{xx}-\sigma_{yy})\cos2\alpha+\sigma_{xy}\sin2\alpha\\
\sigma_{mm}&=\frac{1}{2}(\sigma_{xx}+\sigma_{yy})-\frac{1}{2}(\sigma_{xx}-\sigma_{yy})\cos2\alpha-\sigma_{xy}\sin2\alpha\\
\sigma_{lm}&=\sigma_{xy}\cos2\alpha-\frac{1}{2}(\sigma_{xx}-\sigma_{yy})\sin2\alpha
\end{aligned}\tag{5-23}$$

In establishing these equations, the x, y and l, m axes are taken to have the same sense of "handedness", and the angle α is measured from the x to the l axis, in a sense that corresponds to the "handedness" of the transformation. There is no inference of clockwise or anticlockwise rotation of axes in establishing these transformation equations. However, the way in which the order of the terms is specified in the equations, and related to the sense of measurement of the rotation angle α, should be examined closely.

Now consider the determination of the magnitudes and orientations of the plane principal stresses for a plane problem in the x, y plane. In this case, the σ_{zz}, σ_{yz}, σ_{zx} stress components vanish, the third stress invariant vanishes, and the characteristic Eq. (5. 18), becomes:

$$\sigma_p^2-(\sigma_{xx}+\sigma_{yy})\sigma_p+\sigma_{xx}\sigma_{yy}-\sigma_{xy}^2=0$$

Solution of this quadratic equation yields the magnitudes of the plane principal stresses as:

$$\sigma_{1,2}=\frac{1}{2}(\sigma_{xx}+\sigma_{yy})\pm\left[\frac{1}{4}(\sigma_{xx}-\sigma_{yy})^2+\sigma_{xy}^2\right]^{1/2} \quad (5\text{-}24a)$$

The orientations of the respective principal stress axes are obtained by establishing the direction of the outward normal to a plane which is free of shear stress. Suppose ab, shown in Fig. 5-5, represents such a plane. The outward normal to ab is Ol, and therefore defines the direction of a principal stress, σ_p. Considering static equilibrium of the element aOb under forces operating in the x direction:

$$\sigma_p ab\cos\alpha-\sigma_{xx}ab\cos\alpha-\sigma_{xy}\sin\alpha=0$$

Or:

$$\tan\alpha=\frac{\sigma_p-\sigma_{xx}}{\sigma_{xy}}$$

So:

$$\alpha=\tan\alpha^{-1}\frac{\sigma_p-\sigma_{xx}}{\sigma_{xy}} \quad (5\text{-}24b)$$

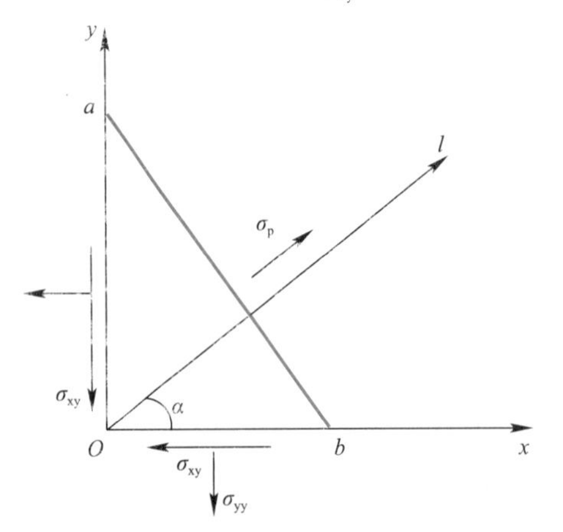

Fig. 5-5 Problem geometry for determination of plane principal stresses and their orientations

Substitution of the magnitudes σ_1, σ_2, determined from Eq. (5-24a), in Eq. (5-24b) yields the orientations σ_1, σ_2, of the principal stress axes relative to the positive direction of the x axis. Calculation of the orientations of the major and minor plane principal stresses in this way associates a principal stress axis unambiguously with a principal stress magnitude. This is not the case with other methods, which employ the last of Eq. (5-23) to determine the orientation of a principal stress axis.

It is to be noted that in specifying the state of stress in a body, there has been no reference to any mechanical properties of thebody which is subject to applied load.

The only concept invoked has been that of static equilibrium of all elements of the body.

5. 1. 7 Displacement and Strain

Application of a set of forces to a body, or change in its temperature, changes the relative positions of points within it. The change in loading conditions from the initial state to the final state causes a displacement of each point relative to all other points.

If the applied loads constitute a self-equilibrating set, the problem is to determine the equilibrium displacement field induced in the body by the loading. A particular difficulty is presented in the analysis of displacements for a loaded body where boundary conditions are specified completely in terms of surface tractions. In this case, unique determination of the absolute displacement field is impossible, since any set of rigid-body displacements can be superimposed on a particular solution, and still satisfy the equilibrium condition. Difficulties of this type are avoided in analysis by employing displacement gradients as the field variables. The related concept of strain is therefore introduced to make basically indeterminate problems tractable.

Fig. 5-6 shows the original positions of two adjacent particles $P(x, y, z)$ and $Q(x+dx, y+dy, z+dz)$ in a body. Under the action of a set of applied loads, P moves to the point

$P^*(x+u_x, y+u_y, z+u_z)$, and Q moves to the point $Q^*(x+dx+u_x^*, y+dy+u_y^*, z+dz+u_z^*)$. If $u_x = u_x^*$, etc., the relative displacement between P and Q under the applied load is zero, i. e., the body has been subject to a rigid-body displacement. The problem of interest involves the case where $u_x \neq u_x^*$, etc. The line element joining P and Q then changes length in the process of load application, and the body is said to be in a state of strain.

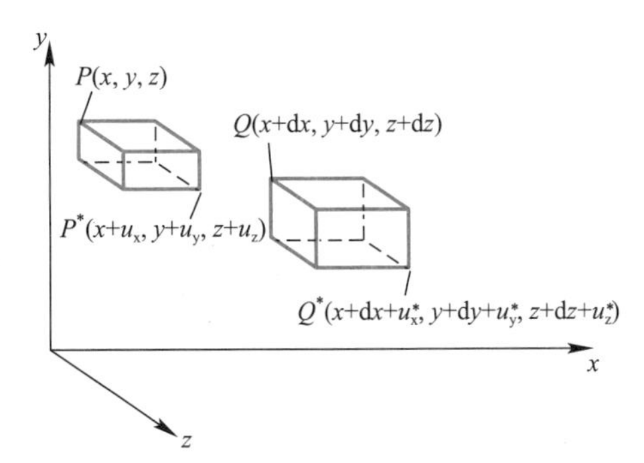

Fig. 5-6 Initial and final positions of points P, Q, in a body subjected to strain

In specifying the state of strain in a body, the objective is to describe the changes in the sizes and shapes of infinitesimal elements in the loaded medium. This is done by considering the displacement components u_x, u_y, u_z of a particle P, and u_x^*, u_y^*, u_z^* of the adjacent particle Q. Since:

$$u_x^* = u_x + du_x, \quad \text{where } du_x = \frac{\partial u_x}{\partial x}dx + \frac{\partial u_x}{\partial y}dy + \frac{\partial u_x}{\partial z}dz$$

And,

$$u_y^* = u_y + du_y, \quad \text{where } du_y = \frac{\partial u_y}{\partial x}dx + \frac{\partial u_y}{\partial y}dy + \frac{\partial u_y}{\partial z}dz$$

$$u_z^* = u_z + du_z, \quad \text{where } du_z = \frac{\partial u_z}{\partial x}dx + \frac{\partial u_z}{\partial y}dy + \frac{\partial u_z}{\partial z}dz$$

The incremental displacements may be expressed by:

$$\begin{bmatrix} du_x \\ du_y \\ du_z \end{bmatrix} = \begin{bmatrix} \frac{\partial u_x}{\partial x} & \frac{\partial u_x}{\partial y} & \frac{\partial u_x}{\partial z} \\ \frac{\partial u_y}{\partial x} & \frac{\partial u_y}{\partial y} & \frac{\partial u_y}{\partial z} \\ \frac{\partial u_z}{\partial x} & \frac{\partial u_z}{\partial y} & \frac{\partial u_z}{\partial z} \end{bmatrix} \begin{bmatrix} dx \\ dy \\ dz \end{bmatrix} \tag{5-25a}$$

Or,

$$[d\delta] = [D][dr] \tag{5-25b}$$

In this expression, $[dr]$ represents the original length of the line element PQ, while $[d\delta]$ represents the relative displacement of the ends of the line element in deforming from the unstrained to the strained state.

The infinitesimal relative displacement defined by Eq. (5-25) can arise from both deformation of the element of which PQ is the diagonal, and a rigid-body rotation of the element. It is needed to define explicitly the quantities related to deformation of the body. Fig. 5-7 shows the projection of the element, with diagonal PQ, on to the yz plane, and subject to a rigid body rotation Ω_x about the x axis. Since the side dimensions of the element are dy and dz, the relative displacement components of Q relative to P are:

$$\begin{aligned} du_y &= -\Omega_x dz \\ du_z &= \Omega_x dy \end{aligned} \tag{5-26}$$

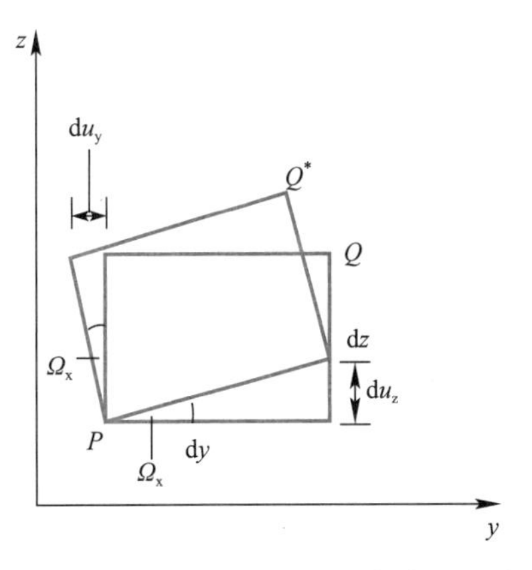

Fig. 5-7 Component displacements of adjacent points resulting from rigid-body rotation of an element

Considering rigid-body rotations Ω_y and Ω_z around the y and z axes, the respective displacements are:

$$\begin{aligned} du_z &= -\Omega_y dx \\ du_x &= \Omega_y dz \end{aligned} \tag{5-27}$$

And,

$$\begin{aligned} du_x &= -\Omega_z dz \\ du_y &= \Omega_z dx \end{aligned} \tag{5-28}$$

The total displacement due to the various rigid-body rotations is obtained by the addition of Eq. (5-26), Eq. (5-27) and Eq. (5-28), i. e. :

$$du_x = -\Omega_z dy + \Omega_y dz$$
$$du_y = \Omega_z dx - \Omega_x dz$$
$$du_z = -\Omega_y dx + \Omega_x dy$$

These equations may be written in the form:

$$\begin{bmatrix} du_x \\ du_y \\ du_z \end{bmatrix} = \begin{bmatrix} 0 & -\Omega_z & \Omega_y \\ \Omega_z & 0 & -\Omega_x \\ -\Omega_z & \Omega_x & 0 \end{bmatrix} \begin{bmatrix} dx \\ dy \\ dz \end{bmatrix} \tag{5-29a}$$

Or,

$$[d\delta'] = [\Omega][dr] \tag{5-29b}$$

The contribution of deformation to the relative displacement $[d\delta]$ is determined by considering elongation and distortion of the element. Fig. 5-8 represents the elongation of the block in the x direction. The element of length dx is assumed to be homogeneously strained, in extension, and the normal strain component is therefore defined by:

$$\varepsilon_{xx} = \frac{du_x}{dx}$$

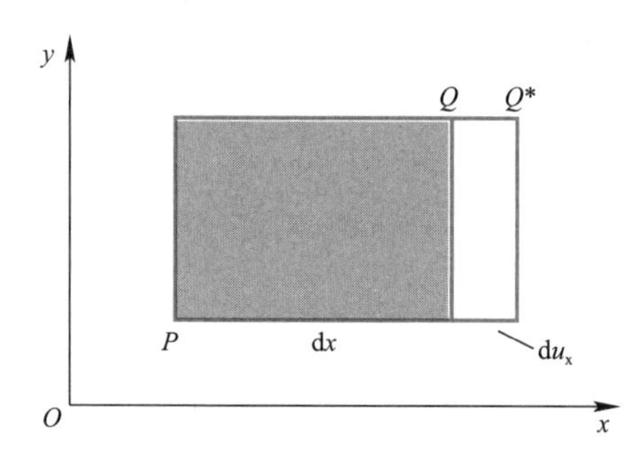

Fig. 5-8 Displacement components produced by pure longitudinal strain

Considering the y and z components of elongation of the element in a similar way, the components of relative displacement due to normal strain can be derived as:

$$du_x = \varepsilon_{xx} dx$$
$$du_y = \varepsilon_{yy} dy$$
$$du_z = \varepsilon_{zz} dz \tag{5-30}$$

The components of relative displacement arising from distortion of the element are derived by considering an element subject to various modes of pure shear strain. Fig. 5-9 shows such an element strained in the x, y plane. Since the angle α is small, pure shear of the element results in the displacement components is:

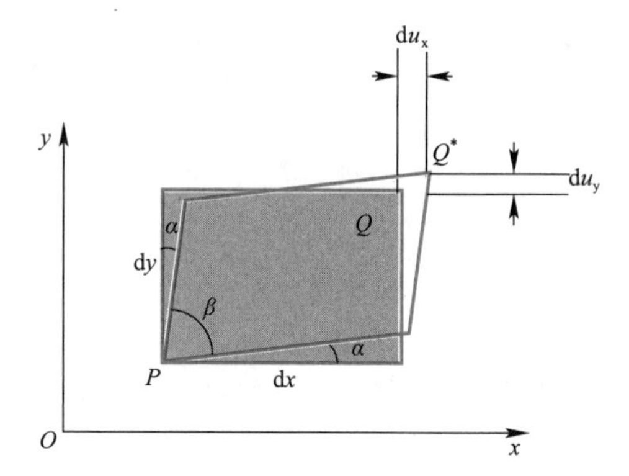

Fig. 5-9 Displacement produced by pure shear strain

$$du_x = \alpha dx$$
$$du_y = \alpha dy$$

Since shear strain magnitude is defined by:

$$\gamma_{xy} = \frac{\pi}{2} - \beta = 2\alpha$$

Then,

$$du_x = \frac{1}{2}\gamma_{xy} dy$$
$$du_y = \frac{1}{2}\gamma_{xy} dx \tag{5-31}$$

Similarly, displacements due to pure shear of the element in the y, z and z, x planes are given by:

$$du_y = \frac{1}{2}\gamma_{yz} dz$$
$$du_z = \frac{1}{2}\gamma_{yz} dy \tag{5-32}$$

And,

$$du_z = \frac{1}{2}\gamma_{zx} dx$$
$$du_x = \frac{1}{2}\gamma_{zx} dz \tag{5-33}$$

The total displacement components due to all modes of infinitesimal strain are obtained by the addition of Eq. (5-30), Eq. (5-31), Eq. (5-32) and Eq. (5-33), i. e. :

$$du_x = \varepsilon_{xx} dx + \frac{1}{2}\gamma_{xy} dy + \frac{1}{2}\gamma_{zx} dz$$
$$du_y = \frac{1}{2}\gamma_{xy} dx + \varepsilon_{yy} dy + \frac{1}{2}\gamma_{yz} dz$$
$$du_z = \frac{1}{2}\gamma_{zx} dx + \frac{1}{2}\gamma_{yz} dy + \varepsilon_{zz} dz$$

These equations may be written in the form:

$$\begin{bmatrix} du_x \\ du_y \\ du_z \end{bmatrix} = \begin{bmatrix} \varepsilon_{xx} & \frac{1}{2}\gamma_{xy} & \frac{1}{2}\gamma_{zx} \\ \frac{1}{2}\gamma_{xy} & \varepsilon_{yy} & \frac{1}{2}\gamma_{yz} \\ \frac{1}{2}\gamma_{zx} & \frac{1}{2}\gamma_{yz} & \varepsilon_{zz} \end{bmatrix} \begin{bmatrix} dx \\ dy \\ dz \end{bmatrix} \tag{5-34a}$$

Or,

$$[d\delta''] = [\varepsilon][dr] \tag{5-34b}$$

Where,

$[\varepsilon]$ —— the strain matrix.

Since,

$$[d\delta] = [d\delta'] + [d\delta'']$$

Eq. (5-25a), Eq. (5-29a) and Eq. (5-34a) yield:

$$\begin{bmatrix} \frac{\partial u_x}{\partial x} & \frac{\partial u_x}{\partial y} & \frac{\partial u_x}{\partial z} \\ \frac{\partial u_y}{\partial x} & \frac{\partial u_y}{\partial y} & \frac{\partial u_y}{\partial z} \\ \frac{\partial u_z}{\partial x} & \frac{\partial u_z}{\partial y} & \frac{\partial u_z}{\partial z} \end{bmatrix} = \begin{bmatrix} \varepsilon_{xx} & \frac{1}{2}\gamma_{xy} & \frac{1}{2}\gamma_{zx} \\ \frac{1}{2}\gamma_{xy} & \varepsilon_{yy} & \frac{1}{2}\gamma_{yz} \\ \frac{1}{2}\gamma_{zx} & \frac{1}{2}\gamma_{yz} & \varepsilon_{zz} \end{bmatrix} + \begin{bmatrix} 0 & -\Omega_z & \Omega_y \\ \Omega_z & 0 & -\Omega_x \\ -\Omega_z & \Omega_x & 0 \end{bmatrix}$$

Equating corresponding terms on the left-hand and right-hand sides of this equation, gives for the normal strain components:

$$\varepsilon_{xx} = \frac{\partial u_x}{\partial x}, \quad \varepsilon_{yy} = \frac{\partial u_y}{\partial y},$$

$$\varepsilon_{zz} = \frac{\partial u_z}{\partial z} \tag{5-35}$$

And,

$$\frac{\partial u_x}{\partial y} = \frac{1}{2}\gamma_{xy} - \Omega_z$$

$$\frac{\partial u_y}{\partial x} = \frac{1}{2}\gamma_{xy} + \Omega_z$$

Thus, expressions for shear strain and rotation are given by:

$$\gamma_{xy} = \frac{\partial u_x}{\partial y} + \frac{\partial u_y}{\partial x}, \quad \Omega_z = \frac{1}{2}\left(\frac{\partial u_y}{\partial x} - \frac{\partial u_x}{\partial y}\right)$$

And similarly,

$$\gamma_{yz} = \frac{\partial u_y}{\partial z} + \frac{\partial u_z}{\partial y}, \quad \Omega_x = \frac{1}{2}\left(\frac{\partial u_z}{\partial y} - \frac{\partial u_y}{\partial z}\right)$$

$$\gamma_{zx} = \frac{\partial u_z}{\partial x} + \frac{\partial u_x}{\partial z}, \quad \Omega_y = \frac{1}{2}\left(\frac{\partial u_x}{\partial z} - \frac{\partial u_z}{\partial x}\right) \tag{5-36}$$

Eq. (5-35) and Eq. (5-36) indicate that the state of strain at a point in a body is completely defined by six independent components, and that these are related simply to the displacement gradients at the point. The form of Eq. (5-34a) indicates that a state of strain is specified by a second-order tensor.

5.1.8 Principal Strains

Since a state of strain is defined by a strain matrix or second-order tensor, determination of principal strains, and other manipulations of strain quantities, are completely analogous to the processes employed in relation to stress. Thus principal strains and principal strain directions are determined by the eigenvalues and associated eigenvectors of the strain matrix. Strain transformation under a rotation of axes is defined, analogously to Eq. (5-13), by:

$$[\varepsilon^*] = [R][\varepsilon][R]^T$$

Where,

$[\varepsilon]$, $[\varepsilon^*]$ —— the strain matrices expressed relative to the old and new sets of co-ordinate axes.

The volumetric strain Δ is defined by:

$$\Delta = \varepsilon_{xx} + \varepsilon_{yy} + \varepsilon_{zz}$$

The deviator strain matrix is defined in terms of the strain matrix and the volumetric strain by:

$$[\varepsilon] = \begin{bmatrix} \varepsilon_{xx} - \Delta/3 & \gamma_{xy} & \gamma_{zx} \\ \gamma_{xy} & \varepsilon_{yy} - \Delta/3 & \gamma_{yz} \\ \gamma_{zx} & \gamma_{yz} & \varepsilon_{zz} - \Delta/3 \end{bmatrix}$$

Plane geometric problems, subject to biaxial strain in the *xy* plane, for example, are described in terms of three strain components ε_{xx}, ε_{yy}, γ_{xy}.

5.1.9 Strain Compatibility Equations

Eq. (5-35) and Eq. (5-36), which define the components of strain at a point, suggest that the strains are mutually independent. The requirement of physical continuity of the displacement field throughout a continuous body leads automatically to analytical relations between the displacement gradients, restricting the degree of independence of strains. A set of six identities can be established readily from Eq. (5-35) and Eq. (5-36). Three of these identities are of the form:

$$\frac{\partial^2 \varepsilon_{xx}}{\partial y^2}+\frac{\partial^2 \varepsilon_{yy}}{\partial x^2}=\frac{\partial^2 \gamma_{xy}}{\partial x \partial y}$$

And three are of the form:

$$\frac{\partial^2 \varepsilon_{xx}}{\partial y \partial z}=\frac{\partial}{\partial x}\left(-\frac{\partial \gamma_{yz}}{\partial x}+\frac{\partial \gamma_{zx}}{\partial y}+\frac{\partial \gamma_{xy}}{\partial z}\right)$$

These expressions play a basic role in the development of analytical solutions to problems in deformable body mechanics.

5.1.10 Stress-strain Relations

It was noted previously that an admissible solution to any problem in solid mechanics must satisfy both the differential equations of static equilibrium and the equations of strain compatibility. It will be recalled that in the development of analytical descriptions for the states of stress and strain at a point in a body, there was no reference to, nor exploitation of, any mechanical property of the solid. The way in which stress and strain are related in a material under load is described qualitatively by its constitutive behaviour. A variety of idealized constitutive models has been formulated for various engineering materials, which describe both the time-independent and time-dependent responses of the material to applied load. These models describe responses in terms of elasticity, plasticity, viscosity and creep, and combinations of these modes. For any constitutive model, stress and strain, or some derived quantities, such as stress and strain rates, are related through a set of constitutive equations. Elasticity represents the most common constitutive behaviour of engineering materials, including many rocks, and it forms a useful basis for the description of more complex behaviour.

In formulating constitutive equations, it is useful to construct column vectors from the elements of the stress and strain matrices, i.e., stress and strain vectors are defined by:

$$[\sigma]=\begin{bmatrix}\sigma_{xx}\\ \sigma_{yy}\\ \sigma_{zz}\\ \sigma_{xy}\\ \sigma_{yz}\\ \sigma_{zx}\end{bmatrix},\ [\varepsilon]=\begin{bmatrix}\varepsilon_{xx}\\ \varepsilon_{yy}\\ \varepsilon_{zz}\\ \gamma_{xy}\\ \gamma_{yz}\\ \gamma_{zx}\end{bmatrix}$$

The most general statement of linear elastic constitutive behaviour is a generalized form of Hooke's Law, in which any strain component is a linear function of all the stress components, i.e.:

$$\begin{bmatrix}\varepsilon_{xx}\\ \varepsilon_{yy}\\ \varepsilon_{zz}\\ \gamma_{xy}\\ \gamma_{yz}\\ \gamma_{zx}\end{bmatrix}=\begin{bmatrix}S_{11} & S_{12} & S_{13} & S_{14} & S_{15} & S_{16}\\ S_{21} & S_{22} & S_{23} & S_{24} & S_{25} & S_{26}\\ S_{31} & S_{32} & S_{33} & S_{34} & S_{35} & S_{36}\\ S_{41} & S_{42} & S_{43} & S_{44} & S_{45} & S_{46}\\ S_{51} & S_{52} & S_{53} & S_{54} & S_{55} & S_{56}\\ S_{61} & S_{62} & S_{63} & S_{64} & S_{65} & S_{66}\end{bmatrix}\begin{bmatrix}\sigma_{xx}\\ \sigma_{yy}\\ \sigma_{zz}\\ \sigma_{xy}\\ \sigma_{yz}\\ \sigma_{zx}\end{bmatrix} \quad (5\text{-}37a)$$

Or,

$$[\varepsilon]=[S][\sigma] \quad (5\text{-}37b)$$

Each of the elements S_{ij} of the matrix $[S]$ is called a compliance or an elastic modulus. Although Eq. (5-37a) suggests that there are 36 independent compliances, a reciprocal theorem, such as that due to Maxwell (1864), may be used to demonstrate that the compliance matrix is symmetric. The matrix therefore contains only 21 independent constants.

In some cases, it is more convenient to apply Eq. (5-37) in inverse form, i.e.:

$$[\sigma]=[D][\varepsilon] \quad (5\text{-}38)$$

The matrix $[D]$ is called the elasticity matrix or the matrix of elastic stiffnesses. For general anisotropic elasticity there are 21 independent stiffnesses.

Eq. (5-37a) indicates complete coupling between all stress and strain components. The existence of axes of elastic symmetry in a body de-couple some of the stress-strain relations, and reduces the number of independent constants required to define the material elasticity. In the case of isotropic elasticity, any arbitrarily or isotropic axis in the medium is an axis of elastic symmetry. Eq. (5-37a), for isotropic elastic materials, reduces to:

$$\begin{bmatrix}\varepsilon_{xx}\\ \varepsilon_{yy}\\ \varepsilon_{zz}\\ \gamma_{xy}\\ \gamma_{yz}\\ \gamma_{zx}\end{bmatrix}=\frac{1}{E}\begin{bmatrix}1 & -v & -v & 0 & 0 & 0\\ -v & 1 & -v & 0 & 0 & 0\\ -v & -v & 1 & 0 & 0 & 0\\ 0 & 0 & 0 & 2(1+v) & 0 & 0\\ 0 & 0 & 0 & 0 & 2(1+v) & 0\\ 0 & 0 & 0 & 0 & 0 & 2(1+v)\end{bmatrix}\begin{bmatrix}\sigma_{xx}\\ \sigma_{yy}\\ \sigma_{zz}\\ \sigma_{xy}\\ \sigma_{yz}\\ \sigma_{zx}\end{bmatrix} \tag{5-39}$$

The more common statements of Hooke's Law for isotropic elasticity are readily recovered from Eq. (5-39), i. e.:

$$\varepsilon_{xx}=\frac{1}{E}[\sigma_{xx}-v(\sigma_{yy}+\sigma_{zz})]$$

$$\gamma_{xy}=\frac{1}{G}\sigma_{xy} \tag{5-40}$$

Where,

$$G=\frac{E}{2(1+v)}$$

The quantities E, G, and u are Young's modulus, the modulus of rigidity (or shear modulus) and Poisson's ratio. Isotropic elasticity is a two-constant theory, so that determination of any two of the elastic constants characterizes completely the elasticity of an isotropic medium.

The inverse form of the stress-strain Eq. (5-39), for isotropic elasticity, is given by:

$$\begin{bmatrix}\sigma_{xx}\\ \sigma_{yy}\\ \sigma_{zz}\\ \sigma_{xy}\\ \sigma_{yz}\\ \sigma_{zx}\end{bmatrix}=\frac{E(1-v)}{(1+v)(1-2v)}\begin{bmatrix}1 & v/(1-v) & v/(1-v) & 0 & 0 & 0\\ v/(1-v) & 1 & v/(1-v) & 0 & 0 & 0\\ v/(1-v) & v/(1-v) & 1 & 0 & 0 & 0\\ 0 & 0 & 0 & \frac{(1-2v)}{2(1-v)} & 0 & 0\\ 0 & 0 & 0 & 0 & \frac{(1-2v)}{2(1-v)} & 0\\ 0 & 0 & 0 & 0 & 0 & \frac{(1-2v)}{2(1-v)}\end{bmatrix}\begin{bmatrix}\varepsilon_{xx}\\ \varepsilon_{yy}\\ \varepsilon_{zz}\\ \gamma_{xy}\\ \gamma_{yz}\\ \gamma_{zx}\end{bmatrix} \tag{5-41}$$

The inverse forms of Eq. (5-40), usually called Lamé's equations, are obtained from Eq. (5-41), i. e.:

$$\sigma_{xx}=\lambda\Delta+2G\sigma_{xx}$$

$$\sigma_{xy}=G\gamma_{xy}$$

Where λ is Lamé's constant, defined by:

$$\lambda=\frac{2vG}{(1-2V)}=\frac{vE}{(1+v)(1-2V)}$$

Δ is the volumetric strain.

Transverse isotropic elasticity ranks second to isotropic elasticity in the degree of expression of elastic symmetry in the material behaviour. Media exhibiting transverse isotropy include artificially laminated materials and stratified rocks, such as shales. In the latter case, all lines lying in the plane of bedding are axes of elastic symmetry. The only other axis of elastic symmetry is the normal to the plane of isotropy. In Fig. 5-10, illustrating a stratified rock mass, the plane of isotropy of the material coincides with the x, y plane. The elastic constitutive equations for this material are given by:

$$\begin{bmatrix}\varepsilon_{xx}\\ \varepsilon_{yy}\\ \varepsilon_{zz}\\ \gamma_{xy}\\ \gamma_{yz}\\ \gamma_{zx}\end{bmatrix}=\frac{1}{E_1}\begin{bmatrix}1 & -v_1 & -v_2 & 0 & 0 & 0\\ -v_1 & 1 & -v_2 & 0 & 0 & 0\\ -v_2 & -v_2 & E_1/E_2 & 0 & 0 & 0\\ 0 & 0 & 0 & 2(1+v_1) & 0 & 0\\ 0 & 0 & 0 & 0 & E_1/G_2 & 0\\ 0 & 0 & 0 & 0 & 0 & E_1/G_2\end{bmatrix}\begin{bmatrix}\sigma_{xx}\\ \sigma_{yy}\\ \sigma_{zz}\\ \sigma_{xy}\\ \sigma_{yz}\\ \sigma_{zx}\end{bmatrix} \tag{5-42}$$

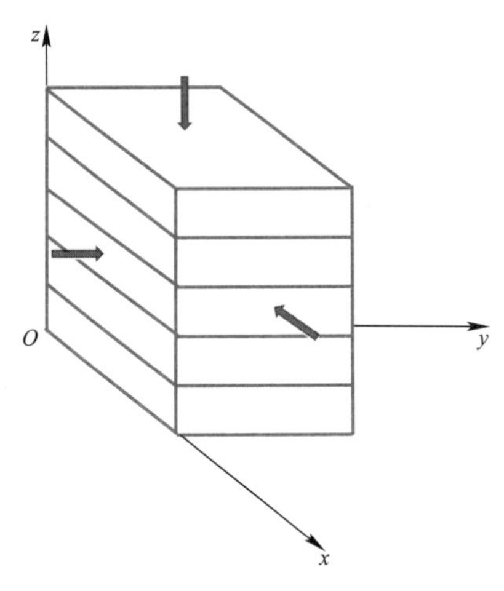

Fig. 5-10 A transversely isotropic body for which the x, y plane is the plane of isotropy

It appears from Eq. (5-42) those five independent elastic constants are required to characterize the elasticity of a transversely isotropic medium: E_1 and v_1 define properties in the plane of isotropy, and E_2, v_2, G_2 properties in a plane containing the normal to, and any line in, the plane of isotropy. Inversion of the compliance matrix in Eq. (5-42), and putting $E_1/E_2=n$, $G_2/E_2=m$, produces the elasticity matrix given by:

$$[D]=\frac{E_2}{(1+v_1)(1-v_1-2nv_2^2)}\begin{bmatrix} n(1-nv_2^2) & n(v_1+nv_2^2) & n^2v_2(1+v_1) & 0 & 0 & 0 \\ & n(1-nv_2^2) & n^2v_2(1+v_1) & 0 & 0 & 0 \\ & & (1-v_1^2) & 0 & 0 & 0 \\ & & & 0.5n(1-v_1-2nv_2^2) & 0 & 0 \\ & \text{symmetric} & & & m(1+v_1)(1-v_1-2nv_2^2) & 0 \\ & & & & & m(1+v_1)(1-v_1-2nv_2^2) \end{bmatrix}$$

Although it might be expected that the modulus ratios n and m, and Poisson's ratios v_1 and v_2, may be virtually independent, such is not the case. The requirement for positive definiteness of the elasticity matrix, needed to assure a stable continuum, restricts the range of possible elastic ratios.

5.2 Graphical Representation of Biaxial Stress

Analytical procedures for plane problems subject to biaxial stress have been discussed above. Where equations or relations appropriate to the two-dimensional case have not been proposed explicitly, they can be established from the three-dimensional equations by deleting any terms or expressions related to the third co-ordinate direction. For example, for biaxial stress in the x, y plane, the differential equations of static equilibrium, in the geomechanics convention, reduce to:

$$\frac{\partial\sigma_{xx}}{\partial x}+\frac{\partial\sigma_{xy}}{\partial y}-X=0$$

$$\frac{\partial\sigma_{xy}}{\partial x}+\frac{\partial\sigma_{yy}}{\partial y}-Y=0$$

One aspect of biaxial stress that requires careful treatment is graphical representation of the state of stress at a point, using the Mohr circle diagram. In particular, the geomechanics convention for the sense of positive stresses introduces some subtle difficulties which must be overcome if the diagram is to provide correct determination of the sense of shear stress acting on a surface.

Correct construction of the Mohr circle diagram is illustrated in Fig. 5-11. The state of stress in a small element *abcd* is specified, relative to the x, y co-ordinate axes, by known values of σ_{xx}, σ_{yy}, σ_{xy}. A set of reference axes for the circle diagram construction is defined by directions σ_n and τ, with the sense of the positive τ axis directed downwards. If O is the origin of the σ_n-τ co-ordinate system, a set of quantities related to the stress components is calculated from:

$$OC=\frac{1}{2}(\sigma_{xx}+\sigma_{yy})$$

$$CD=\frac{1}{2}(\sigma_{xx}-\sigma_{yy})$$

$$DF=-\sigma_{xy}$$

Points corresponding to C, D, F are plotted in the σ, τ plane as shown in Fig. 5-11, using some convenient scale. In the circle diagram construction, if σ_{xy} is positive, the point F plots above the σ_n axis. Construction of the line FDF' returns values of $\tau=\sigma_{xy}$ and $\sigma_n=\sigma_{xx}$ which are the shear and normal stress components acting on the surface *cb* of the element. Suppose the surface *ed* in Fig. 5-11 is inclined at an angle θ to the negative direction of the y axis, or, alternatively, its outward normal is inclined at an angle θ to the x axis. In the circle diagram, the ray FG is constructed at an angle θ to FDF', and the normal GH constructed. The scaled distances OH and HG then represent the normal and shear stress components on the plane *ed*.

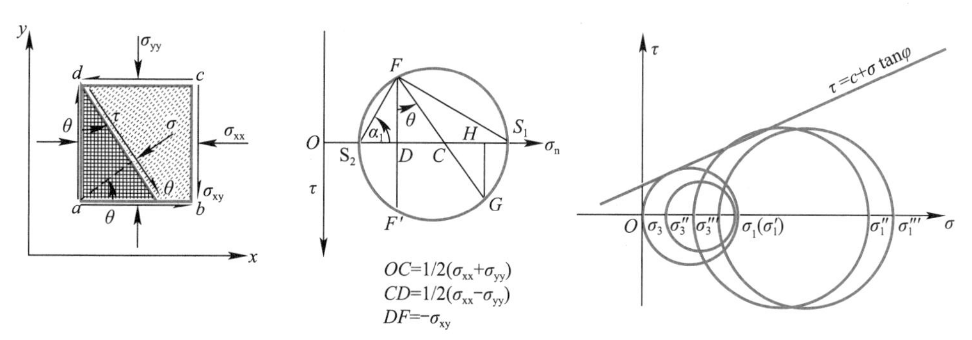

Fig. 5-11 Construction of a Mohr circle diagram, appropriate to the geomechanics convention of stresses

A number of useful results can be obtained or verified using the circle diagram. For example, OS_1 and OS_2 represent the magnitudes of the major and minor principal stresses σ_1, σ_2. From the geometry of the circle diagram, they are given by:

$$\sigma_{1,2}=OC\pm CF=\frac{1}{2}(\sigma_{xx}+\sigma_{yy})\pm\left[\sigma_{xy}^2+\frac{1}{4}(\sigma_{xx}-\sigma_{yy})^2\right]^{1/2}$$

Confirming the solution given in Eq. (5-24a). The ray FS_1 defines the orientation of the major principal plane, so FS_2, normal to FS_1, represents the orientation of the major principal axis. If this axis is inclined at an angle α_1, to the x axis, the geometry of the circle diagram yields:

$$\tan\alpha_1=\frac{(OS_1-OD)}{DF}=\frac{(\sigma_1-\sigma_{xx})}{\sigma_{xy}}$$

This expression is completely consistent with that for orientations of principal axes established analytically (Eq. 5-24b).

Sometimes the facilities required to prepare specimens and carry out uniaxial compression tests to the standard described above are not available. In other cases, the number of tests required to determine the properties of the range of rock types encountered on a project may become prohibitive. There may be still further cases, in which the uniaxial compressive strength and the associated stress-strain behaviour need not be studied in detail, with only an approximate measure of peak strength being required. In all of these instances, the point load test may be used to provide an indirect estimate of uniaxial compressive strength. This account is based on the ISRM Suggested Method for determining point load strength (ISRM Commission, 1985).

In this test, rock specimens in the form of core (the diametral and axial tests), cut blocks (the block test) or irregular lumps (the irregular lump test) are broken by a concentrated load applied through a pair of spherically truncated, conical platens.

The test can be performed in the field with portable equipment or in the laboratory using apparatus such as that shown in Fig. 5-12. The load should be applied at least $0.5D$ from the ends of the specimen in diametral tests, where D is the core diameter, and equivalent distances in other tests as specified by the ISRM Commission. From the measured value of the force P, at which the test specimen breaks, an Uncorrected Point Load Index I_s, is calculated as:

$$I_s = \frac{P}{D_e^2}$$

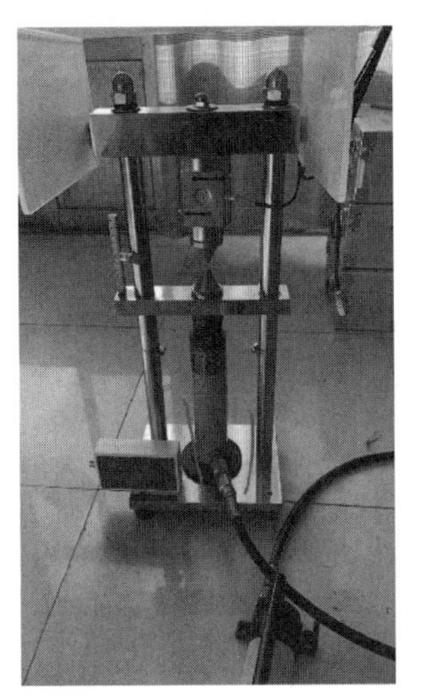

Fig. 5-12 Point load test apparatus

Where D_e, the equivalent core diameter, is given by the core diameter, D, for diametral tests, and by $4A/\pi$ for axial, block and lump tests, where A is the minimum cross-sectional area of a plane through the specimen and the platen contact points.

The index, I_s, varies with D_e and so size correction must be applied in order to obtain a unique point load strength index for a particular rock sample for use for strength classification. Wherever possible, it is preferable to carry out diametral tests on 50—55mm diameter specimens. The size-corrected Point Load Strength Index, $I_{s(50)}$, is defined as the value of I_s that would have been measured in a diametral test with $D = 50$mm. The results of several series of tests carried out by a number of investigators show that the value of Is determined in a test of equivalent diameter, D_e, may be converted to an $I_{s(50)}$ value by the relation:

$$I_{s(50)} = I_s \times \left(\frac{D_e}{50}\right)^{0.45}$$

Beginning with Broch and Franklin (1972), a number of investigators have developed correlations of the point load index with the uniaxial compressive strength, σ_c. The most commonly used correlation is:

$$\sigma_c \approx (22—24) I_{s(50)}$$

Caution must be exercised in carrying out point load tests and in interpreting the results. The test is one in which fracture is caused by induced tension, and it is essential that a consistent mode of failure be produced if the results obtained from different specimens are to be comparable. Very soft rocks, and highly anisotropic rocks or rocks containing marked planes of weakness such as bedding planes, are likely to give spurious results. A high degree of scatter is a general feature of point load test results and large numbers of individual determinations (often in excess of 100) are required in order to obtain reliable indices. For anisotropic rocks, it is usual to determine a strength anisotropy index $I_{a(50)}$, defined as the ratio of mean $I_{a(50)}$ values measured perpendicular and parallel to the planes of weakness.

5.3 Shear Behaviour of Discontinuities

The shear behavior of discontinuities plays a significant role in the overall stability and behavior of rock masses. Discontinuities refer to planes or surfaces of weakness within the rock mass, such as joints, faults, bedding planes, and fractures. These structural features can have a profound influence on the mechanical response of a rock mass when subjected to external forces or shearing stresses. Here are some key aspects of the shear behavior of discontinuities.

Shear strength: discontinuities possess their own shear strength characteristics, which may differ from the intact rock material. The shear strength of a discontinuity is often quantified using parameters such as cohesion (interfacial bond strength) and friction (angle of shearing resistance). These parameters can vary depending on factors such as roughness, alteration, mineralogy, and filling materials within the discontinuities.

Shear displacement: when subjected to shear forces, discontinuities can undergo displacement or sliding along the plane. The amount of shear displacement exhibited by a discontinuity depends on various factors, including the applied stress level, normal stress, surface roughness, persistence of the joint, and presence of interlocking blocks along the interface.

Shear stiffness: discontinuities exhibit their own stiffness or deformability characteristics during shear deformation. The stiffness of a discontinuity refers to its ability to resist shear deformation and can be influenced by factors such as joint roughness, infilling materials, asperity contact, and rock mass properties.

Shear behavior modes: discontinuities can exhibit different modes of shear behavior, which include pure sliding, dilation (opening of the joint), or contraction (closure of the joint). The mode of shear behavior is dependent on factors like the orientation of the discontinuity surface, boundary conditions, and stress paths.

Shear strength anisotropy: discontinuities often exhibit anisotropic shear strength behavior, meaning that their shear strength can vary depending on the direction of shear loading or the alignment of the discontinuity surface relative to the principal stress directions. This shear strength anisotropy is influenced by factors like joint roughness, mineralogy, and orientation.

Understanding the shear behavior of discontinuities is crucial for geotechnical engineers, geologists, and rock mechanics specialists involved in various applications such as slope stability analysis, underground excavations, rock support system design, and rock slope reinforcement. It helps in assessing the potential failure mechanisms, predicting deformation patterns, optimizing engineering designs, and implementing appropriate measures to ensure the stability and safety of rock masses. The most commonly used method for the shear testing of discontinuities in rock is the direct shear test.

As shown in Fig. 5-13, the discontinuity surface is aligned parallel to the direction of the applied shear force. The two halves of the specimen are fixed inside the shear box using a suitable encapsulating material, generally epoxy resin or plaster. This type of test is commonly carried out in the laboratory, but it may also be carried out in the field, using a portable shear box to test discontinuities contained in pieces of drill core or as an in situ test on samples of larger size.

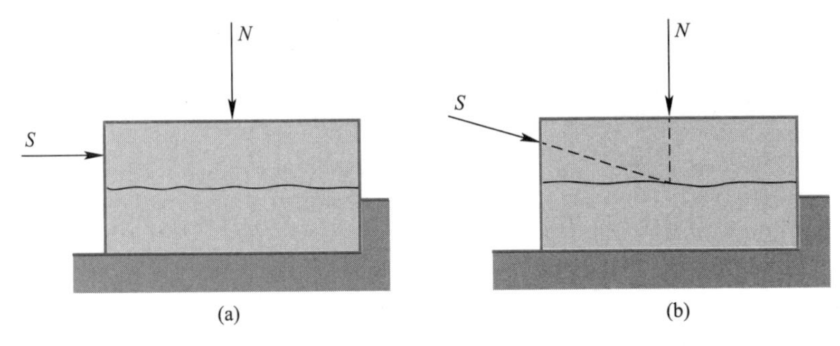

Fig. 5-13 Direct shear test
(a) The shear force applied parallel to the discontinuity; (b) An inclined shear force

Test arrangements of the type shown in Fig. 5-13(a) can cause a moment to be applied around the lateral axis on a discontinuity surface. This produces relative rotation of the two halves of the specimen and a non-uniform distribution of stress over the discontinuity surface. To minimise these effects, the shear force may be inclined at an angle (usually 10°—15°) to the shearing direction as shown in Fig. 5-13(b). This is almost always done in the case of large-scale in situ tests. Because the mean normal stress on the shear plane increases with the applied shear force up to peak strength, it is not possible to carry out tests in this configuration at very low normal stresses.

Direct shear tests in the configuration of Fig. 5-13(a) are usually carried out at constant normal force or constant normal stress. Tests are most frequently carried out on dry specimens, but many shear boxes permit specimens to be submerged and drained shear tests to be carried out with excess joint water pressures being assumed to be fully dissipated throughout the test. Undrained testing with the measurement of induced joint water pressures, is generally not practicable using the shear box.

The triaxial cell is sometimes used to investigate the shear behaviour of discontinuities. Specimens are prepared from cores containing discontinuities inclined at 25°—40° to the specimen axis. A specimen is set up in the triaxial cell as shown in Fig. 5-14(a) for the case of anisotropic rocks, and the cell pressure and the axial load are applied successively. The triaxial cell is well suited to testing discontinuities in the presence of water. Tests may be either drained or undrained, preferably with a known level of joint water pressure being imposed and maintained throughout the test.

It is assumed that slip on the discontinuity will occur. Mohr circle plots are made of the total or effective stresses at slip at a number of values of σ_3, and the points on these circles giving the stresses on the plane of the discontinuity are identified. The required shear strength envelope is then drawn through these points. This requires that a number of tests be carried out on similar discontinuities.

In an attempt to overcome the need to obtain, prepare and set up several specimens containing similar discontinuities, a stage testing procedure is sometimes used. A specimen is tested at a low confining pressure as outlined above. When it appears that slip on the discontinuity has just been initiated (represented by a flattening of the axial load-axial displacement curve that must be continuously recorded throughout each test), loading is stopped, the cell pressure is increased to a new value, and loading is recommenced. By repeating this process several times, a number of points on the peak strength envelope of the discontinuity can be obtained from the one specimen. However, this approach exacerbates the major difficulty involved in using the triaxial test to determine discontinuity shear strengths, namely the progressive change in the geometry of the cell-specimen system that accompanies shear displacement on the discontinuity.

The problem is illustrated by Fig. 5-14. It is clear from Fig. 5-14(a) that, if relative shear displacement of the two parts of the specimen is to occur, there must be lateral as well as axial relative translation. If, as is often the case, one spherical seat is used in the system, axial displacement causes the configuration to change to that of Fig. 4-36(b), which is clearly unsatisfactory. As shown in Fig. 5-14(c), the use of two spherical seats allows full contact to be maintained over the sliding surfaces, but the area of contact changes and frictional and lateral forces are introduced at the seats. Fig. 5-14(d) illustrates the most satisfactory method of ensuring that the lateral component of translation can occur freely and that contact of the discontinuity surfaces is maintained. Pairs of hardened steel discs are inserted between the platens and either end of the specimen. No spherical seats are used. The surfaces forming the interfaces between the discs are polished and lubricated with a molybdenum disulphide grease. In this way, the coefficient of friction between the plates can be reduced to the order of 0.005 which allows large amounts of lateral displacement to be accommodated at the interface with little resistance.

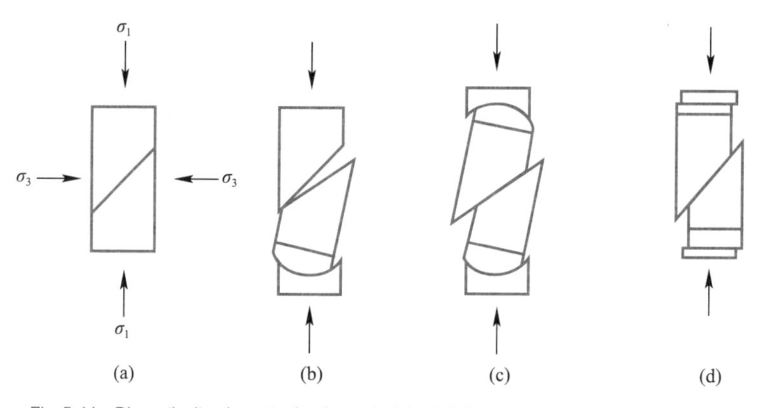

Fig. 5-14 Discontinuity shear testing in a triaxial cell (after Jaeger and Rosengren, 1969)

The roughness of joints or discontinuities has a significant effect on their shear behavior and the overall stability of rock masses. Joint roughness refers to the unevenness and irregularity of the joint surfaces, which can range from smooth and flat to rough and undulating. Here are some key effects of joint roughness on shear behavior.

Shear strength: the roughness of joint surfaces directly affects the shear strength of the discontinuity. Rough joint surfaces provide more interlocking points or asperities, resulting in increased frictional resistance along the joint plane. This leads to higher shear strength compared to smooth joint surfaces. The increased contact area between the joint surfaces also contributes to enhanced cohesion along the interface.

Shear displacement: joint roughness influences the propensity for shear displacement along the discontinuity. Rough joint surfaces tend to impede sliding due to the presence of irregularities and asperities that need to be overcome during shear deformation. As a result, rough joints generally require higher shear stresses to initiate and sustain shear displacement compared to smoother joints.

Dilatancy behavior: joint roughness can affect the dilatant behavior of the discontinuity during shear deformation. Rough joints with irregular surfaces tend to close or contract under shear loading, resulting in a decrease in aperture or dilation. This behavior can influence fluid flow, permeability, and the occurrence of localized strain patterns within the rock mass.

Shear stiffness: the roughness of joint surfaces influences the stiffness or deformability of the discontinuity when subjected to shear forces. Rough joints with high asperity contact tend to exhibit higher shear stiffness compared to smooth joints. This difference in shear stiffness affects the distribution of shear stress within the rock mass and can impact deformation patterns and stress redistribution.

Joints persistence: joint roughness can affect the persistence or continuity of joint sets within a rock mass. Rough joints often exhibit greater persistence, meaning they extend over longer distances before terminating orintersecting with other discontinuities. Persistent joint sets can act as planes of weakness and play a significant role in determining the overall stability and mechanical behavior of the rock mass.

Understanding the effect of joint roughness on shear behavior is crucial for assessing the stability of rock slopes, designing support systems for underground excavations, and predicting the behavior of rock masses under various loading conditions. Rock engineers and geologists consider joint roughness as one of the key parameters in characterizing the shear behavior of discontinuities and incorporating it into numerical models and engineering designs.

5.4 Anisotropic Behaviour of Discontinuous Rock Masses

Anisotropy is one of the most distinct features of stratified rock mass and it must be considered in engineering design and stability analysis. The way in which stress and strain are related in a material under load is described qualitatively by its constitutive behavior. A variety of idealized constitutive models has been formulated for various engineering materials. Elasticity represents the most common constitutive behavior of engineering materials, including many rocks, and it forms a useful basis for the description of more complex behavior. Rock materials exhibiting transverse isotropy include stratified rocks, such as shale or slate rocks. In Fig. 5-15, illustrating a stratified rock mass, the plane of isotropy of the material coincides with the x, y plane. The elastic constitutive equations for this material are given by Eq. (5-43) according to Brady and Brown (1993).

(a)

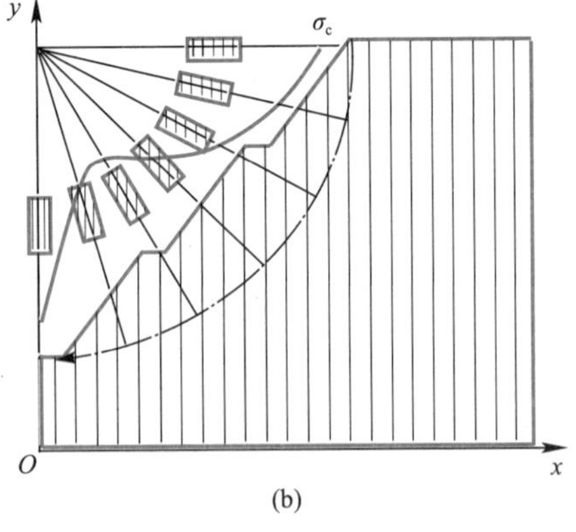

(b)

Fig. 5-15 Stratified rock mass in engineering scale and the related mechanical problem in core-drilling

(a) Stratified rock mass in geological engineering; (b) Diagram of rock samples cored from rock mass

$$\begin{bmatrix}\varepsilon_x\\ \varepsilon_y\\ \varepsilon_z\\ \gamma_{yz}\\ \gamma_{zx}\\ \gamma_{xy}\end{bmatrix}=\begin{bmatrix}\frac{1}{E} & -\frac{v'}{E'} & -\frac{v}{E} & 0 & 0 & 0\\ -\frac{v'}{E'} & \frac{1}{E'} & -\frac{v'}{E'} & 0 & 0 & 0\\ -\frac{v}{E} & -\frac{v'}{E'} & \frac{1}{E} & 0 & 0 & 0\\ & & & \frac{1}{G'} & & 0\\ & & & & \frac{2(1+v)}{E} & 0\\ & & & & & \frac{1}{G'}\end{bmatrix}\begin{bmatrix}\sigma_x\\ \sigma_y\\ \sigma_z\\ \tau_{yz}\\ \tau_{zx}\\ \tau_{xy}\end{bmatrix} \tag{5-43}$$

The quantities E, G, and v are Young's modulus, the shear modulus and Poisson's ratio. It appears from Eq. (5-43) that five independent elastic constants are required to characterize the elasticity of a transversely isotropic medium: E and v define properties in the plane of isotropy, and E', v', G' properties in a plane containing the normal to the plane of isotropy. The shear modulus G' could be obtained according to Saint-Venant empirical equation (Saint-Venant, 1863) by Eq. (5-44).

$$\frac{1}{G'}=\frac{1}{E}+\frac{1}{E'}+2\frac{v'}{E'} \tag{5-44}$$

In experimental tests for transversely isotropic rocks, the elasticity constitutive relationship can be simplified as a plane problem by Eq. (5-45). Once the five elastic parameters are obtained, the mechanical behavior of transversely isotropic rocks can be characterized. For this type of rocks, many researchers have performed laboratory tests and got the corresponding elastic parameters. However, only certain parameters are obtained such as rock specimens with certain dip angles under compression tests.

$$\begin{bmatrix}\varepsilon_1\\ \varepsilon_3\\ \varepsilon_{13}\end{bmatrix}=\begin{bmatrix}\frac{1}{E_1} & -\frac{v_{31}}{E_3} & 0\\ -\frac{v_{31}}{E_3} & \frac{1}{E_3} & 0\\ 0 & 0 & \frac{1}{G_{13}}\end{bmatrix}\begin{bmatrix}\sigma_1\\ \sigma_3\\ \sigma_{13}\end{bmatrix}, \tag{5-45}$$

Some researchers discuss the rotated constitutive relationships based on the rotating coordinate system, by experimentally or numerically analyzing the principal elastic parameters which are in the directions parallel and perpendicular to the structural planes. Though it is an efficient technique to apply the constitutive equations in some codes like COMSOL Multiphysics or FLAC, the actual mechanical behaviors are not represented. The influences of mechanical and geometrical properties of rock matrix as well as joint elements are not studied. In this section, a constitutive relationship is proposed by considering the mechanical and geometrical properties of rock matrix as well as joint. And the influence on elastic modulus and Poisson's ratio are also discussed. Furthermore, the theoretical equations are compared with experimental test results and the feasibility is discussed. Moreover, the proposed relationships can be used to estimate elastic parameters of stratified rocks under compressive test.

5.4.1 The Constitutive Relationship

A typical transversely isotropic rock with layered discontinuities is shown in Fig. 5-16. Relevant parameters for the problem are also illustrated in this figure, i. e. the stress conditions σ_1 and σ_3, the inclination angle of the layers φ, the spacing between layers s_m, the elastic parameters of rock E, the stiffness of joint elements k_n and k_s, and the normal stress and shear stress perpendicular and parallel to the layered planes.

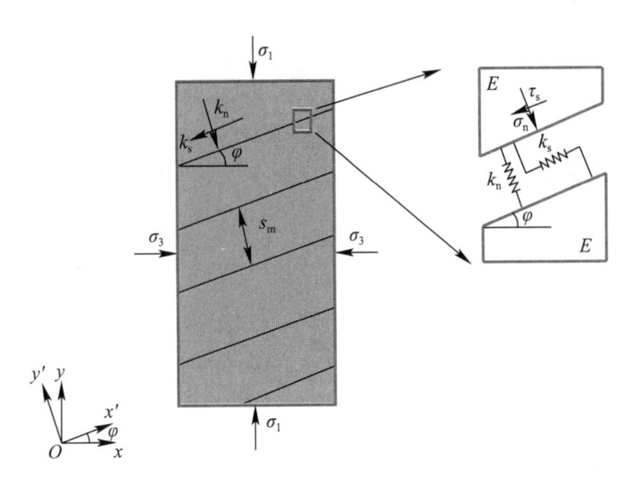

Fig. 5-16 The model of rock mass with one set of stratified joints

Assuming that the rock matrix is isotropic and linearly elastic material, and the joint planes are smooth and have no thickness. For plane problems, the strain is related to the stress conditions and elastic modulus through Eq. (5-46).

$$\varepsilon_i=\frac{1}{E_i}(\sigma_1-v_i\sigma_3) \tag{5-46}$$

Where,

ε_i—— the axial strain of the rock matrix;

v_i—— the Poisson's ratio.

The i subscript denotes the i-th rock matrix layer or rock bridge. Meanwhile, the displacements of the joint elements are controlled by the contact stiffnesses and the contact stress in the normal and shear directions are shown in Eq. (5-47) and Eq. (5-48).

$$\delta_n=\frac{\sigma_n}{k_n} \tag{5-47}$$

$$\delta_s=\frac{\tau_s}{k_s} \tag{5-48}$$

Where,

δ_n and δ_s—— the normal and shear displacement of joint element, respectively.

5.4.2 Equivalent Deformation Parameters in Axial Direction

Let δ_t be the total axial displacement of rock and joint, δ_i and δ_j is the axial displacement of rock matrix and joint, respectively. Therefore:

$$\delta_t=\delta_i+\delta_j \tag{5-49}$$

For the axial displacement of rock matrix, δ_i can be expressed by Eq. (5-50).

$$\delta_i=h\varepsilon_i=\frac{h}{E_i}(\sigma_1-v_i\sigma_3) \tag{5-50}$$

Where,

h —— the height of the model.

We adopt a global coordinate system, XOY, and a rotating coordinate system, $X'OY'$, which has the same direction as joint planes. φ is the angle of XOY and $X'OY'$ and the relationship is listed in Table 5-1. Then, the normal stress σ_n, and shear stress τ_s, which is perpendicular and normal to joint plane, respectively, can be obtained by Eq. (5-51), Eq. (5-52) and Eq. (5-53).

Relation of joint orientation and local coordinate Table 5-1

	X	Y
X'	$\cos\varphi$	$\sin\varphi$
Y'	$-\sin\varphi$	$\cos\varphi$

$$\begin{pmatrix}\sigma_n & 0\\ 0 & \tau_s\end{pmatrix}=\begin{pmatrix}\cos\varphi & \sin\varphi\\ -\sin\varphi & \cos\varphi\end{pmatrix}\begin{pmatrix}\sigma_1 & 0\\ 0 & \sigma_3\end{pmatrix}\begin{pmatrix}\cos\varphi & -\sin\varphi\\ \sin\varphi & \cos\varphi\end{pmatrix} \tag{5-51}$$

$$\sigma_n=\sigma_1\cos^2\varphi+\sigma_3\sin^2\varphi \tag{5-52}$$

$$\tau_s=(\sigma_1-\sigma_3)\sin\varphi\cos\varphi \tag{5-53}$$

We assigned the joint a normal stiffness k_n equal to the slope of the joint compression stress versus compressive displacement. Thus, the compressive displacement can be obtained by Eq. (5-54).

$$\delta_n=\frac{\sigma_n}{k_n}=\frac{\sigma_1\cos^2\varphi+\sigma_3\sin^2\varphi}{k_n} \tag{5-54}$$

Similarly, the shear displacement along joint planes can be obtained by Eq. (5-55).

$$\delta_s=\frac{\tau_s}{k_s}=\frac{(\sigma_1-\sigma_3)\sin\varphi\cos\varphi}{k_s} \tag{5-55}$$

Then, the axial displacement of single joint plane can be expressed as Eq. (5-56).

$$\delta_v(\varphi)=\delta_n\cos\varphi+\delta_s\sin\varphi=\frac{(\sigma_1\cos^2\varphi+\sigma_3\sin^2\varphi)\cos\varphi}{k_n}+\frac{(\sigma_1-\sigma_3)\sin^2\varphi\cos\varphi}{k_s} \tag{5-56}$$

Then the total axial displacement of joint elements can be obtained by Eq. (5-57) considering the spacing of joint planes.

$$\delta_j(\varphi)=\frac{h}{s_m/\cos\varphi}\delta_v(\varphi)=\frac{h(\sigma_1\cos^2\varphi+\sigma_3\sin^2\varphi)\cos^2\varphi}{s_m k_n}+\frac{h(\sigma_1-\sigma_3)\sin^2\varphi\cos^2\varphi}{s_m k_s} \tag{5-57}$$

Substituting Eq. (5-50) and Eq. (5-57) into Eq. (5-49) results in:

$$\delta_t=\delta_i+\delta_j=\left(\frac{1}{E_i}+\frac{\cos^4\varphi}{s_m k_n}+\frac{\sin^2\varphi\cos^2\varphi}{s_m k_s}\right)h\sigma_1-\left[\frac{v_i}{E_i}+\frac{(k_n-k_s)\sin^2\varphi\cos^2\varphi}{s_m k_n k_s}\right]h\sigma_3 \tag{5-58}$$

Then the axial strain ε_t, can be expressed in Eq. (5-59).

$$\varepsilon_t=\frac{\delta_t}{h}=\left(\frac{1}{E_i}+\frac{\cos^4\varphi}{s_m k_n}+\frac{\sin^2\varphi\cos^2\varphi}{s_m k_s}\right)\sigma_1-\left[\frac{v_i}{E_i}+\frac{(k_n-k_s)\sin^2\varphi\cos^2\varphi}{s_m k_s k_n}\right]\sigma_3 \tag{5-59}$$

We define E_1 and E_3 as the equivalent axial and lateral elastic modulus, respectively. ε_1 and ε_3 is the equivalent axial and lateral strain, respectively. v_{13} and v_{31} is the equivalent Poisson's ratio in axial and lateral directions, respectively. Then, ε_1 can be expressed as follows:

$$\varepsilon_1=\frac{1}{E_1}\sigma_1-\frac{v_{31}}{E_3}\sigma_3 \tag{5-60}$$

According to Eq. (5-59) and Eq. (5-60), the following equations can be established.

$$\frac{1}{E_1}=\frac{1}{E_i}+\frac{\cos^4\varphi}{S_m k_n}+\frac{\sin^2\varphi\cos^2\varphi}{S_m k_s} \tag{5-61}$$

$$\frac{v_{31}}{E_3}=\frac{v_i}{E_i}+\frac{(k_n-k_s)\sin^2\varphi\cos^2\varphi}{S_m k_s k_n} \tag{5-62}$$

As shown in Eq. (5-61) and Eq. (5-62), three unknown parameters E_1, E_3, and v_{31} are listed and cannot be obtained here. More equations about these parameters are needed.

5.4.3 Equivalent Deformation Parameters in Lateral Direction

Similarly, the lateral equivalent strain ε_c, can be expressed as Eq. (5-63).

$$\varepsilon_c=-\left[\frac{v_i}{E_i}+\frac{(k_n-k_s)\sin^2\varphi\cos^2\varphi}{s_m k_s k_n}\right]\sigma_1+\left(\frac{1}{E_i}+\frac{\sin^4\varphi}{s_m k_n}+\frac{\sin^2\varphi\cos^2\varphi}{s_m k_s}\right)\sigma_3 \tag{5-63}$$

Then, ε_3 can be expressed as follows:

$$\varepsilon_3=-\frac{v_{13}}{E_1}\sigma_1+\frac{1}{E_3}\sigma_3 \tag{5-64}$$

According to Eq. (5-63) and Eq. (5-64), the following equations can be established.

$$\frac{1}{E_3}=\frac{1}{E_i}+\frac{\sin^4\varphi}{s_m k_n}+\frac{\sin^2\varphi\cos^2\varphi}{s_m k_s} \tag{5-65}$$

$$\frac{v_{13}}{E_1}=\frac{v_i}{E_i}+\frac{(k_n-k_s)\sin^2\varphi\cos^2\varphi}{s_m k_s k_n} \tag{5-66}$$

Finally, the equivalent deformation parameters of the transversely isotropic rock are quantifiedas Eq. (5-67) by solving the four simultaneous equations of Eq. (5-61), Eq. (5-62), Eq. (5-65) and Eq. (5-66).

$$\begin{cases} E_1 = \dfrac{E_i k_n k_s s_m}{s_m k_n k_s + E_i k_s \cos^4\varphi + E_i k_n \sin^2\varphi\cos^2\varphi} \\ v_{13} = \dfrac{v_i k_n k_s s_m + E_i(k_n - k_s)\sin^2\varphi\cos^2\varphi}{s_m k_n k_s + E_i k_s \cos^4\varphi + E_i k_n \sin^2\varphi\cos^2\varphi} \\ E_3 = \dfrac{E_i k_n k_s s_m}{s_m k_n k_s + E_i k_s \sin^4\varphi + E_i k_n \sin^2\varphi\cos^2\varphi} \\ v_{31} = \dfrac{v_i k_n k_s s_m + E_i(k_n - k_s)\sin^2\varphi\cos^2\varphi}{s_m k_n k_s + E_i k_s \sin^4\varphi + E_i k_n \sin^2\varphi\cos^2\varphi} \end{cases} \tag{5-67}$$

5.4.4 Equivalent Elastic Modulus of Rock Mass

Han et al. (2011) have theoretically discussed axial deformation properties of layered rocks. However, the effect of lateral deformation on axial deformation is not considered in their study. To verify the reliability of the above equations, the geometrical and mechanical parameters are used in this section. In this case, the elastic modulus and Poisson's ratio of rock matrix is 30GPa and 0.2, respectively. Joint angle is 30° and the spacing between joint planes is 0.03m. The shear and normal stiffness of joint element are 20GPa/m and 100GPa/m, respectively. Except in special cases, the micro parameters in different scenarios are the same to analyze the influence on macro properties.

Fig. 5-17 shows the relationship between equivalent modulus of deformation and rock matrix modulus based on the proposed equations. In this section, the values of rock matrix modulus vary from 10 to 110GPa with an interval of 5GPa. Other parameters like joint spacing, stiffness, and dip angle, etc., remain unchanged. As shown in Fig. 5-17(a), the equivalent modulus in both axial and lateral directions have similarly increasing trends with the increase of elastic modulus of rock matrix. The values of equivalent modulus in lateral direction are all higher than those in axial directions. However, the ratios of equivalent modulus of jointed rock mass to that of the rock matrix decrease with the increase of rock modulus according to Fig. 5-17(b). The trends are sharp when the elastic modulus of rock matrix is lower than 40GPa.

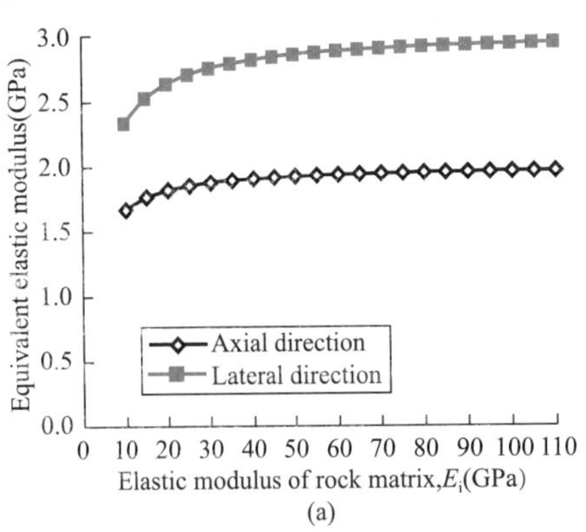

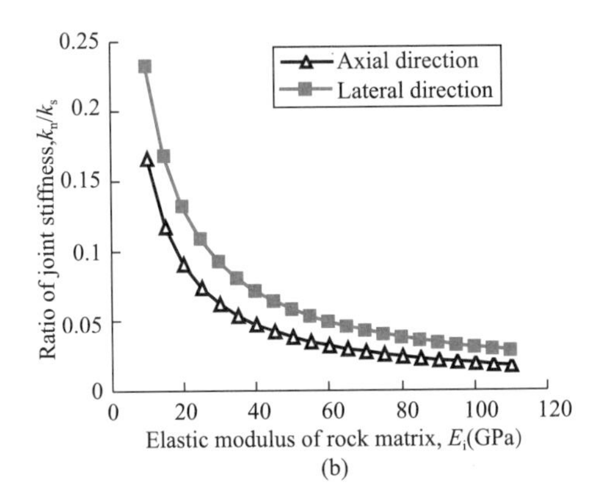

Fig. 5-17 Relationship curves between equivalent elastic modulus and elastic modulus E_i of rock matrix

(a) Relationship with the elastic modulus of rock matrix; (b) Ratio of elastic modulus of rock mass to rock matrix

The spacing of discontinuities determines the sizes of the blocks making up the jointed rock mass. The mechanism of deformation can vary with the joints spacing. To verify the effect of joint space, the oretical analysis is carried out with different spacings of bedding plane for the same bedding plane orientation of 45°. Fig. 5-18(b) depicts the relationship of axial and lateral modulus of deformation against joint spacings. Both the values of equivalent modulus increase significantly with increasing bedding plane orientation. However, the increasing trends gradually decrease with the increase of spacings. The difference between the normal stiffness and shear stiffness of joint elements determines the spatial distinction of mechanical properties of rock mass. Fig. 5-18(c) shows the relationship of axial and lateral equivalent modulus of deformation against the ratio of normal to shear stiffness. The shear stiffness is 20GPa/m and the ratio increased with the increasing normal stiffness. It is shown that the elastic modulus are significantly influenced by the ratios. The lateral modulus decrease with the increase of ratios, however, the axial modulus in-crease with the increase of ratios. In this scenario, it is observed that both of values of axial and lateral modulus tended to be a certain value after the ratio reaches 8.

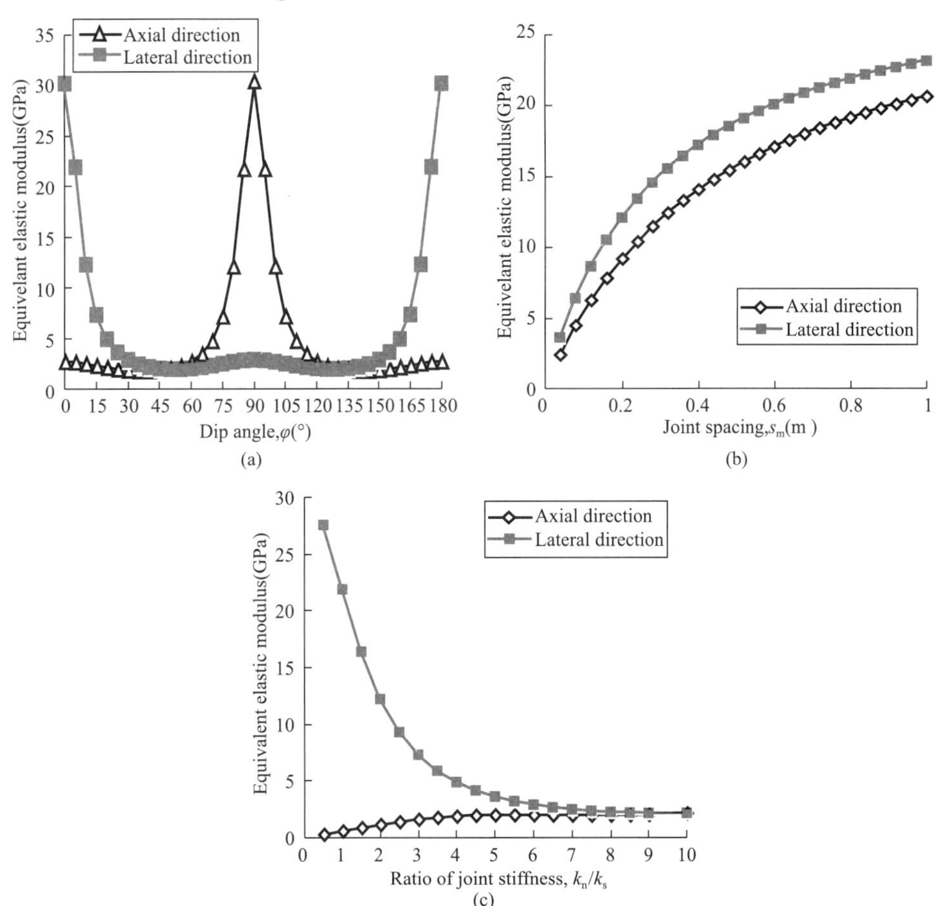

Fig. 5-18 Relationship curves between equivalent elastic modulus and geometric and mechanical properties of rock mass

(a) Relationship with dip angles; (b) Relationship with spacing of bedding planes; (c) Relationship with the joint stiffness ratio

5.4.5 Equivalent Poisson's Ratio of Rock Mass

Similarly, the relationship of equivalent Poisson's ratio with geometric and mechanical properties of rock mass is studied. Fig. 5-19(a) shows the development of equivalent Poisson's ratio with increasing elastic modulus of rock matrix. In this study, both of the axial and lateral Poisson's ratio increase slightly. It should be noticed that the value of lateral Poisson's ratio is higher than 0. 5 in this certain scenario. In this case, the normal and shear stiffness of joint element is 100GPa/m and 20GPa/m, respectively. The ratio of stiffness is 5. 0 and the Poisson's ratio exceeds 0. 5 according to Fig. 5-19(d). Due to the existence of joint elements, the response to the uncoordinated deformation appears. The curves of equivalent Poisson's ratio in both axial and lateral directions were symmetrical at $\varphi = 90°$ as is expected. Meanwhile, the values of two curves equal when $\varphi = 45°$ and 135° as shown in Fig. 5-19(b). Fig. 5-19(c) depicts the relationship with joint spacings. Both the axial and lateral values of Poisson's ratio show gradually decreasing trend with increasing joint spacings and approach to certain values.

The relationship of axial and lateral equivalent Poisson's ratio against the ratio of normal to shear stiffness is depicted in Fig. 5-19(d). In this scenario, the shear stiffness is 20GPa/m and the ratio increases with the increasing normal stiffness. It is shown that the equivalent Poisson's ratios in the two directions are significantly influenced by the ratio. All the values of Poisson's ratio increase with the increase of ratios. It is observed that all the values of Poisson's ratio are lower than 0. 5 when the ratio of stiffness is lower than 2. 5. For Poisson's ratio in axial directions, the value exceeds 0. 5 after the ratio of stiffness is higher than 5. 5. Both of values of axial and lateral Poisson's ratio tend to be a certain value after the ratio reaches 9. According to the results, Poisson's ratio can exceed 0. 5 in this scenario based on the proposed theoretical equations. The properties in other combinations of geometrical and mechanical parameters need a further study and will be discussed in the following research.

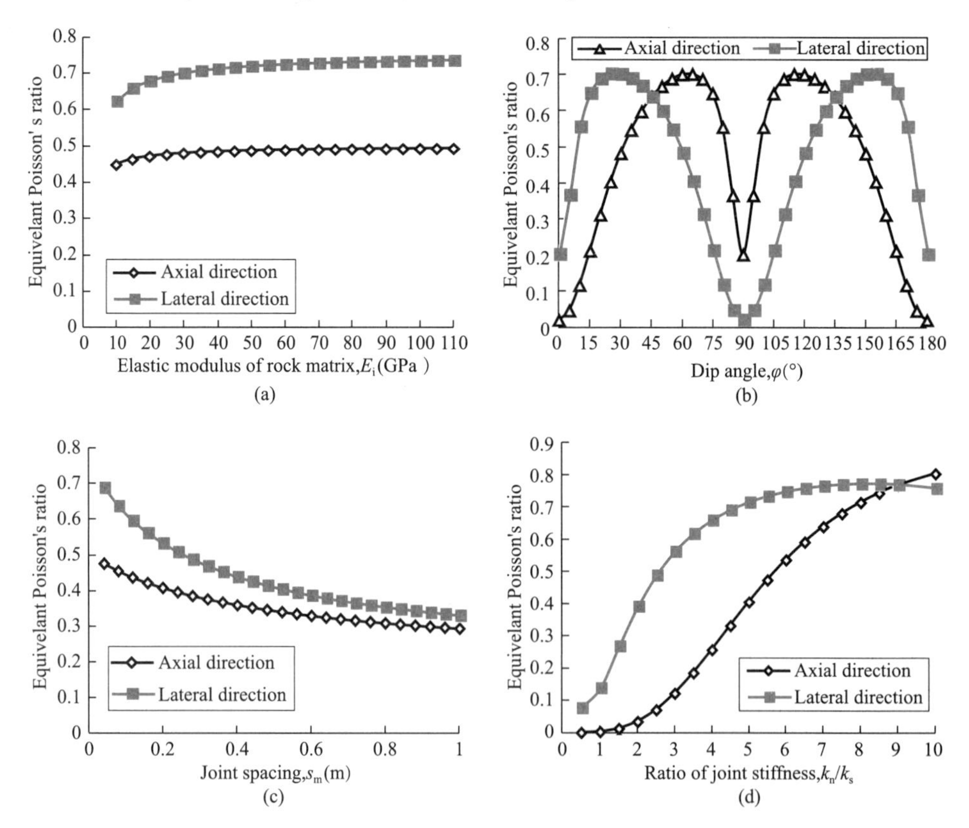

Fig. 5-19 Relationship curves between equivalent Poisson's ratio and geometric and mechanical properties of rock mass
(a) Relationship with elastic modulus of rock; (b) Relationship with dip angle of joint; (c) Relationship with joint spacing;
(d) Relationship with joint stiffness ratio

5.4.6 Verifications and Discussions

To verify the reliability of the proposed theoretical model, experimental tests are also performed on stratified rocks as shown in Fig. 5-20(a). Rock specimens with varied dip angles of bedding planes are cored and tested under uniaxial compression. And the experimental results are listed in Table 5-2. Moreover, the spacing of bedding planes is analyzed using the microphotograph of thin sections of rock as shown in Fig. 5-20(b). According to the laboratory test, the average elastic modulus of rock matrix is 12.0GPa, and the Poisson's ratio is 0.24. The average spacing between bedding planes is about 1.0mm.

The Eq. (5-68) and Eq. (5-69) show that the relationship between the properties of discontinuity, rock matrix and anisotropic rock materials can be described as follows (Goodman, 1989).

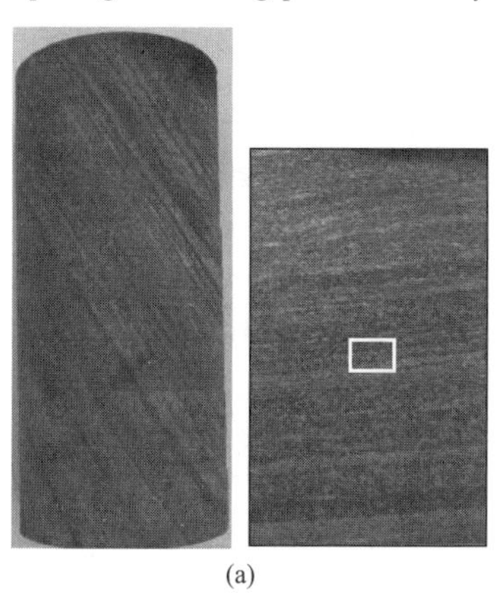

(a)

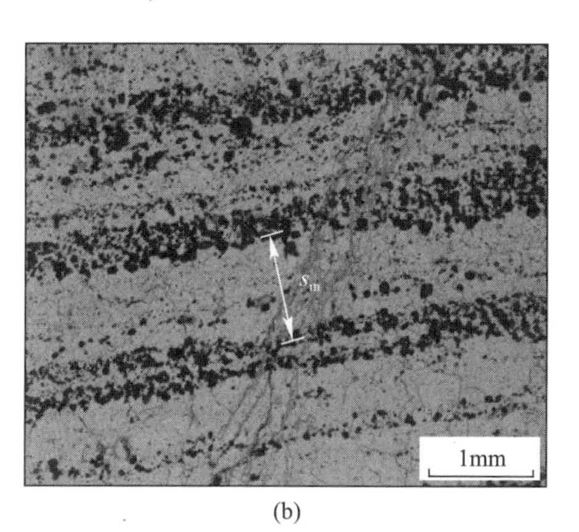

(b)

Fig. 5-20 The spacing analysis of the stratified rock specimen
(a) The layered rock sample; (b) Microphotograph of thin sections of rock

Deformation parameters in experimental tests Table 5-2

Mechanical properties		0°	15°	30°	45°	60°	75°	90°
Elastic modulus (GPa)	Lab test values	13.00	12.80	12.20	10.70	14.10	12.80	10.90
		11.40	10.80	9.42	10.00	7.50	11.80	13.60
		11.00	10.20	8.73	9.85	7.34	7.09	11.50
	Average	11.80	11.27	10.12	10.18	9.65	10.56	12.00
Poisson's ratio	Lab test values	0.21	0.24	0.30	0.23	0.28	0.26	0.24
		0.21	0.22	0.24	0.22	0.28	0.25	0.24
		0.17	0.18	0.20	0.33	0.25	0.23	0.24
	Average	0.20	0.21	0.25	0.26	0.27	0.25	0.24

$$E_e = \left(\frac{1}{E_r} + \frac{1}{\delta k_n} \right)^{-1} \tag{5-68}$$

$$G_e = \left(\frac{1}{G_r} + \frac{1}{\delta k_s} \right)^{-1} \tag{5-69}$$

Where,

δ —— a mean vertical spacing;

k_n and k_s —— normal and shear stiffness of discontinuities;

E_r and G_r —— Young's modulus and shear modulus in rock matrix;

E_e and G_e —— Young's modulus and shear modulus in transversely anisotropic rock with inclined angle of 0°, respectively.

Generally, the strength of joint elements

are lower than those of rock materials. However, the mechanical parameters of joint layer could also be higher than those of rock matrix, such as some kind of sandstone (Kim et al., 2016). Thus, the stiffness of joint could be expressed by Eq. (5-70) and Eq. (5-71).

$$k_n = \frac{E_r E_e}{\delta \mid E_r - E_e \mid} \tag{5-70}$$

$$k_s = \frac{G_r G_e}{\delta \mid G_r - G_e \mid} \tag{5-71}$$

From these relationships, the normal and shear stiffness of discontinuities can be obtained according to the experimental results. E_r, G_r, E_e, G_e and δ are determined as 12.0GPa, 4.84GPa, 11.80GPa, 4.92GPa and 1mm, respectively. Therefore, k_n and k_s are calculated as 7.08×10^{13}Pa/m and 2.98×10^{13}Pa/m, respectively. Fig. 5-21 (a) shows the comparison of elastic modulus between theoretical analysis and experimental results. The theoretical results fall in the range of the experimental results. Moreover, a similar trend is observed. Fig. 5-21(b) depicts the comparison of Poisson's ratio of stratified rocks. The theoretical results also have similar trend with those of experimental tests. The feasibility of proposed theoretical model can be efficiently used to evaluate the deformational behavior of stratified rocks. Furthermore, the proposed theoretical model considers geometrical and mechanical properties of rock mass, and thus, can also be used to numerically estimate the macro deformation behaviors of stratified rock mass in engineering scale.

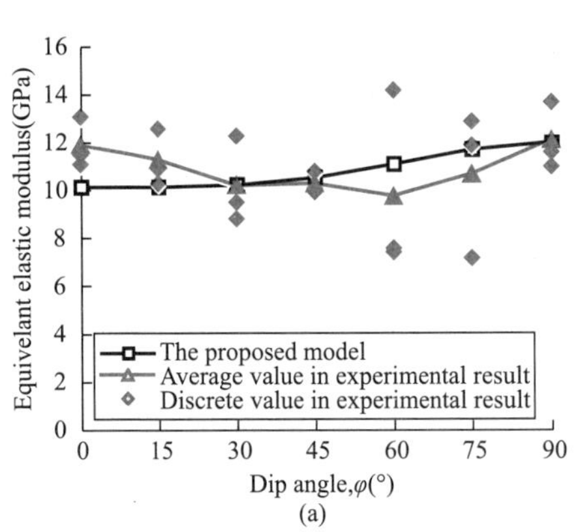

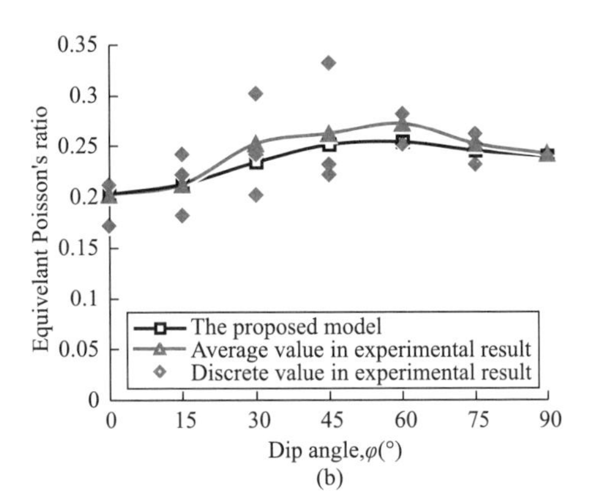

Fig. 5-21 Comparison of deformation characteristics in theoretical results with experimental results

(a) Comparison of elastic modulus; (b) Comparison of Poisson's ration

References

[1] Brady B. H. G., Brown E. T.. Rock Mechanics for Underground Mining [M]. 3rd ed. Boston: Kluwer Academic Publishers, 2004.

[2] Goodman, R. E.. Introduction to Rock Mechanics [M]. 2nd ed. New Jersey: Wiley, 1989.

[3] Jaeger J. C.. Friction of Rocks and Stability of Rock Slopes [J]. Géotechnique, 1971, 21(2): 97-134.

[4] Kim K. Y., Zhuang L., Yang H., et al. Strength Anisotropy of Berea Sandstone: Results of X-Ray Computed Tomography, Compression Tests, and Discrete Modeling [J]. Rock Mechanics and Rock Engineering, 2016, 49(4): 1201-1210.

Chapter 6
Slope Engineering and Stability

6.1 Introduction

An exposed ground surface that stands at an angle with the horizontal is called a slope. The slope can be natural or man-made, as shown in Fig. 6-1. Slope engineering involves the design and analysis of slope excavation and support, and construction. Rock slope engineering including slope excavation design and methods, slope stability analysis, rock slope support design and methods, and slope protection measures, will be covered in this section. It involves assessing and mitigating the risks associated with natural and man-made slopes, such as hillsides, embankments, cliffs, and cuttings.

Fig. 6-1 The natural slope and man-made slope

The primary goal of slope engineering is to ensure the safety and stability of slopes. Slope failures can result in significant damage to infrastructure, property, and even loss of life. Therefore, understanding the factors that influence slope stability and implementing appropriate engineering measures are crucial in slope engineering. Here are some key aspects of slope engineering.

Site investigation: this involves conducting a detailed study of the site, including geological, geomorphological, and hydrogeological characteristics. It helps in understanding the properties and behavior of the slope materials, identifying potential failure mechanisms, and selecting suitable engineering solutions.

Slope stability analysis: slope stability analysis evaluates the equilibrium and safety of slopes under different loading and environmental conditions. Various methods, such as limit equilibrium analysis, finite element analysis, or numerical modeling, are used to assess factors like soil strength, cohesion, internal friction, pore water pressure, and the influence of external forces like gravity and seismic activity.

Slope protection and stabilization: based on the analysis, appropriate slope protection and stabilization measures are designed. These may include retaining walls, ground reinforcement (e. g., anchors, soil nails), drainage systems, surface erosion control (e. g., vegetation, erosion blankets), and slope reinforcement techniques.

Monitoring and maintenance: continuous monitoring of slopes is essential for detecting any signs of instability or changes in slope behavior. Monitoring techniques can include instrumentation (such as inclinometers, piezometers, and ground-based radar), remote sensing technologies, and visual inspections. Regular maintenance and timely repairs help ensure the long-term stability and performance of engineered slopes.

Environmental considerations: slope engineering also considers environmental factors, such as the impact of slopes on ecosystems, water resources, and natural habitats. Appropriate environmental safeguards and mitigation measures are integrated into slope engineering projects.

Slope engineering is crucial in various applications, including transportation infrastructure (roads, railways), mining operations, urban development, and land management. It requires a multidisciplinary approach, integrating geotechnical engineering, geological analysis, hydrology, and environmental considerations to ensure the stability, safety, and sustainability of slopes in both natural and engineered environments.

6.2 Consequences of Slope Failures

This section describes the primary issues that need to be considered in rock slope design for civil projects and open pit mines. The basic difference between these two types of projects is that in civil engineering, a high degree of reliability is required because slope failure, or even rock falls, can rarely be tolerated. In contrast, some movement of open pit slopes is accepted if production is not interrupted, and rock falls are of little consequence. It is of interest to note that there is some correspondence between the steepest and highest stable slopes for both natural and man-made slopes.

The common design requirement for rock cuts is to determine the maximum safe cut face angle compatible with the planned maximum height. The design process is a trade-off between stability and economics. That is, steep cuts are usually less expensive to construct than flat cuts because there is less volume of excavated rock, less acquisition of right-of-way and smaller cut face areas. However, with steep slopes it may be necessary to install extensive stabilization measures such as rock bolts and shotcrete in order to minimize both the risk of overall slope instability and rock falls during the operational life of the project.

Typical open pit slope geometry shows the relationship between overall slope angle, inter-ramp angle and bench geometry (Fig. 6-2). Slope stability is the analysis of soil/rock covered slopes and its potential to undergo movement. Stability is determined by the balance of shear stress and shear strength. A previously stable slope may be initially affected by preparatory factors, making the slope conditionally unstable. Triggering factors of a slope failure can be climate events because they can make a slope actively unstable, leading to mass movements. Mass movements can be caused by increases in shear stress, such as loading, lateral pressure, and transient forces. Alternatively, shear strength may be decreased by weathering, changes in pore water pressure, and organic material. Failures of rock slopes, both man-made and natural, include rock falls, overall slope instability and landslides, as well as slope failures in open pit mines.

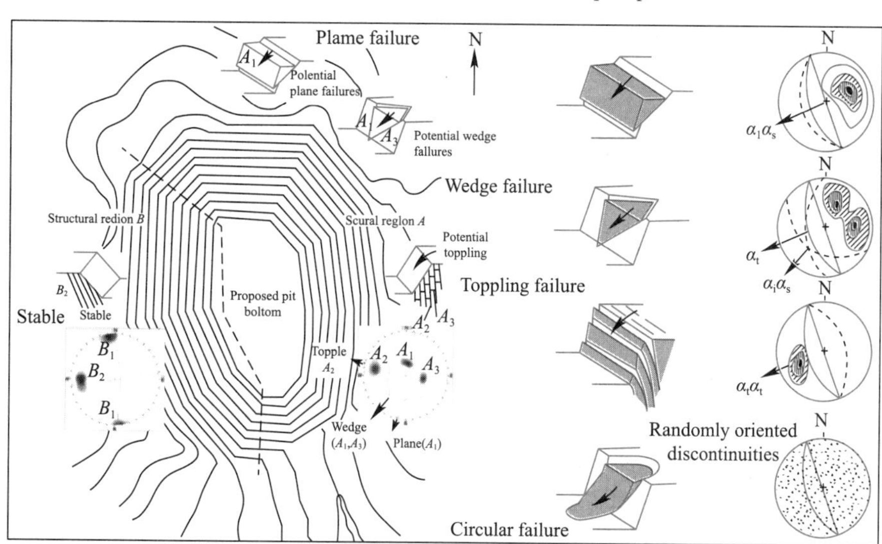

Fig. 6-2 The failure modes of the slopes of different situations (one)

Fig. 6-2 The failure modes of the slopes of different situations (two)

Landslide is a typical slope failure. Many-rock slopes move to varying degrees during the course of their operational lives. Such movement indicates that the slope is in a quasi-stable state, but this condition may continue for many years, or even centuries, without failure occurring. However, in other cases, initial minor slope movement may be a precursor for accelerating movement followed by collapse of the slope.

1. Circular Failure

Usually occurs in waste rock, heavily fractured rock and weak rock, with no identifiable structural pattern (Fig. 6-3). Rock masses are crushed or highly fractured rock masses. A strongly defined structural pattern no longer exists, and the failure surface is free to find a line of least resistance through the slope. The crushed or highly fractured rock masses are assumed to be homogeneous, and the shear strengths are controlled by cohesion and friction.

2. Plane Failure

Occurs in rocks with plane discontinuities, e. g., bedding planes (Fig. 6-4). The sliding plane must strike parallel or near parallel (within ±20°) to the slope plane. The sliding plane must"daylight" in the slope face, i. e., dip angle of sliding plane less than dip angle of slope. Dip angle of sliding plane is more than friction angle of sliding plane. Lateral resistance to sliding is negligible.

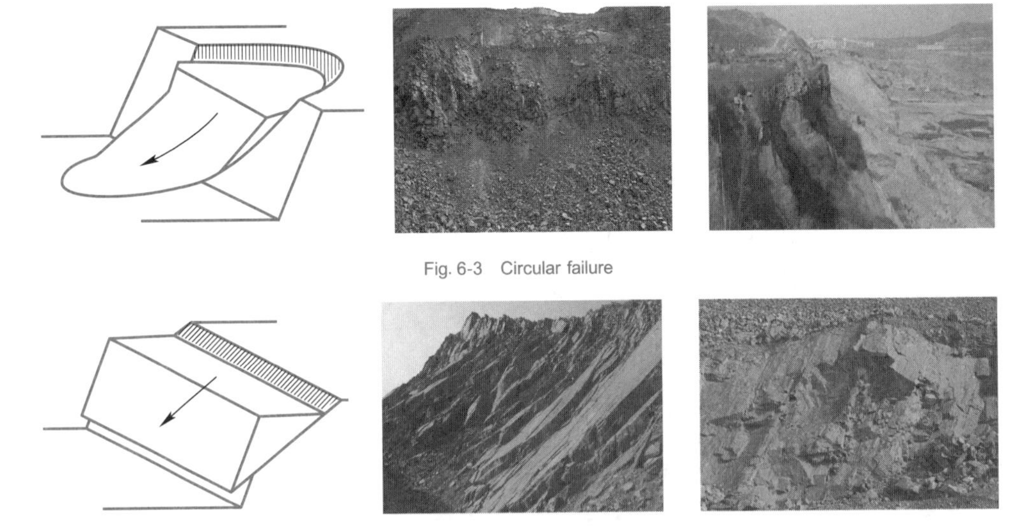

Fig. 6-3 Circular failure

Fig. 6-4 Plane failure

3. **Wedge Failure**

Occurs in rocks with intersecting discontinuities forming wedges (Fig. 6-5). Two joint planes and slope plane cut rock to form a wedge that is "daylight" in the slope face, i. e., plunge of the line of intersection of the joint (sliding) planes less than dip angle of the slope. Plunge of the line of intersection of the two sliding planes is more than angle of friction of the sliding planes.

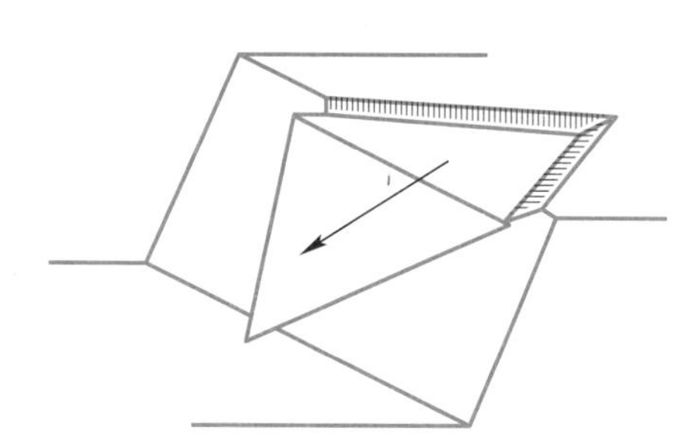

Fig. 6-5 Wedge failure

4. **Toppling failure**

Occurs in rocks with columnar or block structures separated by steeply dipping discontinuities (Fig. 6-6). Rock mass of block structures separated by steeply dipping discontinuities. Rock block width/height is less than the gradient of the toppling plane. When the dip angle of the toppling plane is less than the angle of friction of this plane, toppling only. When the dip angle of the toppling plane is more than the angle of friction of this plane, mixed toppling with sliding.

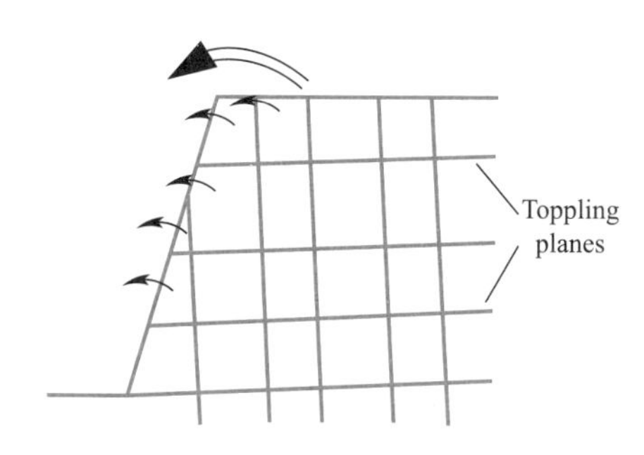

Fig. 6-6 Toppling failure

6.3 Factors Inducing Slope Failure

In addition to these man-made excavations, in mountainous terrain the stability of natural rock slopes may also be of concern. For example, high-ways and railways located in river valleys may be located below such slopes, or cut into the toe, which may be detrimental to stability. One of the factors that may infiuence the stability of natural rock slopes is the regional tectonic setting. Factors of safety may be only slightly greater than unity where there is rapid uplift of the land mass and corresponding down-cutting of the water-courses, together with earthquakes that loosen and displace the slope.

Geological factors include:

Low strength

Joints and fractures

Anisotropy
Discontinuous of seepage
Discontinuous of stiffness

Morphological factors include:

Tectonic or volcanic uplift
Glacial movement
Fluvial, wave, or glacial erosion of slope toe or lateral margins
Subterranean erosion
Deposition loading of slope or its crest
Vegetation removal (by fire, drought)

Thawing
Freeze-and-thaw weathering
Shrink-and-swell weathering

Human activities include:

Excavation of slope or its toe
Loading of slope or itscrest
Deforestation
Irrigation
Mining
Artificial vibration(explode)
Waterleakage from utilities

6.4 Slope Features and Dimensions

Landslides are geological phenomena characterized by the mass movement of rock, soil, or debris downslope. They can occur in various forms and exhibit different features depending on the type of material involved, the slope angle, and the triggering factors. Some common landslide features are shown in Fig. 6-7 and listed in Table 6-1 and Table 6-2. It's important to note that these features may not be present in every landslide and can vary depending on the specific circumstances. Additionally, landslides can occur in a variety of forms, such as rockfalls, debris flows, or rotational slides, each with their own distinct characteristics.

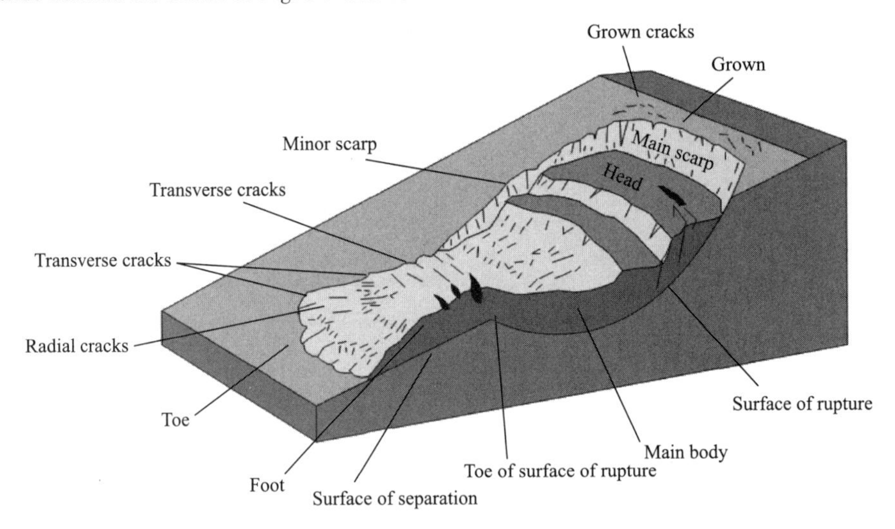

Fig. 6-7 The typical landslide features

Definition of landslide features **Table 6-1**

No.	Name	Definition
1	Crown	Practically undisplaced material adjacent to the highest parts of the main scarp
2	Main scarp	Steep surface on undisturbed ground at upper edge of landslide caused by movement of displaced material away from undisturbed ground; it is visible part of surface rupture

Continued

No.	Name	Definition
3	Top	Highest point of contact between displaced material and main scarp
4	Head	Upper parts of landslide along contact between displaced material and main scarp
5	Minor scarp	Steep surface on displaced material of landslide produced by differential movements within displaced material
6	Main body	Part of displaced material of landslide that overlies surface of rupture between main scarp and toe of surface of rupture
7	Foot	Portion of landslide that has moved beyond toe of surface of rupture and overlies original ground surface
8	Tip	Point on toe farthest from top of landslide
9	Toe	Lower, usually curved margin of displaced material of a landslide, most distant from main scarp
10	Surface of rupture	Surface that forms (or that has formed) lower boundary of displaced material below original ground surface; mechanical idealization of surface of rupture is called sliding surface in stability analysis
11	Toe of surface of rupture	Intersection (usually buried) between lower part of surface of rupture of a landslide and original ground surface
12	Surface of separation	Part of original ground surface now overlain by foot of landslide
13	Displaced material	Material displaced from its original positionon slope by movement in landslide; forms both depleted mass and accumulation
14	Zone of depletion	Area of landslide within which displaced material lies below original ground surface
15	Zone of accumulation	Area of landslide within which displaced material lies above original ground surface
16	Depletion	Volume bounded by main scarp, depleted mass, and original ground surface
17	Depleted mass	Volume of displaced material that overlies surface of rupture but underlies original ground surface
18	Accumulation	Volume of displaced material that lies above original ground surface
19	Flank	Undisplaced material adjacent to sides of surface of rupture; compass directions are preferable in describing flanks, but if left and right are used, they refer to flanks as viewed from crown
20	Original ground surface	Surface of slope that existed before landslide took place

Definitions of landslide dimensions Table 6-2

No.	Name	Definition
1	Width of displaced mass	Maximum breadth of displaced mass perpendicular to length L_d
2	Width of surface of rupture	Maximum width between flanks of landslide perpendicular to length L_r
3	Length of displaced mass	Minimum distance from tip to top
4	Length of surface of rupture	Minimum distance from toe of surface of rupture to crown
5	Depth of displaced mass	Maximum depth of surface of rupture below original ground surface measured perpendicular to plane containing W_d and L_d
6	Depth of surface of rupture	Maximum depth of surface of rupture below original ground surface measured perpendicular to plane containing W_r and L_r
7	Total length	Minimum distance from tip of landslide to crown
8	Length of center line	Distance from crown to tip of landslide through points on original ground surface equidistant from lateral margins of surface of rupture and displaced material

6.4.1 Summary of Design Methods

A basic feature of all slope design methods is that shear takes place along either a discrete sliding surface, or within a zone, behind the face. If the shear force (displacing force) is greater than the shear strength of the rock (resisting force) on this surface, then the slope will be unstable.

Instability could take the form of displacement that may or may not be tolerable, or the slope may collapse either suddenly or progressively. The definition of instability will depend on the application. For example, an open pit slope may undergo several meters of displacement without effecting operations, while a slope supporting a bridge abutment would have little tolerance for movement. Also, a single rock fall from a slope above a highway may be of little consequence if there is an adequate ditch to contain the fall, but failure of a significant portion of the slope that reaches the traveled surface could have serious consequences.

Based upon these concepts of slope stability, the stability of a slope can be expressed in using the factor of safety, FS—Stability quantified by limit equilibrium of the slope, which is stable if FS> 1. The factor of safety is the most common method of slope design, and there is wide experience in its application to all types of geological conditions, for both rock and soil. Furthermore, there are generally accepted factors of safety values for slopes excavated for different purposes, which promotes the preparation of reasonably consistent designs. The ranges of minimum total factors of safety as proposed by Terzaghi and Peck are given in Table 6-3. In Table 6-3, the upper values of the total factors of safety apply to usual loads and service conditions, while the lower values apply to maximum loads and the worst expected geological conditions. For open pit mines, the factor of safety is generally used in the range of 1.2—1.4, using either limit equilibrium analysis to calculate directly the factor of safety, or numerical analysis to calculate the onset of excessive strains in the slope.

Values of minimum total safety factors Table 6-3

Failure type	Category	Safety factor
Shearing	Earth works	1.3—1.5
	Earth retaining structures, excavations	1.5—2.0
	Foundations	2—3

6.4.2 Limit Equilibrium Analysis

Limit equilibrium analysis is a widely used method for assessing the stability of slopes and analyzing the potential for slope failure. It involves evaluating the forces acting on a slope and determining whether the resisting forces exceed the driving forces that could cause the slope to fail. The analysis assumes that the slope failure occurs along a specific failure surface within the soil or rock mass. The failure surface can be assumed to be either circular (for rotational failures) or planar (for translational failures).

In limit equilibrium analysis, the following steps are typicallyperformed.

- Define the geometry of the slope: the slope's geometry, including its height, inclination, and the shape of the potential failure surface, is defined.
- Identify the material properties: the engineering properties of the soil or rock materialsin the slope are determined, such as shear strength parameters and unit weights.
- Determine the external forces: the external forces acting on the slope, such as the weight of the soil or rock mass, groundwater pressure, and additional loads, are considered.
- Assess the internal forces: the internal forces within the slope, including the shear forces along the potential failure surface, are evaluated.
- Conduct a moment equilibrium analysis: the analysis involves establishing the equilibrium of moments around a critical point within the slope to determine the factor of safety against slope failure.
- Calculate the factor of safety: the factor of safety is the ratio of the resisting forces to the driving forces. If the factor of safety is equal to or greater than 1, the slope is considered stable; if it is less than 1, the slope may be prone to failure.
- Sensitivity analysis and design optimization: various analysis and iterative steps may be performed to optimize the slopedesign and assess the effect of different parameters on the factor of safety.

It is important to note that limit equilibrium analysis provides a simplified assessment of slope stability and has certain limitations. Factors such as rainfall infiltration, seismic forces, complex geological conditions, and time-dependent behavior are not explicitly considered in this analysis method. Therefore, professional judgment, field investigations, and additional analysis techniques may be necessary to fully evaluate slope stability and design appropriate mitigation measures.

For all shear type failures, the rock can be assumed to be a Mohr-Coulomb material in which the shear strength is expressed in terms of the cohesion c and friction angle φ. For a sliding surface on which there is an effective normal stress σ' acting, the shear strength τ developed on this surface is given by:

$$\tau = c + \sigma' \tan\varphi \quad (6\text{-}1)$$

Eq. (6-1) is expressed as a straight line on a normal stress—shear stress plot (Fig. 6-8a), in which the cohesion is defined by the intercept on the shear stress axis, and the friction angle is defined by the slope of the line. The effective normal stress is the difference between the stress due to the weight of the rock lying above the sliding plane and the uplift due to any water pressure acting on this surface. Fig. 6-8 (b) shows a slope containing a continuous joint dipping out of the face and forming a sliding block. Calculation of the factor of safety for the block shown in Fig. 6-8 (b) involves the resolution of the force acting on the sliding surface into components acting perpendicular and parallel to this surface.

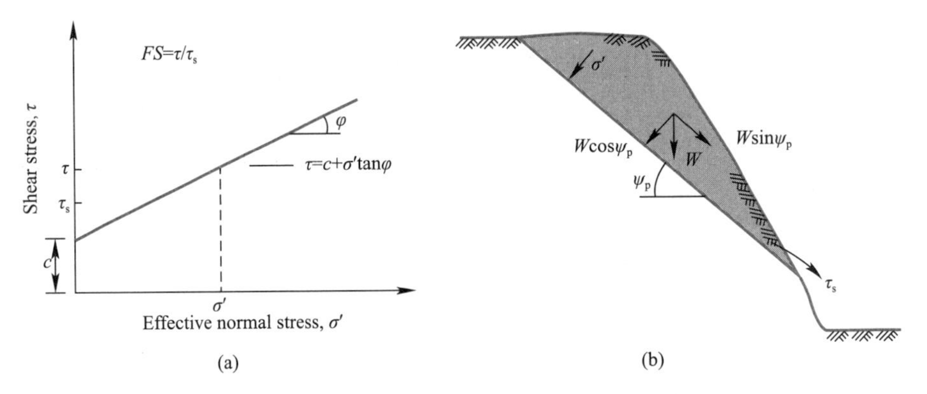

Fig. 6-8 Method of calculating factor of safety of sliding block

(a) Mohr diagram showing shear strength defined by cohesion c and friction angle φ;

(b) Resolution of force W due to weight of block into components parallel and perpendicular to sliding plane (ψ_p)

The factor of safety analysis described above involves selection of a single value for each of the parameters that define the driving and resisting forces in the slope. In reality, each parameter has a range of values, and a method of examining the effect of this variability on the factor of safety is to carry out sensitivity analysis using upper and lower bound values for those parameters considered critical to design. However, to carry out sensitivity analysis for more than three parameters is cumbersome, and it is difficult to examine the relationship between each of the parameters. Consequently, the usual design procedure involves a combination of analysis and judgment in assessing the influence on stability of variability in the design parameters, and then selecting an appropriate factor of safety.

6.5 Factor of Safety

An exposed ground surface that stands at an angle with the horizontal is called an unrestrained slope. The slope can be natural or man-made. It can fail in various modes. This chapter primarily relates to the quantitative analysis that falls under the category of slide. We will discuss in detail the following.

- Definition of factor of safety.
- Stability of infinite slopes.
- Stability of finite slopes with plane and circular failure surfaces.
- Analysis of the stability of finite slopes with steady-state seepage and in rapid drawdown conditions.

The task of the engineer charged with analyzing slope stability is to determine the factor of safety. Generally, the factor of safety is defined as:

$$F_s = \frac{\tau_f}{\tau_d} \tag{6-2}$$

Where,

F_s—— factor of safety with respect to strength;

τ_f—— average shear strength of the soil;

τ_d—— average shear stress developed along the potential failure surface.

The shear strength of a soil consists of two components, cohesion and friction, and may be written as:

$$\tau_f = c' + \sigma' \tan\varphi' \quad (6\text{-}3)$$

Where,

c' —— cohesion;

φ' —— angle of friction;

σ' —— normal stress on the potential failure surface.

In a similarmanner, we can write:

$$\tau_d = c'_d + \sigma' \tan\varphi'_d \quad (6\text{-}4)$$

Where c'_d and φ'_d are, respectively, the cohesion and the angle of friction that develop along the potential failure surface. Substituting Eq. (6-3) and Eq. (6-4) into Eq. (6-2), we get:

$$F_s = \frac{c' + \sigma' \tan\varphi'}{c'_d + \sigma' \tan\varphi'_d} \quad (6\text{-}5)$$

Now we can introduce some other aspects of the factor of safety—that is, the factor of safety with respect to cohesion $F_{c'}$, and the factor of safety with respect to friction $F_{\varphi'}$. They are defined as:

$$F_{c'} = \frac{c'}{c'_d} \quad (6\text{-}6)$$

And,

$$F_{\varphi} = \frac{\tan\varphi'}{\tan\varphi'_d} \quad (6\text{-}7)$$

When we compare Eq. (6-5) through Eq. (6-7), we can see that when $F_{c'}$ becomes equal to $F_{\varphi'}$, it gives the factor of safety with respect to strength. Or, if:

$$\frac{c'}{c'_d} = \frac{\tan\varphi'}{\tan\varphi'_d} \quad (6\text{-}8)$$

Then we can write:

$$F_s = F_{c'} = F_{\varphi'} \quad (6\text{-}9)$$

When F_s is equal to 1, the slope is in a state of impending failure. Generally, a value of 1.5 for the factor of safety with respect to strength is acceptable for the design of a stable slope.

References

[1] Terzaghi, K., Peck, R. B.. Soil Mechanics in Engineering Practice[M]. 3rd ed. New Jersey: Wiley, 1996.

Chapter 7
Underground Engineering

7.1 Introduction

Underground engineering refers to the construction of structures and infrastructure below the ground surface. This type of construction is typically used in urban areas where space is limited, or in cases where surface construction is not feasible due to environmental or other constraints. Underground engineering includes a wide range of structures, such as tunnels, subways, underground parking garages, and underground utility networks.

The design and construction of underground engineering projects presents a unique set of challenges, including the need to excavate and stabilize the ground, manage groundwater and soil pressure, and ensure adequate ventilation and lighting. Underground engineering projects often require specialized equipment and expertise to safely and efficiently construct and maintain. One of the key benefits of underground engineering is the ability to maximize land use in urban areas. By constructing infrastructure underground, valuable surface space can be preserved for other uses, such as parks or commercial development. Underground engineering can help to reduce traffic congestion and improve transportation efficiency by providing alternative routes for vehicles and pedestrians.

Despite the many benefits of underground engineering, there are also potential risks and challenges associated with this type of construction. These include the risk of ground instability or collapse, the potential for water infiltration and flooding, and the need for ongoing maintenance and repair of underground infrastructure. Fig. 7-1 shows an excavated tunnel, also called roadway. The surrounding rock mass may face the instability problems such as roof fall and rib spalling. To ensure the safe and effective design and construction of underground engineering projects, a variety of techniques are used, including geological surveys, geotechnical testing, and computer modeling. These tools can help to assess the feasibility of underground construction, identify potential risks and challenges, and inform decision-making throughout the project lifecycle.

Overall, underground engineering is an important and rapidly growing field, with significant potential to improve urban infrastructure and enhance the quality of life in cities around the world. With the development of deep underground engineering, high in-situ stresses. Rock at engineering scale is discontinuous, inhomogeneous, anisotropic, and non-linearly elastic. Rock mechanics deals with the response of rock when the boundary conditions are disturbed by engineering. Tunnel engineering involves design and analysis of tunnel excavation, support and construction.

Fig. 7-1 The underground tunnel based on 3D laser scanning (one)

Fig. 7-1 The underground tunnel based on 3D laser scanning (two)

7.2 Types of Underground Openings

Rock tunnel is a general term. It includes tunnels, caverns, and shafts, and for various applications (Fig. 7-2).

(a)

(b)

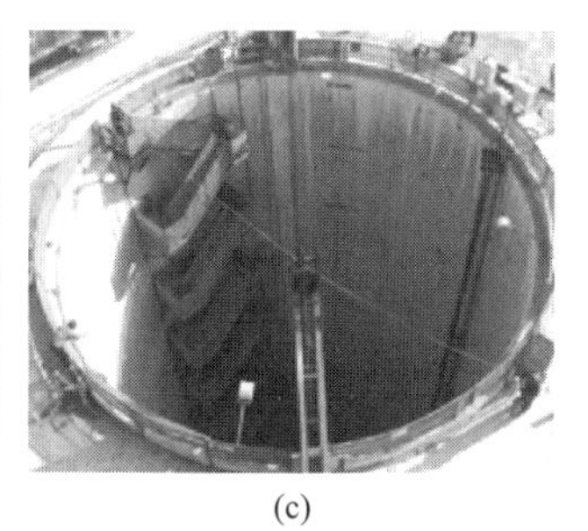

(c)

Fig. 7-2 Three types of underground openings
(a) Tunnel; (b) Cavern; (c) Shaft

7.3 Principal Failure Patterns

The principal failure patterns of tunnels can vary depending on factors such as geological conditions, construction methods, and environmental factors. Here are some common failure patterns that can occur in tunnels.

- **Tunnel collapse**: this is a severe failure pattern where the tunnel structure fails and caves in, typically due to inadequate support or unstable ground conditions. It can pose a significant risk to personnel and equipment inside the tunnel.
- **Ground settlement**: in some cases, tunnels can experience ground settlement, leading to deformation and sinking of the tunnel structure. This can occur due to the excavation process, soil consolidation, or changes in groundwater levels.
- **Water ingress**: tunnels may encounter water ingress from sources such as groundwater, surface water runoff, or leaking pipes. Excessive water infiltration can weaken the surrounding rock or soil, causing instability and potential collapse.
- **Rockbursts**: in situations where the tunnel passes through rock masses under high stress, sudden and violent failures known as

rockbursts can occur. This can result in the ejection of rock fragments and pose a severe hazard to personnel and infrastructure.

- **Slope instability**: tunnels constructed through slopes or mountainsides can be prone to slope instability. Landslides or slope failures adjacent to the tunnel can lead to deformation or collapse of the tunnel structure.
- **Foundation failure**: if the tunnel's foundation is poorly designed or constructed, it can result in the settlement or shifting of the tunnel structure. Foundation failure can compromise the stability and integrity of the tunnel.
- **Construction-induced damage**: sometimes, failures can occur during the construction phase of the tunnel. These can include accidental collapses, equipment malfunctions, or errors in construction techniques that result in structural damage.

It's worth noting that proper geotechnical investigations, engineering design, and appropriate construction techniques are essential to minimize the risk of these failure patterns. Continuous monitoring and maintenance of tunnels are crucial to identify and address any emerging issues promptly. Some engineering problems met during the excavation of tunnel, include squeeze, water inrush, and mixed faces. In summary, there are four principal failure patterns for underground engineering.

7.3.1 Overburden Soils and Heavily Weathered Rock

At shallow depth in overburden soil or heavily weathered poor-quality rock, excavation problems are generally associated with squeezing or flowing ground and very short stand-up times. Either cut or cover of soft ground tunnelling techniques have to be used and adequate support has to be provided immediately behind the advancing face, as shown in Fig. 7-3(a).

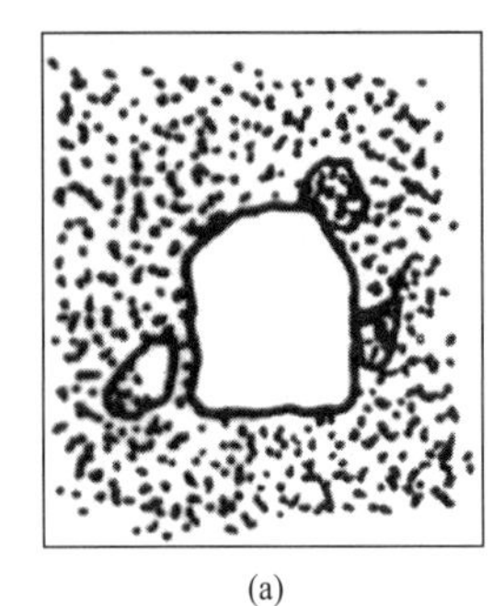
(a)

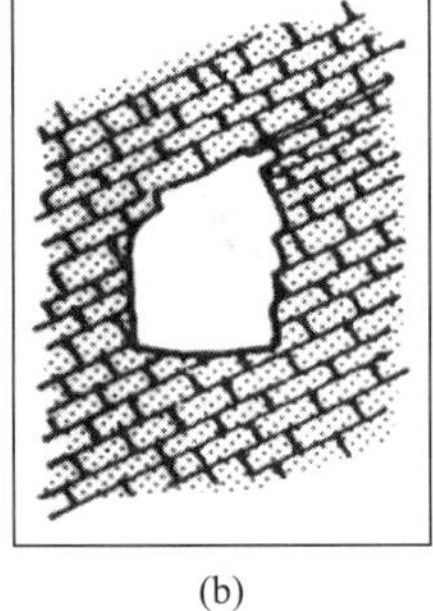
(b)

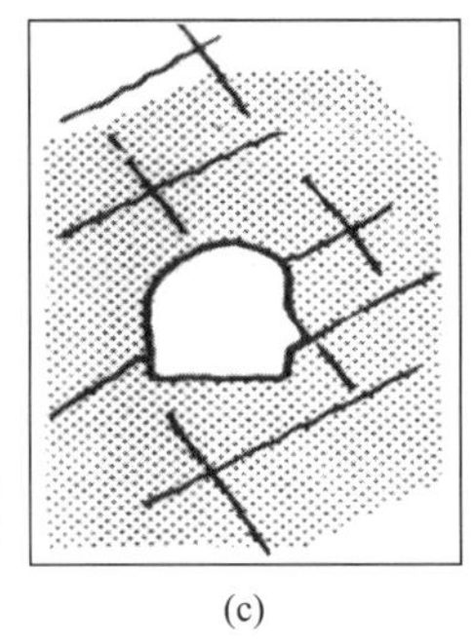
(c)

(d)

Fig. 7-3 Four principal failure patterns
(a) Overburden soils and heavily weathered rock; (b) Blocky jointed and partially weathered rock; (c) Massive rock with few unweathered joints; (d) Massive rock at great depth

7.3.2 Blocky Jointed and Partially Weathered Rock

Stability problems in blocky jointed rock mass are generally associated with gravity falling and sliding of blocks from roof and sidewalls. Rock stress at shallow depth is generally low that does not control the failure mechanism. Structurally controlled failure can be analyzed by projection method. The optimum orientation and shape of an excavation should have the smallest volume of potentially unstable wedges, as shown in Fig. 7-3(b).

7.3.3 Massive Rock with Few Unweathered Joints

Excavation in unweathered massive rock mass with few joints do not usually suffer from serious stability problems when the stresses in the rock surrounding the excavations are less than approximately 1/5 of the uniaxial compressive strength of the rock material, as shown in Fig. 7-3(c).

7.3.4 Massive Rock at Great Depth

At depth, rock stress increases to a level at which failure is induced in the rock surrounding the excavation. This stress induced failure can range from minor spalling or slabbing in the surface rock to major rock bursts. Shape of the excavation open can be optimized to minimize the potential of stress induced failure, as shown in Fig. 7-3(d).

Falling or sliding of wedges or blocks released by intersecting discontinuities, as shown in Fig. 7-4. This type of failure is structurally controlled failure, generally occurs in hard rock at relatively shallow depth. Spalling, popping and rock bursts of rock caused by high in situ stresses. This type of failure is stress induced failure, and occurs in highly stressed brittle rock. Squeezing is the large deformation resulting from "plastic" failure of relatively ductile rock masses when the ratio of rock mass strength to in-situ stress falls below about 30%.

Fig. 7-4 Failure mechanism of rock tunnels

7.4 Excavation Design for Massive Elastic Rock

Rock is used as a structural material, i. e., often rock masses are not supported but primarily reinforced. Support design is based on rock mass quality, i. e., empirical methods, with deformation as design control/criterion. Numerical methods are used to predict problem areas and to extrapolate experience. Monitoring is used to verify and modify support. When designing the excavation of tunnels or underground openings in massive elastic rock, several key considerations should be taken into account. Here are some important factors to consider.

- **Geotechnical site investigation**: conduct a thorough geotechnical site investigation to gather information about the rock mass properties, such as strength, deformability, and bedding characteristics. This data is essential for determining the appropriate excavation method and support system.
- **Excavation method**: select an appropriate excavation method based on the geotechnical properties of the rock mass. Common methods include drilling and blasting, mechanical excavation (such as tunnel boring machines), or a combination of both. The selected method should consider factors such as rock strength, stability, anticipated stress redistribution, and environmental impact.
- **Excavation geometry and support**

design: determine the optimal tunnel size and shape considering the geotechnical conditions and intended use of the tunnel. The design should ensure proper stability and support during and after excavation. This includes selecting suitable ground reinforcement measures, such as rock bolts, shotcrete, steel supports, or other support systems based on the anticipated rock behavior.

- **Ground reinforcement and stabilization**: develop a comprehensive ground reinforcement plan to enhance the stability of the tunnel. This may involve installing rock bolts, steel arches, shotcrete lining, or other reinforcement measures to control potential rock deformation and ensure structural integrity.
- **Rock mass behavior analysis**: analyze the behavior of the rock mass during excavation using numerical modeling techniques. This can help predict potential ground movement, stress redistribution, and potential failure modes. It allows for evaluating different scenarios and optimizing the support design accordingly.
- **Monitoring system**: implement a monitoring system to continuously assess the performance of the excavated tunnel. This can include monitoring devices such as convergence measurements, strain gauges, or displacement sensors. Regular monitoring helps detect potential issues and allows for timely corrective actions.
- **Construction sequence and phasing**: develop a construction sequence and phasing plan to minimize potential impacts on the rock mass stability. This involves considering factors such as stress redistribution, support installation order, and excavation rates to mitigate risks associated with ground movements.

It's important to note that each project and site will have unique characteristics, and a qualified geotechnical engineer or tunneling expert should be involved in the design process. Their expertise and experience will ensure that the excavation design adequately addresses the specific conditions of the massive elastic rock mass. Mining excavations are of two types—service openings and production openings. Service openings include mine accesses, ore haulage drives, airways, crusher chambers and underground workshop space. They are characterized by a duty life approaching the mining life of the orebody. It is therefore necessary to design these openings so that their operational functions can be assured and maintained at low cost over a relatively long operational life. Mine production openings have a temporary function in the operation of the mine. These openings include the ore sources, or stopes, and related excavations such as drill headings, stope accesses, and ore extraction and service ways. In these cases, it is necessary to assure control of the rock around the excavation boundary only for the life of the stope, which may be as short as a few months.

7.4.1 Zone of Influence of an Excavation

The most common examples of such problems are stresses around boreholes, or around long tunnels, as shown in Fig. 7-5. Many other problems are idealized as being two dimensional so as to take advantage of the relative ease of solving two-dimensional elasticity problems as compared to three-dimensional problems. Hence, it is worthwhile to study the properties of two-dimensional stress tensors. The concept of a zone of influence is important in mine design, since it may provide considerable simplification of a design problem. The essential idea of a zone of influence is that it defines a domain of significant disturbance of the pre-mining stress field by an excavation. It differentiates between the near field and far field of an opening. The extent of an opening's effective near-field domain can be explained by the following example. The stress distribution around a circular hole in a hydrostatic stress field, of magnitude p, is given by Eq. (7-1) as:

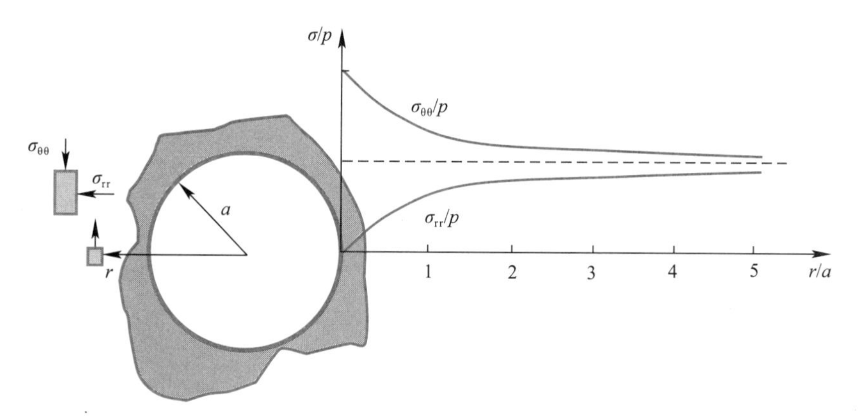

Fig. 7-5 Axisymmetric stress distribution around a circular opening in a hydrostatic stress field

$$\sigma_{rr}=p\left(1-\frac{a^2}{r^2}\right)$$
$$\sigma_{\theta\theta}=p\left(1+\frac{a^2}{r^2}\right) \quad (7\text{-}1)$$
$$\sigma_{r\theta}=0$$

Eq. (7-2) indicates that the stress distribution is axisymmetric, and this is illustrated in Fig. 7-5. Using Eq. (7-2), it is readily calculated that for $r = 5a$, $\sigma_{\theta\theta} = 1.04p$ and $\sigma_{rr} = 0.96p$, i.e. on the surface defined by $r = 5a$, the state of stress is not significantly different (within ±5%) from the field stresses.

If a second excavation (Ⅱ) is generated outside the surface described by $r = 5a$ for the excavation I, as shown in Fig. 7-6, the pre-mining stress field will not be significantly different from the virgin stress field. The boundary stresses for excavation Ⅱ are thus those for an isolated excavation. Similarly, if excavation I is outside the zone of influence of excavation Ⅱ, the boundary stresses around excavation I are effectively those for an isolated opening. The general rule is that openings lying outside one another's zones of influence can be designed by ignoring the presence of all others. For example, for circular openings of the same radius a, in a hydrostatic stress field, the mechanical interaction between the openings is insignificant if the distance $D_{\mathrm{I},\mathrm{II}}$ between their centres is:

$$D_{\mathrm{I},\mathrm{II}} \geqslant 6a \quad (7\text{-}2)$$

It is important to note that, in general, the zone of influence of an opening is related to both excavation shape and pre-mining stresses. Other issues related to the notion of zone of influence include the state of stress in a medium containing a number of excavations, and interaction between different-sized excavations. Fig. 7-6 illustrates the overlap of the zones of influence of two circular openings. In the overlap region, the state of stress is produced by the pre-mining stresses and the stress increments induced by each of the excavations Ⅰ and Ⅱ. In the other sections of each zone of influence, the state of stress is that due to the particular excavation.

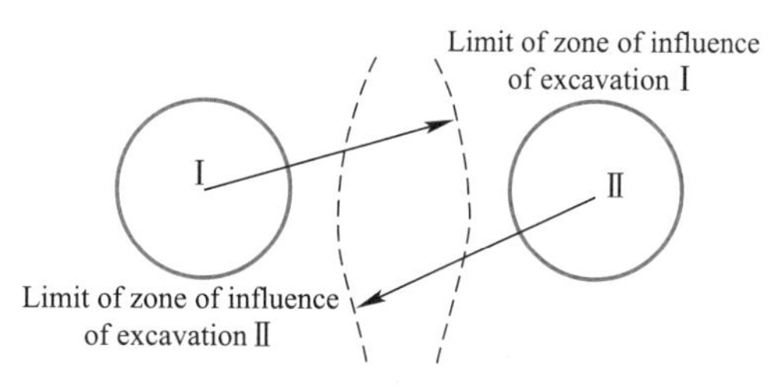

Fig. 7-6 Circular openings in a hydrostatic stress field, effectively isolated by virtue of their exclusion from each other's zone of influence

7.4.2 Zones of Rock Failure

In assessing the performance of excavations and rock structures, it is useful to distinguish between failure of the structure, and failure or fracture of the rock mass. Failure of a structure implies that it is unable to fulfil the designed duty requirement. Failure of a rock structure in massive rock is synonymous with extensive rock fracture, since the stable performance of the structure under these conditions cannot be assured. In a mine structure, control of displacements in a fractured rock mass may require the installation of designed support elements, or implementation of a mining sequence which limits the adverse consequences of an extensive fracture domain. On the other hand, limited fractured rock zones may pose no mining problem, and a structure or opening may completely satisfy the design duty requirements. A simple method of estimating the extent of fracture zones provides a basis for the prediction of rock mass performance, modification of excavation designs, or assessing support and reinforcement requirements.

This indicates that for any value of the minimum principal stress at a point, a major principal stress value can be determined which, if reached, leads to local failure of the rock. When assessing the possibility of rock failure at an excavation boundary, the applicable compressive strength parameter of the rock mass is the crack initiation stress, σ_{ci}. For design purposes, the tensile strength of the rock mass is taken to be universally zero. Prediction of the extent of boundary failure may be illustrated by reference to a circular excavation in a biaxial stress field, as shown in Fig. 7-7. Boundary stresses are given by the expression:

$$\sigma_{\theta\theta}=p[1+K+2(1-K)\cos2\theta] \quad (7\text{-}3)$$

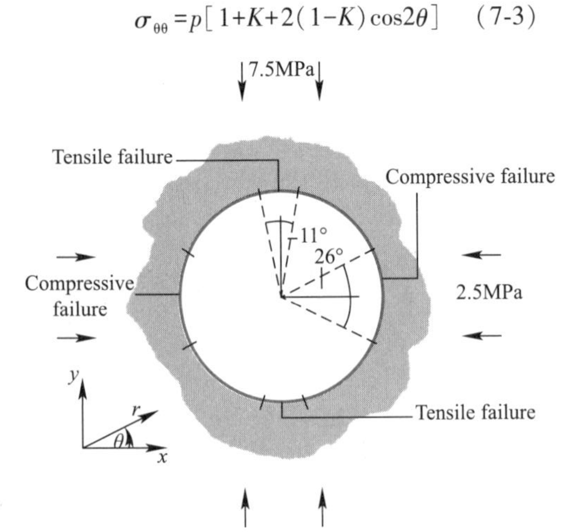

Fig. 7-7 Prediction of the extent of boundary failure around a circular excavation, using the rock mass failure criterion and the elastic stress distribution

7.4.3 Support and Reinforcement of Massive Rock

Mining activity frequently takes place under conditions sufficient to induce extensive failure around mine access and production openings. An understanding of the mechanics of the techniques exploited to control the performance of fractured rock is therefore basic to effective mining practice under these conditions. In evaluating the possible mechanical roles of a support system, it is instructive to consider the effect of support on the elastic stress distribution in the rock medium. One function of support is taken to be the application or development of a support load at the excavation surface.

Rock-support interaction illustrates the interaction between the rock mass surrounding the tunnel and the support material, as shown in Fig. 7-8. It is characterized by the load deformation curve of a tunnel and available support curve of the support material. Tunnel deforms after excavation, at different rates for different rock mass quality. Support pressure required to limit the deformation changes with deformation. And it decreases with further deformation. Load-deformation curve can be produced for a particular tunnel.

Available support curve is a load-deformation curve of the support material. It is a property of the reinforcement or support material, e. g., steel and concrete. In general, steel deforms elastically and after yielding, plastically.

Load-deformation curve and support pressure curve are analyzed together. Support pressure required to limit deformation is to be provided by the available support of the support material, i. e., equilibrium.

(a) Stiff support
(b) Medium support
(c) Yielding support
(d) Soft support
(e) Insufficient support

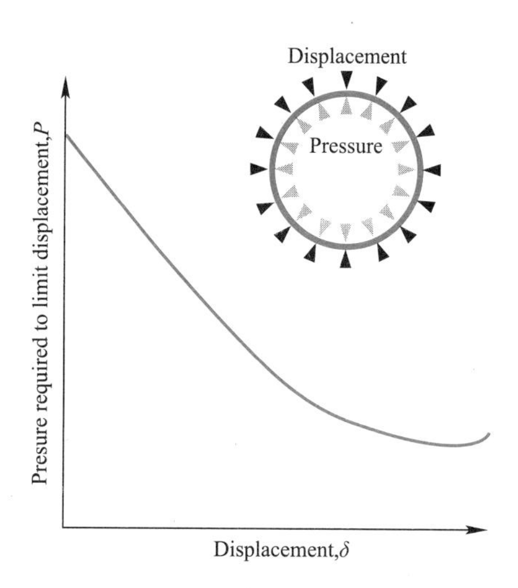

Fig. 7-8 Rock-support interaction

Ideal support:

A good engineering practice is to allow for deformation, but to control further displacement beyond necessary. An ideal support is to match the rock load-deformation curve withthe pressure yielded by the support.

References

[1] Brady B. H. G., Brown E. T.. Rock Mechanics for Underground Mining[M]. 3rd ed. Boston: Kluwer Academic Publishers, 2004.

Chapter 8
Numerical Methods for Geotechnical Engineering

8.1 Introduction

Numerical methods play a vital role in geotechnical engineering by providing valuable tools for analyzing and solving complex problems encountered in soil mechanics and foundation engineering. These methods involve the use of mathematical models and computer algorithms to simulate and predict the behavior of soil and structures. One common application of numerical methods in geotechnical engineering is the analysis of soil deformation and stress distribution. By discretizing the soil domain into smaller elements or grids, finite element analysis (FEA) can be used to solve the governing equations of soil mechanics, such as equilibrium and constitutive relations. FEA allows engineers to assess the response of soil under different loading conditions, determine potential failure mechanisms, and evaluate the stability of foundations and slopes. Another widely used numerical method is finite difference analysis (FDA). It approximates the differential equations governing soil behavior through discretizing the domain into a grid of points. Finite differences are then used to approximate the derivatives, enabling the solution of the governing equations. FDA is commonly employed in time-dependent problems, such as consolidation analysis and groundwater seepage. Other numerical techniques utilized in geotechnical engineering include boundary element method (BEM), meshless methods (e.g., radial basis functions and element-free Galerkin method), and discrete element method (DEM). These methods offer advantages in specific scenarios, such as analyzing problems involving large deformations, nonlinear behavior, or interaction between particles.

Numerical models are computer programs that attempt to represent the mechanical response of a rock mass subjected to a set of initial conditions such as in situ stresses and water levels, boundary conditions and induced changes such as slope excavation. The result of a numerical model simulation typically is either equilibrium or collapse. If an equilibrium result is obtained, the resultant stresses and displacements at any point in the rock mass can be compared with measured values. If a collapse result is obtained, the predicted mode of failure is demonstrated.

Numerical models divide the rock mass into zones. Each zone is assigned a material model and properties. The material models are idealized stress/strain relations that describe how the material behaves. The simplest model is a linear elastic model, which uses the elastic properties (Young's modulus and Poisson's ratio) of the material. Elastic-plastic models use strength parameters to limit the shear stress that a zone may sustain. The zones may be connected together, termed a continuum model, or separated by discontinuities, termed a discontinuum model. Discontinuum models allow slip and separation at explicitly located surfaces within the model.

Overall, numerical methods in geotechnical engineering provide a means to analyze complex soil-structure interactions, assess the performance of geotechnical systems, and optimize designs. They aid in decision-making processes and contribute to more reliable and cost-effective solutions in geotechnical engineering projects.

8.2 Numerical Models

All rock slopes involve discontinuities. Representation of these discontinuities in numerical models differs depending on the type of model. There are two basic types of models: discontinuum models and continuum models. Fig. 8-1 shows different numerical models by using several numerical simulation methods.

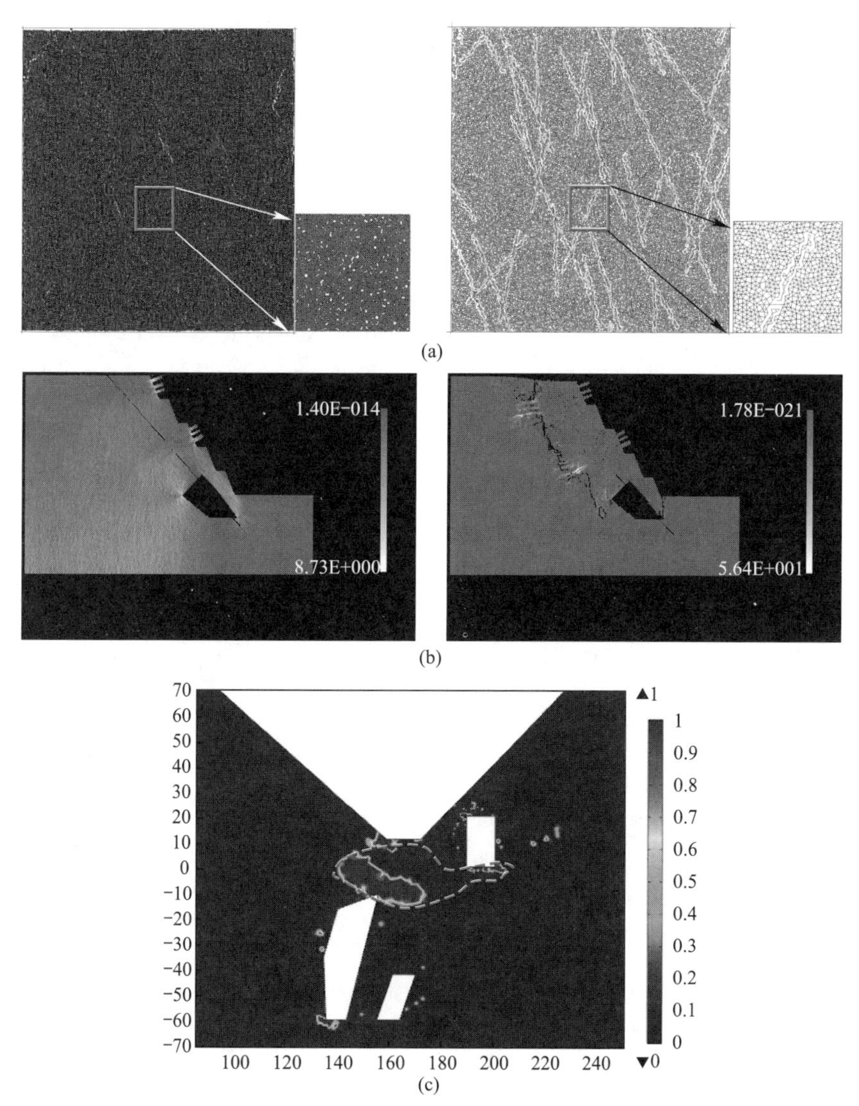

Fig. 8-1 Numerical models by using different methods

(a) Discontinuum model by using the PFC code; (b) Continuum model by using the RFPA code; (c) Continuum model by using the Comsol Multiphysics code

1. Discontinuum Models

Discontinuum models, also known as discontinuous models or particle-based models, are numerical models that represent materials or systems by explicitly considering the interactions and behavior of individual particles or discrete elements. These models are particularly suitable for simulating situations where significant discontinuities, such as fractures, joints, or other types of interfaces, play a crucial role in the system's behavior.

In discontinuum models, the material is represented as a collection of discrete elements, often referred to as particles or blocks, which interact with each other through mechanical contacts or other interaction rules. Each particle has its own properties, such as size, shape, mass, cohesion, friction coefficient, and other material characteristics. The interactions between particles are defined based on contact laws that describe the forces, moments, deformations, and other effects when particles come into contact or separate from each other.

One popular discontinuum modeling method is the discrete element method (DEM), which is widely used to study the behavior of granular materials, rock masses, and fractured solids. DEM tracks the motion and interactions among individual particles, accounting for their geometric and contact properties. It allows for the simulation of particle-scale phenomena like particle rearrangement, breakage, fragmentation, and sliding along interfaces.

Discontinuum models have several advantages. They can capture complex behaviors such as localized failures, brittle fracture propagation, and interaction between intact and fractured materials. They are also suitable for simulating highly deformable or evolving materials, as they accommodate large displacements and deformations without requiring a uniform mesh resolution. Discontinuum models find applications in geotechnical engineering, mining, rock mechanics, earthquake analysis, and other fields involving discontinuous media.

The most widely used discrete element codes for geotechnical studies are PFC2D/3D (two/three-dimensional partical flow code), UDEC (universal distinct element code) and 3DEC (three-dimensional distinct element code). The field of geotechnical engineering relies on several widely used discrete element method (DEM) software for studying granular materials, rock masses, and discontinuous media. Here are some of the most commonly used DEM codes in geotechnical studies.

PFC (particle flow code): PFC is a versatile and widely-used DEM software developed by Itasca Consulting Group. It allows for the simulation of both two-dimensional (2D) and three-dimensional (3D) problems involving particle-based materials. PFC offers various models to simulate soil, rock, and concrete behavior, including contact mechanics, particle breakage, fluid flow, and thermal interactions.

YADE (yet another dynamic engine): YADE is an open-source DEM software primarily used for geotechnical and geomechanical simulations. It provides robust capabilities for modeling granular materials, soils, and cohesive materials. YADE's modular architecture allows users to implement custom models and algorithms. It supports parallel computing, making it suitable for large-scale simulations.

UDEC (universal distinct element code): UDEC, developed by Itasca Consulting Group, focuses on analyzing the response and stability of rock masses and jointed structures using the distinct element method. It specifically addresses problems related to discontinuities, such as fractures, faults, and wedges. UDEC has been widely applied in rock mechanics, mining, geology, geotechnical engineering, and tunneling applications.

PFC3D (particle flow code in 3 dimensions): PFC3D is a commercial software package developed by Itasca Consulting Group, specializing in simulating three-dimensional geotechnical problems. It provides advanced capabilities for analyzing deformations, fragmentation, and failure phenomena in granular media, soils, and rocks. PFC3D is used extensively in studying soil-structure interaction, landslides, environmental geotechnics, and mining applications.

2. Continuum Models

Continuum models, also referred to as continuum-based models or field models, represent materials or systems as continuous entities, where properties and behavior are described in terms of averaged or statistical quantities. These models assume that the material or system can be described by macroscopic quantities, such as density, stress, strain, and velocity, without explicit consideration of individual particles or discrete elements.

In continuum models, the material or system is considered to be homogeneous and continuously distributed in space. The behavior of the material or system is represented by constitutive equations that relate stresses, strains, and other variables. These equations, derived from fundamental principles of mechanics and physics, describe the response of the material to external forces and constraints.

One of the most commonly used continuum models in geotechnical engineering is the finite element method (FEM). FEM divides the material or structure into a finite number of smaller elements and approximates the behavior within each element. The governing equations, typically derived from conservation laws and constitutive relations, are solved using numerical methods to obtain the system's response under specific loading conditions.

Continuum models have several advantages. They offer computational efficiency, allowing for the analysis of large-scale systems and complex geometries. They provide a continuous representation of the material or system, enabling the evaluation of global response and overall structural integrity. Continuum models are widely used in analyzing soil mechanics, fluid dynamics, structural mechanics, heat transfer, and other fields where the behavior of materials can be adequately described by continuum assumptions.

In geotechnical studies, continuum codes are commonly used to simulate the behavior of soil and rock materials as continua. These codes utilize constitutive models based on the principles of continuum mechanics to analyze the deformation, stress distribution, and stability of geotechnical structures. Here are some of the most widely used continuum codes in geotechnical studies:

PLAXIS: PLAXIS is a powerful commercial software package that specializes in finite element analysis for geotechnical applications. It provides comprehensive capabilities for modeling various geotechnical problems, including soil-structure interaction, foundation engineering, slope stability, tunneling, and embankments. PLAXIS offers advanced soil models, such as Mohr-Coulomb, Hardening Soil, Soft Soil Creep, and Consolidation models, allowing engineers to simulate complex geotechnical behavior.

FLAC (fast lagrangian analysis of continua): FLAC, developed by Itasca Consulting Group, is a versatile numerical tool that combines both continuum and discontinuum methods. With its emphasis on the explicit finite difference method, FLAC is particularly suitable for modeling complex geological and geotechnical phenomena, including excavation-induced ground deformations, landslides, dynamic response of soil deposits, and geological hazards.

ABAQUS: ABAQUS is a widely used general-purpose finite element software suite that covers various engineering disciplines, including geotechnical analysis. It provides robust capabilities for simulating nonlinear behaviors in soils, rocks, and structural elements. ABAQUS offers a wide range of material models, contact formulations, and advanced analysis techniques to study geotechnical problems such as pile foundations, retaining walls, soil-structure interaction, and underground excavations.

MIDAS GTS: MIDAS GTS (geotechnical and tunneling software) is a comprehensive software package that offers a complete solution for geotechnical analysis and design. It encompasses various modules, including soil-structure interaction, tunneling, slope stability, foundation engineering, and seismic analysis. MIDAS GTS employs advanced finite element algorithms to simulate complex geotechnical problems accurately.

GeoStudio: GeoStudio is a suite of integrated software tools specifically designed for geotechnical and geo-environmental modeling and analysis. It includes various modules such as SLOPE/W for slope stability analysis, SEEP/W for seepage analysis, SIGMA/W for stress-deformation analysis, and others. GeoStudio combines the power of finite element, finite difference, and limit equilibrium methods to address a wide range of geotechnical challenges.

Both discontinuum and continuum models have their strengths and limitations. The choice between these modeling approaches depends on the characteristics of the problem at hand, such as the presence of discontinuities, the level of detail required, the material behavior, and the available computational resources. By employing appropriate modeling techniques, engineers and scientists can gain insights into the behavior of complex systems and make informed decisions in various fields of study.

8.3 Modeling Issues

Modeling requires that the real problem be idealized, or simplified, in order to fit the constraints imposed by factors such as available material models and computer capacity. Analysis of rock mass response involves different scales. It is impossible and undesirable to include all features, and details of rock mass response mechanisms, into one model. This section discusses the basic issues that must be resolved when setting up a numerical model.

8.3.1 Two-dimensional Analysis Versus Three-dimensional Analysis

Two-dimensional (2D) and three-dimensional (3D) analysis in numerical simulation refer to different approaches for modeling and analyzing engineering problems.

Two-dimensional analysis: in 2D analysis, the problem is simplified by assuming that the behavior of a system occurs only within a two-dimensional plane. This approach reduces the complexity of the problem and computational requirements compared to 3D analysis. 2D analysis is commonly used when the problem can be adequately represented and analyzed within a plane, such as in simple foundation systems, shallow slopes, and certain structural elements like beams or plates. However, it may neglect important effects occurring perpendicular to the analyzed plane, leading to limitations in capturing the full behavior of the system.

Three-dimensional analysis: in 3D analysis, all three dimensions of the problem are considered, allowing for a more comprehensive representation of the actual physical system. It provides a more realistic depiction of the behavior, especially when dealing with complex geometries, irregular boundaries, or systems with significant variations along the third dimension. 3D analysis is often necessary when accurate predictions of stress distribution, deformation, and interaction between different components are required. However, the computational cost and complexity increase significantly compared to 2D analysis.

The choice between 2D and 3D analysis depends on several factors, including the complexity and nature of the problem, availability of data, resources (computational power, time), desired level of accuracy, and specific engineering objectives. In numerical simulations for geotechnical analysis, the choice between two-dimensional (2D) and three-dimensional (3D) analysis depends on several factors, including the complexity of the problem, available data, computational resources, and the desired level of accuracy. Here are some key considerations when deciding between 2D and 3D analysis.

Problem complexity: if the geotechnical problem involves significant out-of-plane effects or three-dimensional behavior, such as complex soil-structure interactions or tunnel excavations, a 3D analysis may be more appropriate. On the other hand, simpler problems with predominantly in-plane behavior, such as stability analysis of slopes or shallow foundations, can often be adequately captured using a 2D analysis.

Available data: the decision between 2D and 3D analysis may also depend on the availability of site-specific data. If the project has detailed information about soil layering, stratigraphy, and geological features in three dimensions, a 3D analysis can better represent the true conditions. However, if the available data is limited or primarily in 2D (e. g., cross-sections), a 2D analysis might be more feasible.

Computational resources: 3D analysis generally requires significantly higher computational resources (processing power, memory, and disk space) compared to 2D analysis. Running 3D simulations can be computationally expensive and time-consuming. Therefore, the available computational resources should be considered when choosing between 2D and 3D analysis. If the available resources are limited or if obtaining results within a reasonable timeframe is crucial, a 2D analysis may be preferred.

Accuracy and detail: 3D analysis potentially provides a more accurate representation of the actual behavior of the system. It can capture complex spatial effects, stress redistributions, and interactions between different components. 2D analysis, although simplified, can still provide useful insights and reasonable approximations, especially when the problem is laterally homogeneous and has limited out-of-plane effects. It's important to assess whether the additional complexity of a 3D analysis justifies the increased computational effort and resources.

Engineering judgment: ultimately, the choice between 2D and 3D analysis relies on engineering judgment and experience. Engineers need to evaluate the specific characteristics and requirements of each geotechnical problem and determine the most suitable approach for their particular application.

It is worth noting that some commercial software packages offer options for performing both 2D and 3D analysis.

8.3.2 Continuum Versus Discontinuum Models

The next step is to decide whether to use a continuum code or a discontinuum code. The choice between continuum and discontinuum models depends on the specific behavior being analyzed.

Deciding whether to use a continuum code or a discontinuum code involves evaluating various factors related to the problem at hand and the desired level of analysis. The following steps can help in making this decision.

Problem understanding: gain a thorough understanding of the nature of the problem and its associated phenomena. Identify whether the behavior can be adequately represented by assuming a continuous medium or if there are distinct discontinuities present.

System geometry: consider the geometry and complexity of the system. If the problem involves regular or smooth boundaries, and the behavior can be reasonably approximated as continuous, a continuum code may be suitable. On the other hand, if there are significant discontinuities, fractures, or interaction between discrete elements, a discontinuum code may be more appropriate.

Available data: examine the availability and quality of data for the problem. Continuum codes often require material properties and boundary conditions that can be obtained from standard laboratory tests or well-established models. Discontinuum codes may require more specific and detailed information about the discontinuities, such as joint orientations, roughness, mechanical properties, and strength parameters.

Computational resources: evaluate the computational resources (computer power, time) available for the analysis. Discontinuum codes typically involve greater computational effort compared to continuum codes due to considerations of individual block interactions and complex fracture propagation. Determine whether the available resources are sufficient for the chosen code.

Accuracy and precision: consider the required accuracy and precision for the analysis. Continuum codes provide a macroscopic view of the system's behavior, while discontinuum codes offer a more detailed representation of localized discontinuities and fractures. Assess whether the desired level of analysis can be achieved with the chosen approach.

Engineering objectives: align the choice of the code with the engineering objectives of the analysis. If the goal is to understand the overall response of the system or evaluate the stability of large-scale structures, a continuum code may be appropriate. Conversely, if the focus is on understanding the behavior of individual blocks, fracture propagation, or rock mass deformations, a discontinuum code might be more suitable.

Validation and experience: consider the availability of validation studies or prior experience with similar problems. Review existing literature, case studies, and research to determine which approach has been effectively used in similar scenarios. Collecting feedback and experiences from experts in the field can provide valuable insights for the decision-making process.

Remember, the choice between continuum and discontinuum codes is not always binary. Hybrid approaches that combine both techniques may also be applicable in certain cases to capture important aspects of the problem. It is crucial to carefully evaluate all relevant factors before making a final decision.

8.3.3 Selecting Appropriate Zone Size

The next step in the process is to select an appropriate zone size. Selecting an appropriate zone size (also known as discretization size or mesh size) is crucial in numerical simulations to balance accuracy and computational efficiency. The following steps can help in determining an appropriate zone size.

Problem understanding: gain a thorough understanding of the nature of the problem, including the physical phenomena and length scales involved. Consider the key features and regions that need to be accurately captured in the simulation.

Mesh type: identify the type of mesh or grid system that will be used for discretization. Common types include structured grids (rectangular or hexahedral), unstructured grids (triangular or tetrahedral), or specialized meshes suitable for specific applications (e. g., boundary-fitted or adaptive meshes).

Desired resolution: determine the desired resolution or level of detail required in the simulation results. This depends on the specific variables of interest, such as stress gradients, fluid flow patterns, or temperature distributions. Consider whether a fine-scale representation is necessary throughout the entire domain or if coarser resolution can be used in certain regions.

Numerical stability: ensure that the chosen zone size satisfies the stability requirements of the numerical method being employed. Some numerical methods have stability conditions that impose limits on the maximum zone size allowable for accurate and reliable simulations. Check the literature or consult experts to determine any specific stability criteria associated with the chosen method.

Computational resources: evaluate the available computational resources, including computing power and time constraints. Smaller zone sizes generally result in more accurate results but require increased computational effort. Consider the trade-off between accuracy and computational efficiency based on the available resources.

Convergence analysis: perform a convergence analysis by conducting preliminary simulations with different zone sizes. Start with a relatively coarse mesh and progressively refine it while monitoring the solution convergence. Assess whether the solution exhibits consistent behavior and converges towards a stable solution as the zone size decreases. This analysis helps identify the zone size at which further refinement does not significantly impact the results.

Validation and benchmarking: compare the simulation results against available experimental data or validated numerical solutions, if applicable. Assess how well the chosen zone size captures the essential features and phenomena observed in the validation data. This step helps provide confidence in the selected zone size and its ability to produce accurate results for the problem at hand.

Sensitivity analysis: perform sensitivity analysis to evaluate the influence of zone size on the simulation results. Vary the zone size within a reasonable range and assess the impact on important outcomes or metrics of interest. This analysis can help identify critical regions or variables that are sensitive to changes.

8.3.4 Initial Conditions

Initial conditions are those conditions that existed prior to mining. The initial conditions of importance at mine sites are the in situ stress field and the ground water conditions. The role of stresses has been traditionally ignored in slope analysis. There are several possible reasons for this.

Limit equilibrium analysis, which are widely used for stability analysis, cannot include the effect of stresses in their analysis. Nevertheless, limit equilibrium analysis are thought to provide reasonable estimates of stability in many cases, particularly where structure is absent, such as soil slopes.

Most stability analysis have traditionally been performed for soils, where the range of possible in situ stresses is more limited than for rocks. Furthermore, many soil analysis have been performed for constructed embankments such as dams, where in situ stresses do not exist.

Most slope failures are gravity driven, and the effects of in situ stress are thought to be minimal.

In situ stresses in rock masses are not routinely measured for slopes, and their effects are largely unknown.

One particular advantage of stress analysis programs such as numerical models is their ability to include pre-mining initial stress states in stability analysis and to evaluate their importance.

8.3.5 Boundary Conditions

Boundary conditions in numerical models refer to the specifications or constraints applied to the boundaries of the modeled domain. They define how the system interacts with its surroundings and can significantly influence the behavior and accuracy of the simulation. Boundaries are either real or artificial. Real boundaries in slope stability problems correspond to the natural or excavated ground surface that is usually stress free.

Artificial boundaries do not exist in reality. All problems in geomechanics, including slope stability problems, require that the infinite extent of a real problem domain be artificially truncated to include only the immediate area of interest. Displacement at the base of the model is always fixed in both the vertical and horizontal directions to inhibit rotation of the model. Two assumptions can be made regarding the displacement boundaries near the toe of any slope. One assumption is that the displacements near the toe are inhibited only in the horizontal direction. This is the mechanically correct condition for a problem that is perfectly symmetric with respect to the plane or axis representing the toe boundary. Strictly speaking, this condition only occurs in slopes of infinite length, which are modeled in two-dimensions assuming plane strain, or in slopes that are axially symmetric in which the pit is a perfect cone. In reality, these conditions are rarely satisfied. Therefore, some models are extended laterally to avoid the need to specify any boundary condition at the toe of the slope. It is important to note that difficulties with the boundary condition near the slope toe are usually a result of the two dimensional assumption. In three-dimensional models, this difficulty generally does not exist.

The far-field boundary location and condition must also be specified in any numerical model for slope stability analysis. The general notion is to select the far-fieldlocation so that it does not significantly influence the results. If this criterion is met, whether the boundary is prescribed-displacement or prescribed-stress is not important. In most slope stability studies, a prescribed-displacement boundary is used. The authors have used a prescribed-stress boundary in a few cases and found no significant differences with respect to the results from a prescribed-displacement boundary. The magnitude of the horizontal stress for the prescribed stress boundary must match the assumptions regarding the initial stress state in order for the model to be in equilibrium. However, following any change in the model, such as an excavation increment, the prescribed-stress boundary causes the far-field boundary to displace toward the excavation while maintaining its original stress value. For this reason, a prescribed-stress boundary is also referred to as a"following" stress, or constant stress boundary, because the stress does not change and follows the displacement of the boundary. However, following stresses are most likely where slopes are cut into areas where the topography rises behind the slope. Even where slopes are excavated into an inclined topography, the stresses will flow around the excavation to some extent, depending on the effective width of the excavation perpendicular to the downhill topographic direction.

A summary of the effects of boundary conditions on analysis results is as follows.

- A fixed boundary causes both stresses and displacements to be underestimated, whereas a stress boundary does the opposite.
- The two types of boundary condition "bracket" the true solution, so that it is possible to conduct tests with smaller models to obtain a reasonable estimate of the true solution by averaging the two results.

A final point to be kept in mind is that all open pit slope stability problems are three-dimensional in reality. This means that the stresses acting in and around the pit are free to flow both beneath and around the sides of the pit. Therefore, it is likely that, unless there are very low strength faults parallel to the analysis plane, a constant stress or following stress boundary will over-predict the stresses acting horizontally.

8.4 Typical Slope Stability Analysis

8.4.1 Geological Conditions

Here, we will introduce an application case in an open pit mine by using the COMSOL Multiphysics code, a FEM method. Sijiaying open pit mine is located in Tangshan city in northeastern China as shown in Fig. 8-2. The length of the open pit is approximate 2870m from north to south and the width is more than 1500m from east to west. The maximum height of the open pit limit is about 560m with the bench height of 12—15m. The angle of a bench is 75° according to the preliminary pit slope design. The cut face in eastern slope coincides with the low friction bedding planes in the rock mass. On Jul 13, 2012, a landslide happened in the eastern slope of Sijiaying open pit mine. According to the in-situ survey result, the weak structural plane which coincides with the dip angle of the slope is the key influencer. The geological drillings were conducted to investigate the geology information as shown in Fig. 8-2. Meanwhile, ShapeMetriX3D system was used for the acquisition of 3D rock mass exposure. Then the geological parameters which can accurately represent rock discontinuities were obtained. The structural plane is found to be relatively smooth and the inclination angle of the dominant joint planes, which have the same dip direction as the slope, is in the range from 50° to 55°. And the density of dominant joint planes is 6.35/m^3.

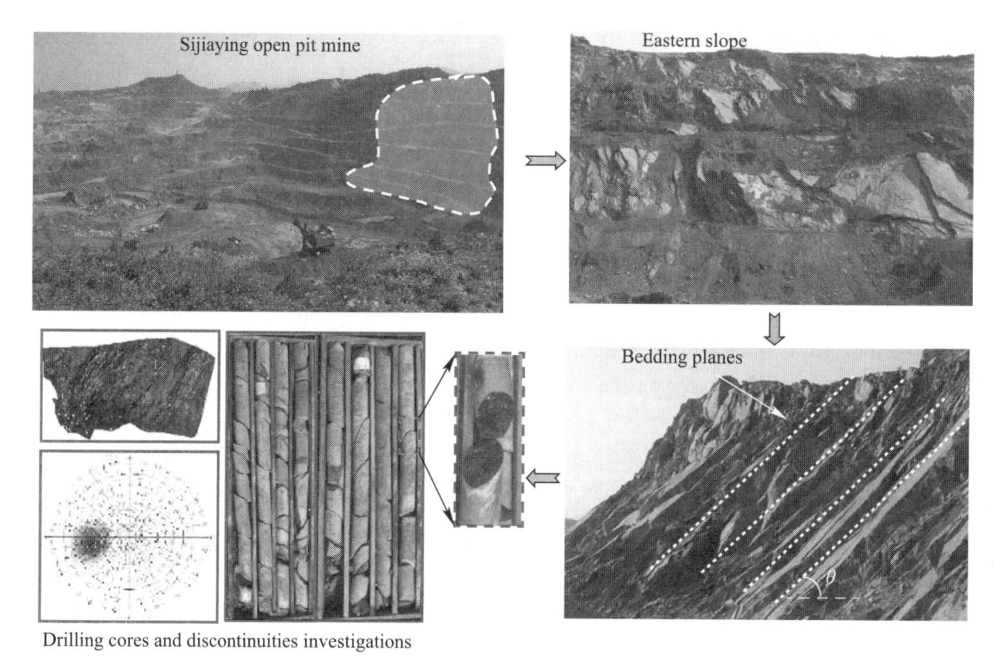

Fig. 8-2 Geological investigations of eastern slope in Sijiaying open pit mine

According to the in-situ survey, the rock type is mainly granulite specimens and the value of material constant m_i, is 33. The Sijiaying iron mine has a large open pit and the slopes suffer significant disturbance due to heavy production blasting. Thus the disturbance factor D, is chosen to be 1.0. According to the structure and rock surface conditions, the geological strength index (GSI) is 34. Then the elastic modulus, compressive strength, tensile strength, cohesive strength and friction angle of rock mass can be obtained. Also, the corresponding anisotropic parameters of rock mass can be determined by using the corresponding anisotropic parameters of intact rocks. The anisotropy of intact rock was also considered and the relevant properties are listed in Table 8-1.

Mechanical parameters of rock mass according to Hoek-Brown criterion

Table 8-1

Inclination (°)	Elastic modulus E(GPa)	Compressive strength σ_c(MPa)	Tensile strength σ_t(MPa)	Cohesive strength c(MPa)	Friction angle φ(°)
0	5.70	0.29	-0.005	0.27	40.91
15	2.86	0.15	-0.002	0.20	35.73

Continued

Inclination (°)	Elastic modulus E(GPa)	Compressive strength σ_c(MPa)	Tensile strength σ_t(MPa)	Cohesive strength c(MPa)	Friction angle φ(°)
30	2.77	0.14	-0.002	0.20	35.50
45	2.23	0.11	-0.002	0.19	33.85
60	1.89	0.10	-0.002	0.18	32.62
75	1.01	0.05	-0.001	0.14	28.06
90	3.39	0.17	-0.003	0.22	37.00

8.4.2 Numerical Model and Boundary Conditions

A numerical model of Shirengou iron ore mine is established to simulate the stability of jointed rock mass taking into account the anisotropic property. As is shown in Fig. 8-2, the crest of the present pit mine and the designed pit limit in this profile are about 100m and 400m, respectively. The sloping plane and foliation planes have the same direction and the principal inclination angle is about 50°. The rock mass of the slope is transversely isotropic and the mechanical properties of the slope rock mass are assumed to be the same. The slope has a length of more than 2800m in the vertical direction. Thus, the model could be assumed to be in a state of plane strain (with no change in elastic strain in the vertical direction) and static mechanical equilibrium. The constitutive relationship of transversely isotropic material can be expressed as Eq. (8-1). The mechanical parameters of the anisotropic rock mass, such as elastic modulus (E, v, E', v'), tensile, compressive and shear strength, have been discussed. The Hoffman anisotropic strength criterion (Hoffman 1967) can be used to assess the damage zone in this numerical model as shown in Eq. (8-2).

$$\begin{bmatrix}\varepsilon_x\\ \varepsilon_y\\ \varepsilon_z\\ \gamma_{yz}\\ \gamma_{zx}\\ \gamma_{xy}\end{bmatrix}=[S][\sigma]=\begin{bmatrix}\frac{1}{E} & -\frac{v'}{E'} & -\frac{v}{E} & 0 & 0 & 0\\ -\frac{v'}{E'} & \frac{1}{E'} & -\frac{v'}{E'} & 0 & 0 & 0\\ -\frac{v}{E} & -\frac{v'}{E'} & \frac{1}{E} & 0 & 0 & 0\\ & & & \frac{1}{G'} & & 0\\ & & & & \frac{2(1+v)}{E} & 0\\ & S & & & & \frac{1}{G'}\end{bmatrix}\begin{bmatrix}\sigma_x\\ \sigma_y\\ \sigma_z\\ \tau_{yz}\\ \tau_{zx}\\ \tau_{xy}\end{bmatrix} \tag{8-1}$$

$$\frac{\sigma_1^2}{X_tX_c}-\frac{\sigma_1\sigma_2}{X_tX_c}+\frac{\sigma_2^2}{Y_tY_c}+\frac{X_c-X_t}{X_tX_c}\sigma_1+\frac{Y_c-Y_t}{Y_tY_c}\sigma_2+\frac{\tau_{12}^2}{S^2}=1 \tag{8-2}$$

Where,

X_t and Y_t—— the tensile strength parallel and perpendicular to joint planes, respectively;

X_c and Y_c—— the compressive strength parallel and perpendicular to joint planes respectively;

S —— the shear strength of the rock mass along joint planes;

σ_1—— the normal stress along the principal direction of elasticity;

σ_2—— the normal stress perpendicular to the principal direction of elasticity;

τ_{12}—— the shear stress.

In the numerical model, the designed pit limit is modeled to study the slope failure patterns. The bottom boundary of the domain is fixed in all directions. Both the left and right boundaries are fixed in the horizontal direction. All the governing equations described above are implemented into COMSOL Multiphysics (CM), a powerful PDE-based multiphysics modeling environment. CM code can supply powerful partial differential equations (PDEs)-based modeling environment. And the specified PDEs may be nonlinear and act on a 2D or 3D geometry with complex boundary values. If the value of Eq. (8-2) reaches 1, then the corresponding region is treated as the damage zone.

8.4.3 Results and Analysis

Until recently, the Hoffman anisotropic criterion has been rarely used in rock engineering. Hoffman criterion is used to study the damage zone of rock slope in Sijiaying open pit mine. The relevant strength parameters used in this criterion have been listed in Table 8-1. The inclination angle is approximately 50° according to the in-situ survey and the slope damage zone by Hoffman criterion in Eq. (8-2) is depicted in Fig. 8-3. Partial failures of the single slope are observed in the eastern slope and no large scale landslide occurs according to the simulation. The numerical result agrees well with the in-situ observation compared with the failure pattern of the natural slope.

8.4.4 Further Discussion

Orientations of weak structures have a dominant effect on the stability of the slope. The influence of anisotropic property on the failure patterns of the slope is analyzed and discussed. Fig. 8-4 shows the failure areas in different models with varied inclination angles. It is observed that the shape and size of the damage zones in eastern slope are significantly influenced by the orientation of the joint planes. Take the model with a joint inclination of 0° as an example, the tensile strength in the direction coinciding with the joint planes is much higher. Thus, the region of damaged zones in both eastern and western slope are in small scale thought damage zone occurs in both the two sides of the

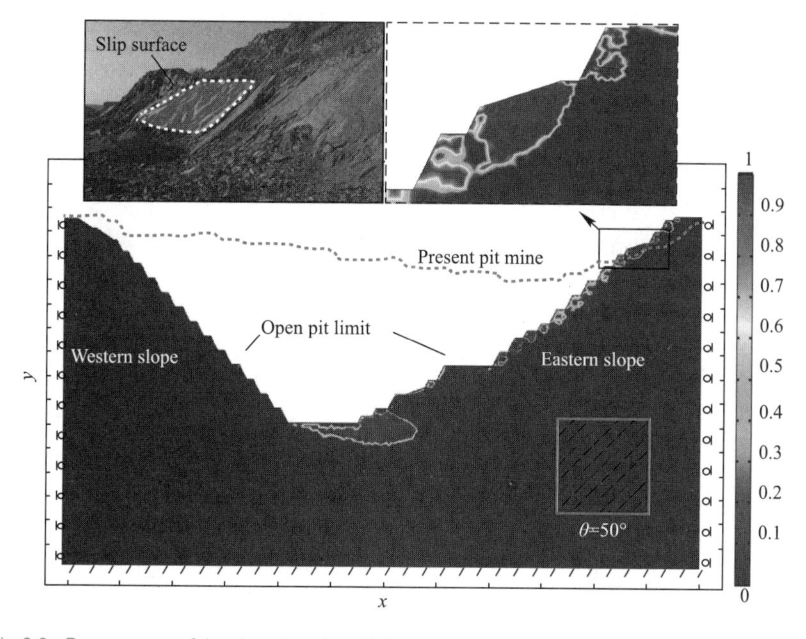

Fig. 8-3 Damage zone of the slope based on Hoffman criterion and comparison with in-situ observation

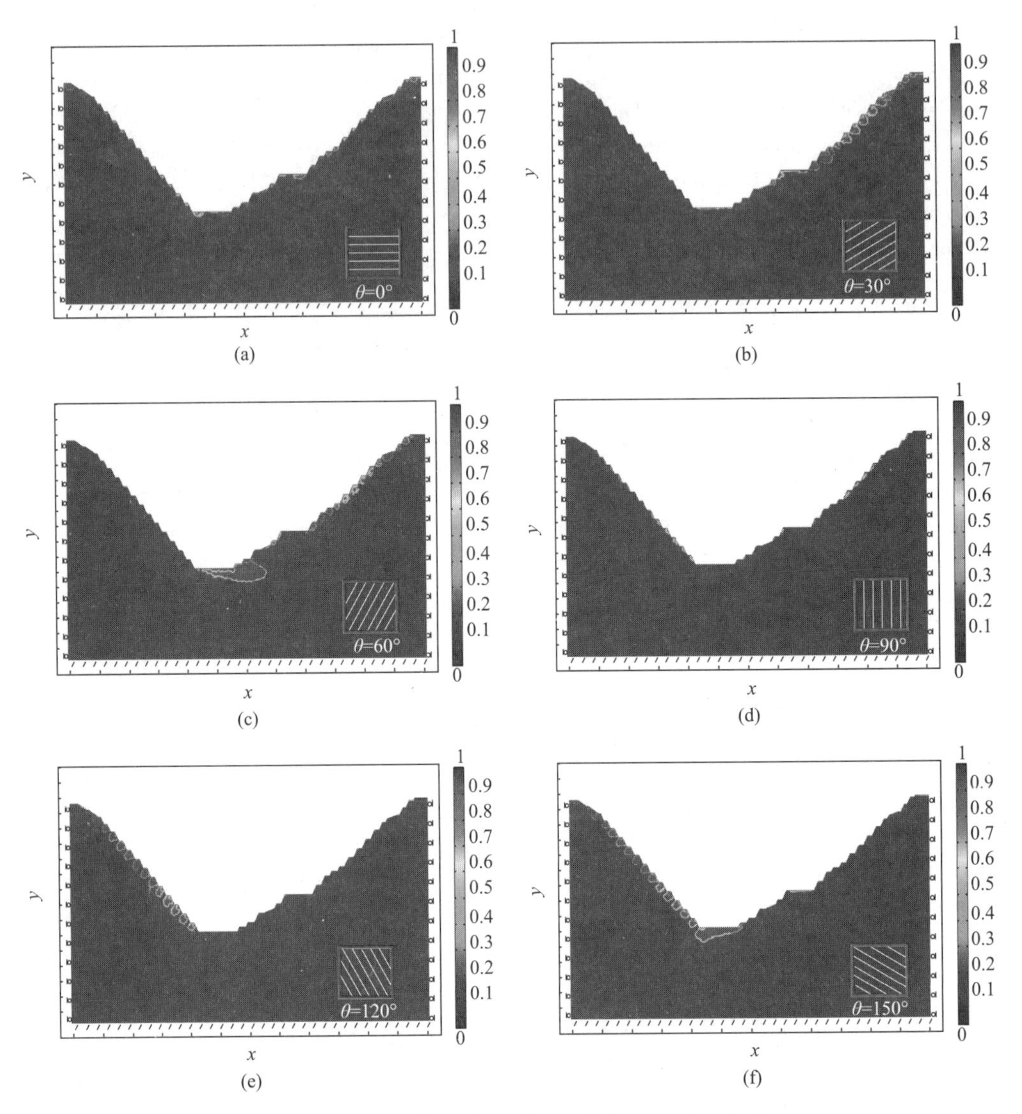

Fig. 8-4 Failure patterns of the slope with respect to the angles of anisotropy

slope. With the increase of inclining angle, the shear strength of the eastern slope is the key factor that determines the fracture patterns. In the cases of θ is 30° or 60° where the dip angle of the structural plane is less than the bench angle, a larger failure area of the single eastern slope is observed. With the increase of inclination angle, the damaged area mainly occurs on the western slope which agrees with the expectations.

Generally, the slope represents a larger rupture when the strength of rock mass is reduced. In the strength reduction methods, the stability factor, which is defined as the ratio of initial strength value to the induced strength value, is always used to evaluate the slope stability conditions.

Fig. 8-5 presents the failure patterns of the anisotropic slope when the strength is 2.0 times reduced, which also means the stability factor is 2.0. It is shown that the failure regions are sensitive to the anisotropic angles. The failure zone in Fig. 8-4(a) shows a minimum area. When the dip of the slope coincides with the bedding planes, the failure area expand (Fig. 8-4b, d). Typical toppling failure pattern is formed when the angle of bedding plane is 90°. It is concluded that the proposed model can efficiently characterize the anisotropic behaviors of layered rock mass. More complex numerical models, such as a 3D numerical model, can then be established to study the failure patterns of rock slopes using the proposed anisotropic constitutive relationship.

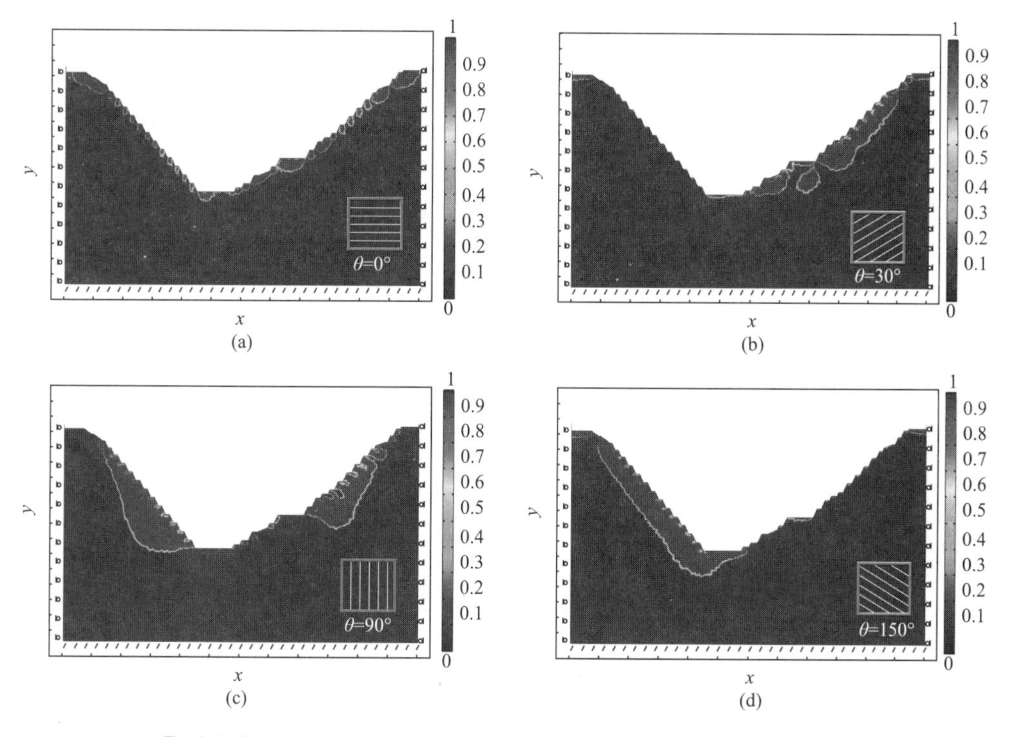

Fig. 8-5 Failure patterns of anisotropic rocky slope considering the strength reduction

References

[1] Hoffman O.. The Brittle Strength of Orthotropic Materials[J]. Journal of Composite Materials, 1967, 1: 200-206.

Chapter 9
Developments of Geotechnical Engineering

Geotechnical engineering is a field of civil engineering that focuses on the behavior of soil and rock in the context of construction and infrastructure development. In recent years, there have been significant advancements in geotechnical engineering, driven by advances in technology, new research, and growing demand for more resilient, sustainable, and cost-effective infrastructure. In this chapter, the development of identification of discontinuities, development of 3D rough discrete fractures network model, application of 3D printing technology in geotechnical engineering, and anisotropic mechanical behavior of jointed rock masses, are introduced.

9.1 Discontinuity Identification Based on 3D Point Clouds

9.1.1 Introduction

Discontinuities are widely distributed in rock mass engineering and have an important impact on both the physical and mechanical properties of rock masses. Finding ways to identify and quantify discontinuity information from three-dimensional point clouds (3DPC) in the rock mass in automatic or semi-automatic manners as an alternative to the use of the conventional approaches based on a geological compass and measuring tape is still a considerable challenge. The accurate quantification of the discontinuity properties has been considered as an initial key step for rock mechanics. We proposed a three-level extraction of rock mass discontinuity characteristics method, named the Cloud-Group-Cluster (CGC) method, from the 3DPC of rock masses based on the MATLAB code. This approach divides the 3DPC into point clouds, point groups, and point clusters. First, the normal vectors of the scattered points of the 3DPC based on the nearest neighbor search method are analyzed. Then, the point group analysis according to the normal vector is conducted according to different values of angle threshold to determine the inclination and orientation of the structural planes. Each point group is separated into several points clusters according to the cluster analysis. The discontinuities information including the orientation (dip angle, strike), area density, spacing of the structure sets are automatically measured based on the CGC method according to the 3DPC of rock masses.

9.1.2 Methodology

9.1.2.1 K-nearest neighbor searching

The normal vector should be the primary task to define the geometry of the discontinuities. To acquire the normal vectors of each point, the *K*-nearest neighbor searching algorithm is used. The principle of the algorithm is shown in Fig. 9-1. The seed p_1, is located in a certain position (x_1, y_1, z_1). Before calculating, the distance d_i between p_1 and other *K* points $p_i(x_i, y_i, z_i)$ will be calculated by Eq. (9-1). The values of distance are then sorted and recorded in one matrix. The *K* nearest points are the target neighbor points for the normal vector calculation.

$$d_i = \| p_i - p_1 \| = \| (x_i + y_i + z_i) - (x_1 + y_1 + z_1) \| \tag{9-1}$$

Eq. (9-2) shows an equation of the plane in the 3D coordinate system. Once the plane parameters (a, b, c) are determined, the fitting plane or rock plane is quantified.

$$z = ax + by + c \tag{9-2}$$

The plane parametersof the (K+1) points, (a, b, c) can be estimated with the least square method by Eq. (9-3) and Eq. (9-4). The normal vector of the fitting plane could be obtained according to the values (a, b, c).

$$z_i = ax_i + by_i + c + e_i \tag{9-3}$$

Where,

(x_i, y_i, z_i) —— the position of p_i in the $(K+1)$ points;

e_i—— the fitting error of each point.

The optimal solution, $Q(a, b)$ could be calculated by solving the minimum value of the quadratic summation of e_i, as shown by Eq. (9-4).

$$Q(a, b) = \sum_{i=1}^{k} e_i^2 = \sum_{i=1}^{k} (z_i - ax_i - by_i - c)^2 \tag{9-4}$$

According to the equation, the corresponding normal vector $(a, b, -1)$ is the normal vector of p_1. The normal vector of any point in the point cloud has two directions, up and down. In this section, the downward normal vector is taken so that the direction of the normal vector is consistent. Take three points (1, 0, 0), (0, 1, 0) and (0, 0, 1), for example, the fitting plane is determined by $z=-x-y+1$. Thus, the normal vector of the three points is (−1, −1, −1).

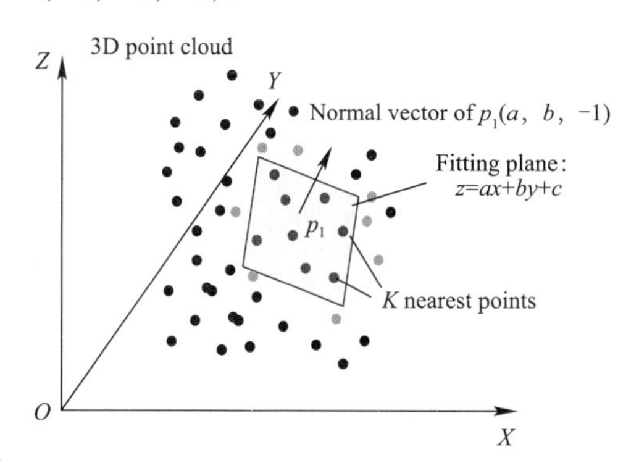

Fig. 9-1 Schematic diagram of K-nearest neighbor searching

9.1.2.2 Identification of coplanar point cloud

This method identifies the structural planes based on the 3DPC by measuring the attitude of the outcrop at each point. The sensitivity of the value K is discussed by identifying the orientation of the coplanar point cloud. As shown in Fig. 9-2, the 123 points are determined by the plane $z=-0.3083-0.0561x+0.0073y$. The normal vector of the plane is (−0.0561, 0.0073, −1). By choosing different K values, several fitting planes will be obtained. Nine different K values, $K=3$, 4, 5, 6, 7, 8, 9, 15, and 30, are selected to verify the sensitivity of the normal vectors. The results are listed in Table 9-1.

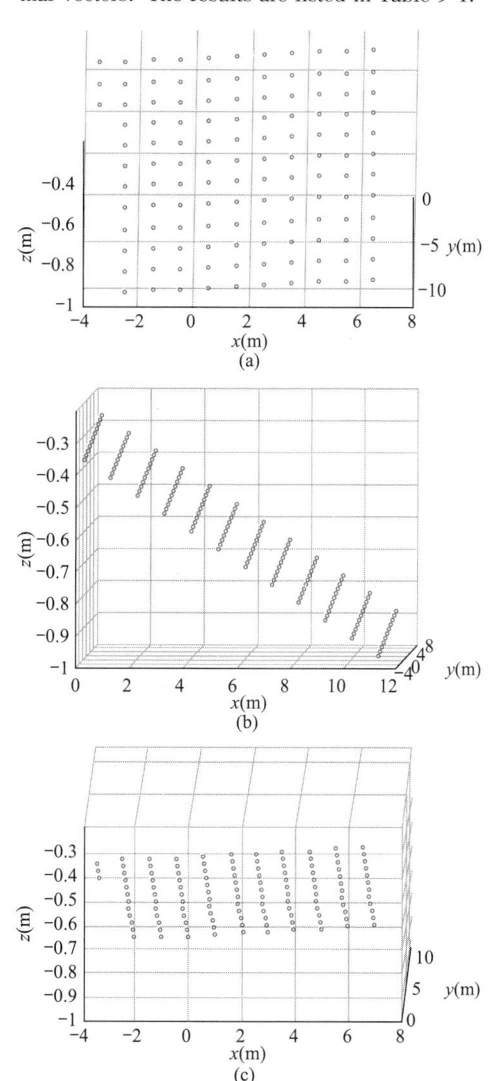

Fig. 9-2 The point cloud distribution of 123 coplanar points (a, b, and c are the different views of the points)

According to Table 9-1, the larger the value of the neighbor number K, the calculated value of the normal vector of each point is closer to the actual value. When K is set to 3, the normal vector calculation error is large, and even very large deviations occur. The number of effective normal vectors is only 6, and the number of other normal vectors is 117. When K is set to

4, the effective normal vector increases rapidly to about 50%. When K is raised to 5, 6, 7, 8, and 9, the number of normal vector calculation values close to the actual value keeps increasing. The normal vector of each point is close to the actual value and the same as the actual value when K= 15 and 30.

List of the normal vectors of varied K values of coplanar points **Table 9-1**

K neighbor points and the proportion ratio in total points	Identification result	
	Normal vector	Number of points
K=3, 2.44%	(−0.0561, 0.0073, −1)	2
	(−0.0562, 0.0072, −1)	1
	(−0.0561, 0.0072, −1)	2
	(−0.0562, 0.0073, −1)	1
	Other vectors	117
K=4, 3.25%	(−0.0561, 0.0073, −1)	30
	(−0.0562, 0.0072, −1)	15
	(−0.0561, 0.0072, −1)	16
	(−0.0562, 0.0073, −1)	14
	Other vectors	48
K=5, 4.07%	(−0.0561, 0.0073, −1)	43
	(−0.0562, 0.0072, −1)	26
	(−0.0561, 0.0072, −1)	13
	(−0.0562, 0.0073, −1)	11
	Other vectors	30
K=6, 4.88%	(−0.0561, 0.0073, −1)	52
	(−0.0562, 0.0072, −1)	23
	(−0.0561, 0.0072, −1)	22
	(−0.0562, 0.0073, −1)	13
	Other vectors	13
K=7, 5.69%	(−0.0561, 0.0073, −1)	55
	(−0.0562, 0.0072, −1)	25
	(−0.0561, 0.0072, −1)	13
	(−0.0562, 0.0073, −1)	12
	Other vectors	18
K=8, 6.50%	(−0.0561, 0.0073, −1)	58
	(−0.0562, 0.0072, −1)	12
	(−0.0561, 0.0072, −1)	25
	(−0.0562, 0.0073, −1)	25
	Other vectors	3

Continued

K neighbor points and the proportion ratio in total points	Identification result	
	Normal vector	Number of points
K=9, 7.32%	(−0.0561, 0.0073, −1)	60
	(−0.0562, 0.0072, −1)	15
	(−0.0561, 0.0072, −1)	16
	(−0.0562, 0.0073, −1)	25
	Other vectors	7
K=15, 12.20%	(−0.0561, 0.0073, −1)	119
	(−0.0562, 0.0072, −1)	1
	(−0.0561, 0.0072, −1)	2
	(−0.0562, 0.0073, −1)	1
	Other vectors	0
K=30, 24.39%	(−0.0561, 0.0073, −1)	122
	(−0.0562, 0.0072, −1)	1
	(−0.0561, 0.0072, −1)	0
	(−0.0562, 0.0073, −1)	0
	Other vectors	0

Fig. 9-3 shows the proportion of the effective normal vector in the total normal vectors with varied neighbor values K. With the increase of the neighbor value, the effective normal vector recognition ratio increases. When K=4, the recognition effectiveness is about 60%. In this section, we can treat the result as an ideal result when the ratio of the effective recognition of normal vector is more than 80%. For this coplanar 3DPC, the recognition effect is ideal when K>6. The identification result is better when K is larger than 8 with the ratio of the effective 3DPC normal vector is more than 90%. The results show that the combination of the K-nearest neighbor searching method and the least square method to determine the normal vectors of the 3DPC is effective. It is also indicated that the accuracy of the solution is significantly affected by the number of K-nearest neighbors.

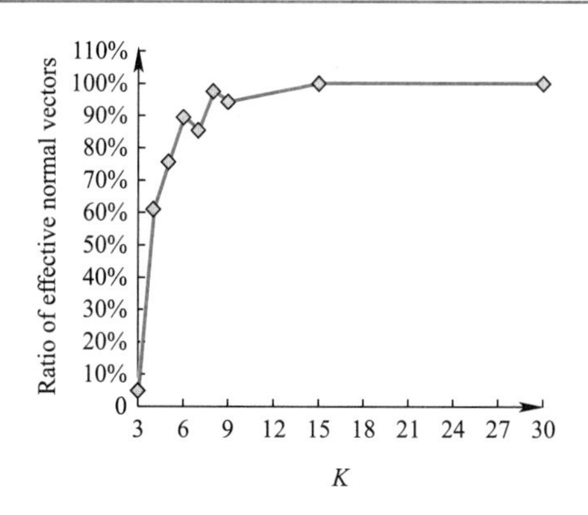

Fig. 9-3 Relationship of the effective normal vectors with varied K values

9.1.2.3 Identification of non-coplanar 3DPC

Seven non-coplanar points (1, 0, 0), (0, 1, 0), (0, 0, 0), (1, 1, 0), (1, 1, 1), (0.5, 0.5, 0) and (0, 0, 1), are listed to discuss the influence of K value on the recognition of normal vectors. Fig. 9-4 shows the distribution of the 3DPC. Table 9-2 shows the recognition results of normal vectors when the K is 3, 4, 5, and 6. The results show that the value of K has a significant influence on the normal vector calculation of non-coplanar 3DPCs. When K=3, the identification scattered points are divided into 2 groups, among which the normal vector of (0, 0, −1) contains 6 points. According to the distribution of normal vectors, the normal vector identification is more concentrated. Three groups of coplanar points are identified when K increases to 4. And the highest group contains 5 points with the same normal vector value. When the K value is 5, the 3DPC is divided into 4 groups according to the normal vector recognition. The highest number of points in the group is 3, and the normal vector is (0, 0, −1). There are 4 groups of coplanar points are identified when the K value is 6. The normal vectors are very discrete according to the distribution. For the 7 non-coplanar points, the recognized 3DPC normal vectors are relatively concentrated, and the recognized normal vectors are mainly (0, 0, −1) when k is set to be 3—5. Comparing with the results of 129 coplanar 3DPC normal vectors, a smaller value of K can make an effective recognition when the number of 3DPCs is large and the difference in surface fluctuations is small. When the number of 3DPCs is small and the fluctuation is large, the result obtained by taking a small value of K would be more ideal. The above results indicate that when conducting relevant research on 3DPCs, sampling discussion and analysis should be carried out in advance to determine the optimal number of neighbors K.

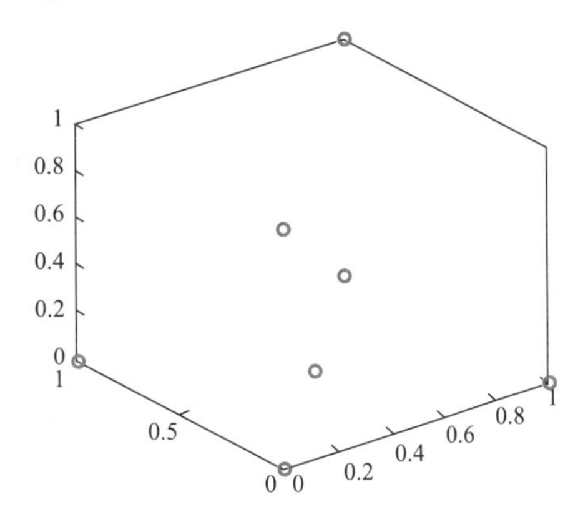

Fig. 9-4 The point cloud distribution of the non-coplanar points

Result of the normal vectors of varied k values of non-coplanar points Table 9-2

K neighbor points and the proportion ratio in total points	Identification result		Distribution of normal vectors
	Normal vector	Number of points	
K=3, 42.86%	(0, 0, −1)	6	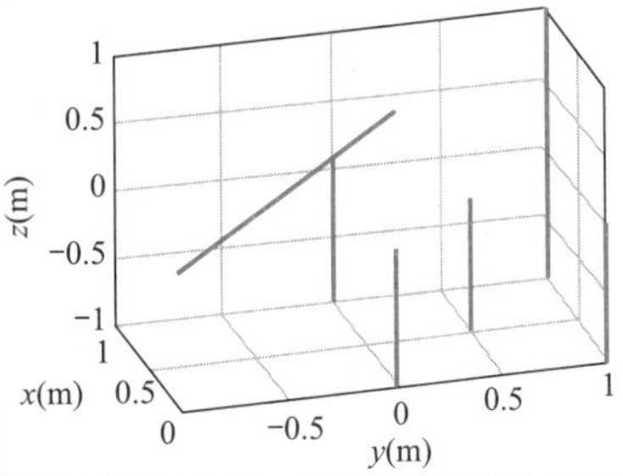
	(−1, 0, −1)	1	
K=4, 57.14%	(0, 0, −1)	5	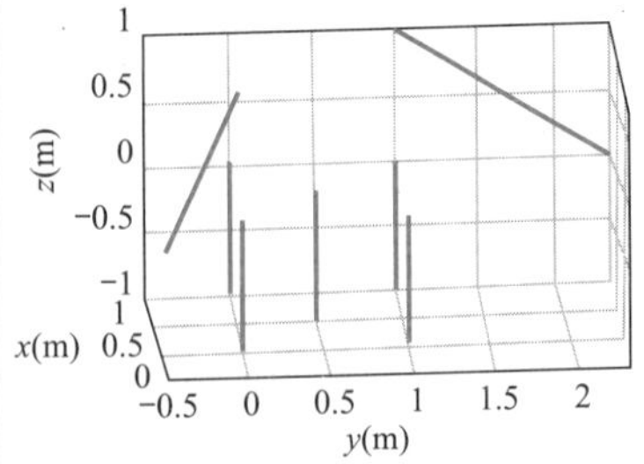
	(1.333334, 0, −1)	1	
	(−0.5, 0.5, −1)	1	

Continued

K neighbor points and the proportion ratio in total points	Identification result		Distribution of normal vectors
	Normal vector	Number of points	
$K=5$, 71.43%	(0, 0, −1)	3	
	(−0.5, −0.5, −1)	2	
	(0.666667, 0.666667, −1)	1	
	(0.5, 0.5, −1)	1	
$K=6$, 85.71%	(0.294118, 0.294118, −1)	3	
	(0.666667, 0.666667, −1)	2	
	(−0.29412, −0.29412, −1)	1	
	(0.117647, 0.117647, −1)	1	

9.1.3 Results

9.1.3.1 Identification of discontinuity groups

The determination of normal vectors of each point is the first step for the recognition of rock discontinuities. Orientation of the structure plane could be determined according to the normal. Angle threshold θ, associated with the difference of the normal vectors between the adjacent points, has a significant influence on the identification of rock discontinuity groups. The angle of two normal vectors of adjacent points can be calculated by Eq. (9-5).

$$O_{ij}=\frac{(a_i,\ b_i,\ -1)\cdot(a_j,\ b_j,\ -1)}{\sqrt{\|a_i,\ b_i,\ -1\|}\sqrt{\|a_j,\ b_j,\ -1\|}} \tag{9-5}$$

Where,

O_{ij}—— the angle of normal between point i and point j;

$(a_i,\ b_i,\ -1)$ and $(a_j,\ b_j,\ -1)$ —— the normal vector of point i and point j, respectively.

The point i and j could be treated as in the same group when O_{ij} is equal to or smaller than threshold θ. It means the two points belong to the same discontinuity group. By determining a suitable threshold value for the recognition, point groups can be realized by obeying the criteria: a normal difference smaller than the angle threshold. Therefore, the determinations of reasonable values for the angel threshold are an essential task.

To verify the sensitivity of K and θ, a 3DPC including 729 points generated from a tetrahedron model (Fig. 9-5) is discussed. Considering $K=5$, 6, 7, 10, 12, 15; $\theta=5°$, 10°, 15°, 20°, 30°, and 45°, the algorithm proposed in this paper is performed to identify the planes from the 3DPC as shown in Fig. 9-6. Points belong to the same group (or plane) are displayed by the same color. There are 4 individual planes and edges in a tetrahedron model. Thus, a reasonable result of point group number S_n should be 4.

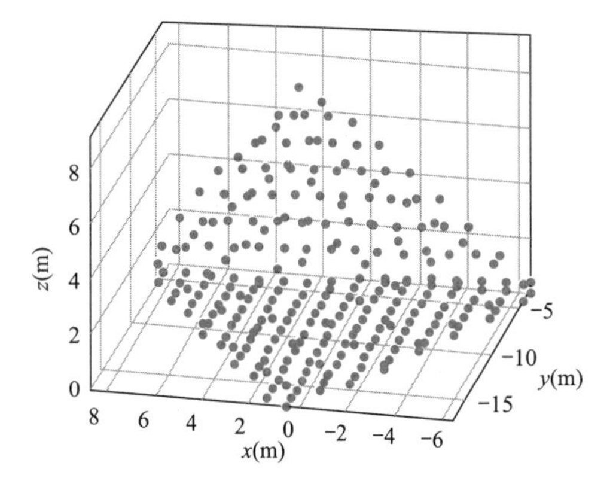

Fig. 9-5 Point clouds distribution of the tetrahedron model

As seen from Fig. 9-6, smaller θ values represent stricter criteria on the detection of difference in normal vectors between seed and neighboring points, resulting in biases and only partial points of a discontinuity can be distinguished. In the case of $K=6$, threshold $\theta=5°$, 58 groups of discontinuity including 4 main planes can be detected from the 3DPC. In this case, the algorithm does not provide the adequate ability to separate points especially the 4 edge points between adjacent discontinuities, leading to increases in sets of discontinuities. As the θ value grows, more points tend to be marked out within the same group of discontinuity. The number of point groups S_n, tends to decrease. In the case of $K=12$, 15, threshold $\theta=30°$, 45°, the point group number is 4, which means reasonable combination parameters for K and θ are obtained. According to Fig. 9-6, the recognition of point grouping of the edge points is more complicated. The calculation of the normal vector of the points near the edges is greatly affected by the value of K. With the increase of the value of K, the number of point groups of the edge points is reduced.

Fig. 9-7 shows the relationship of the recognized number of point groups S_n with the K values and threshold value θ. When increasing the K value, the recognized group's number S_n changes in a small range, except in the case of $\theta=5°$. All the recognition result of point groups number is higher than 40 when the threshold value is 5°, which means a smaller threshold is not suitable for the tetrahedron 3DPC. When the angle threshold is higher than 30°, the number of point groups is gradually lower than 9. When the threshold is 45°, the number of identified structural groups is 4, which is consistent with the actual case. Herein, the optimal value for the threshold θ is suggested to be specified as approximately 45°.

9.1.3.2 Elimination of noise points

According to the above analysis, the threshold selection is important for the measurement of structural planes. However, no standard threshold for all 3DPCs could be determined to reasonably group all the 3D points. Consequently, it is suggested that a filter factor f, is employed to eliminate the noise points. When the number of points in one certain point group is lower than the value of the filter factor, then the points are treated as noise points which will be eliminated in the discontinuities analysis.

The identification algorithm is discussed based on a 3DPC of a hexahedron model (Fig. 9-8). The 3DPC contains 706 points, including 8 edge points. An identification number of 3 point groups should be reasonable. Considering $K=10$, and 20; $\theta=5°$, 10°, 20°, 30°, and 45°; $f=0$, 10, 20, 30, and 50, the algorithm proposed here is conducted to identify the planes from the 3DPC as listed in Table 9-3. According to the results, it can be seen that the filter factor f can greatly reduce the number of point groups and eliminate some invalid point groups in edges. Take the case of $K=10$ and threshold $\theta=5°$, for example, the number of identified point groups is 62 when filter $f=0$. According to the point group identification result, it can be seen that the result for recognizing planes is not reasonable. When the filter $f=10$, the influence of boundary points is effectively eliminated and the number of point groups is optimized to 5. With the filter f increasing to 20, the number of recognized point

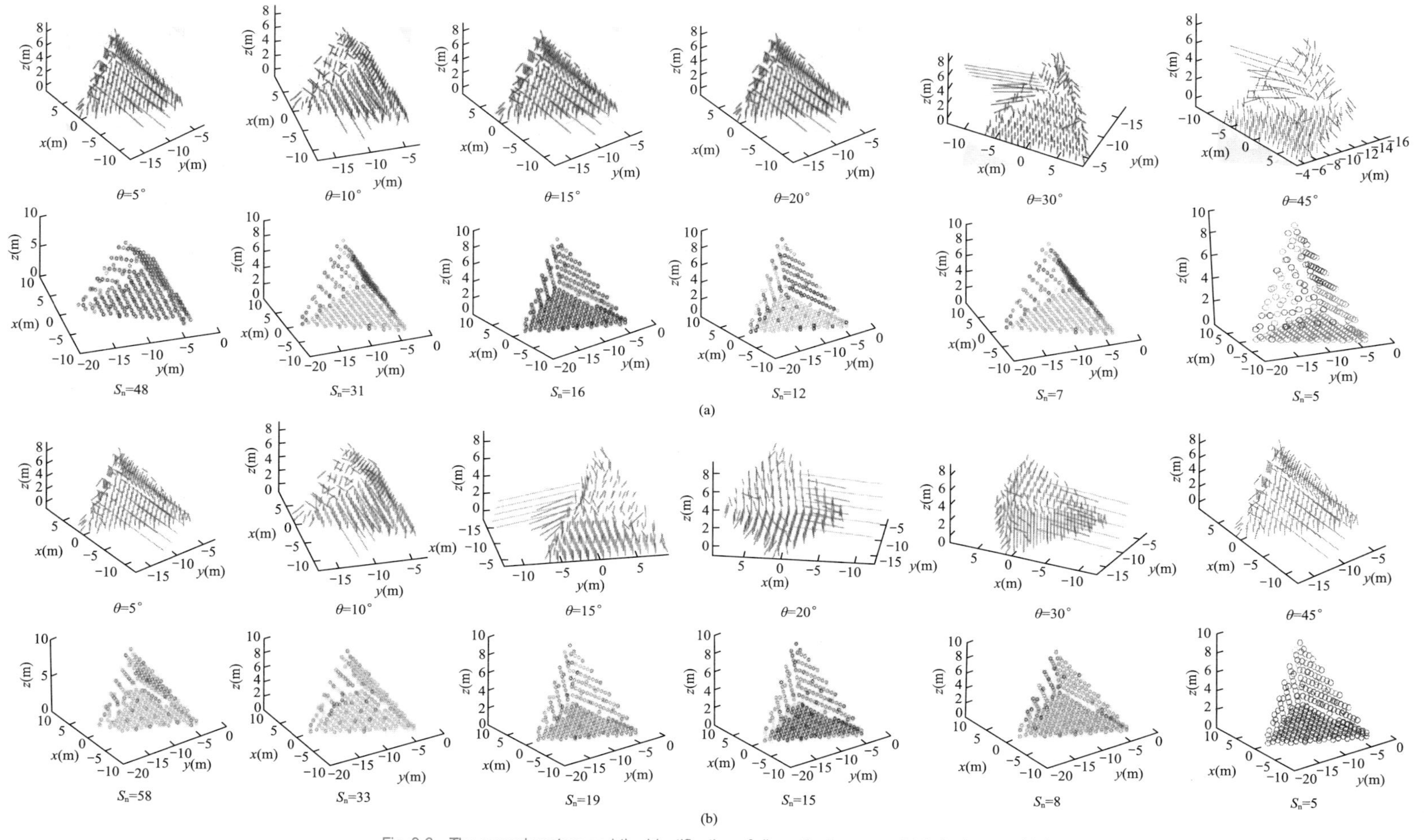

Fig. 9-6 The normal vectors and the identification of discontinuity group of tetrahedron model (one)

(a) K = 5; (b) K = 6;

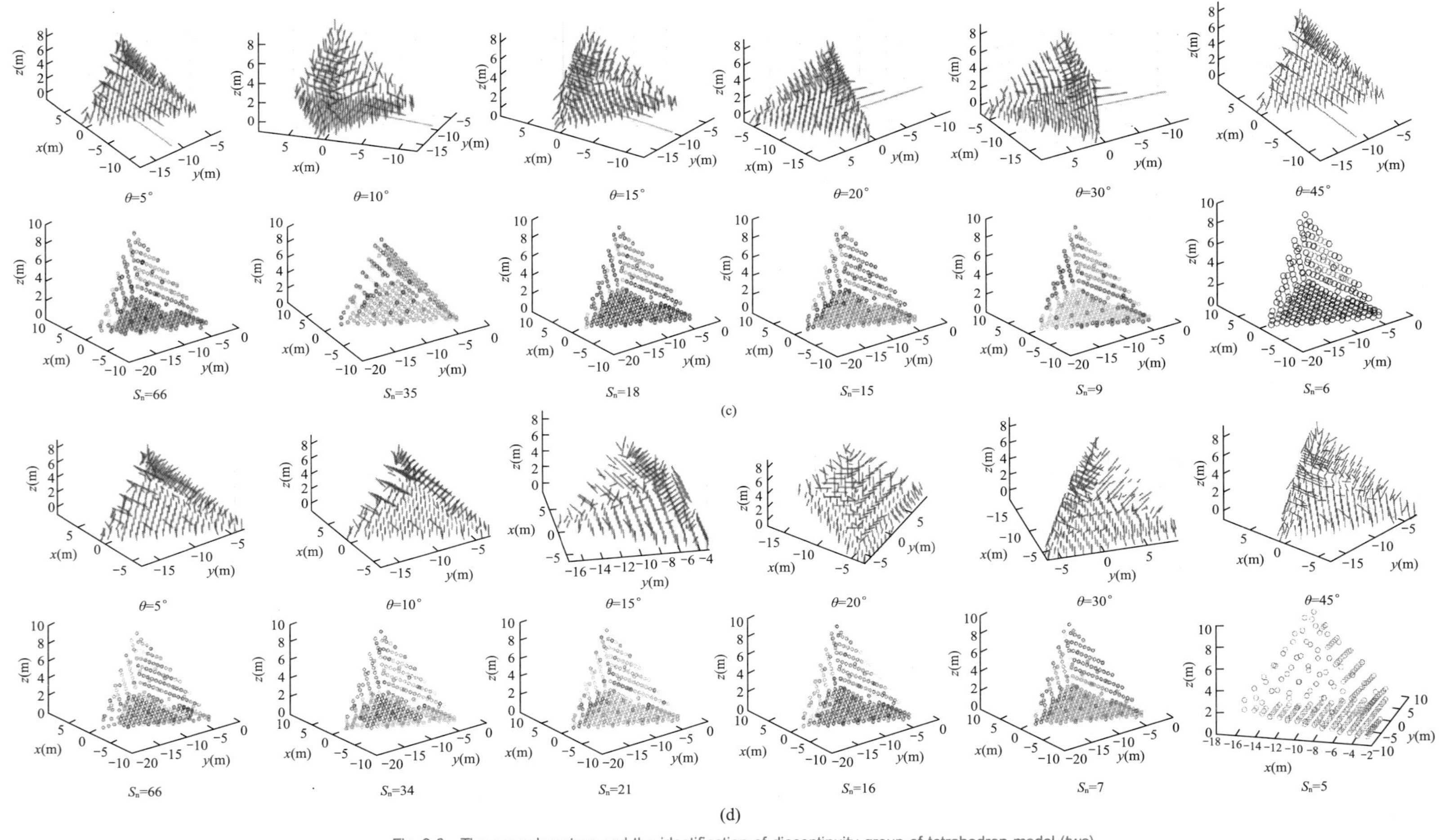

Fig. 9-6 The normal vectors and the identification of discontinuity group of tetrahedron model (two)

(c) $K=7$; (d) $K=10$;

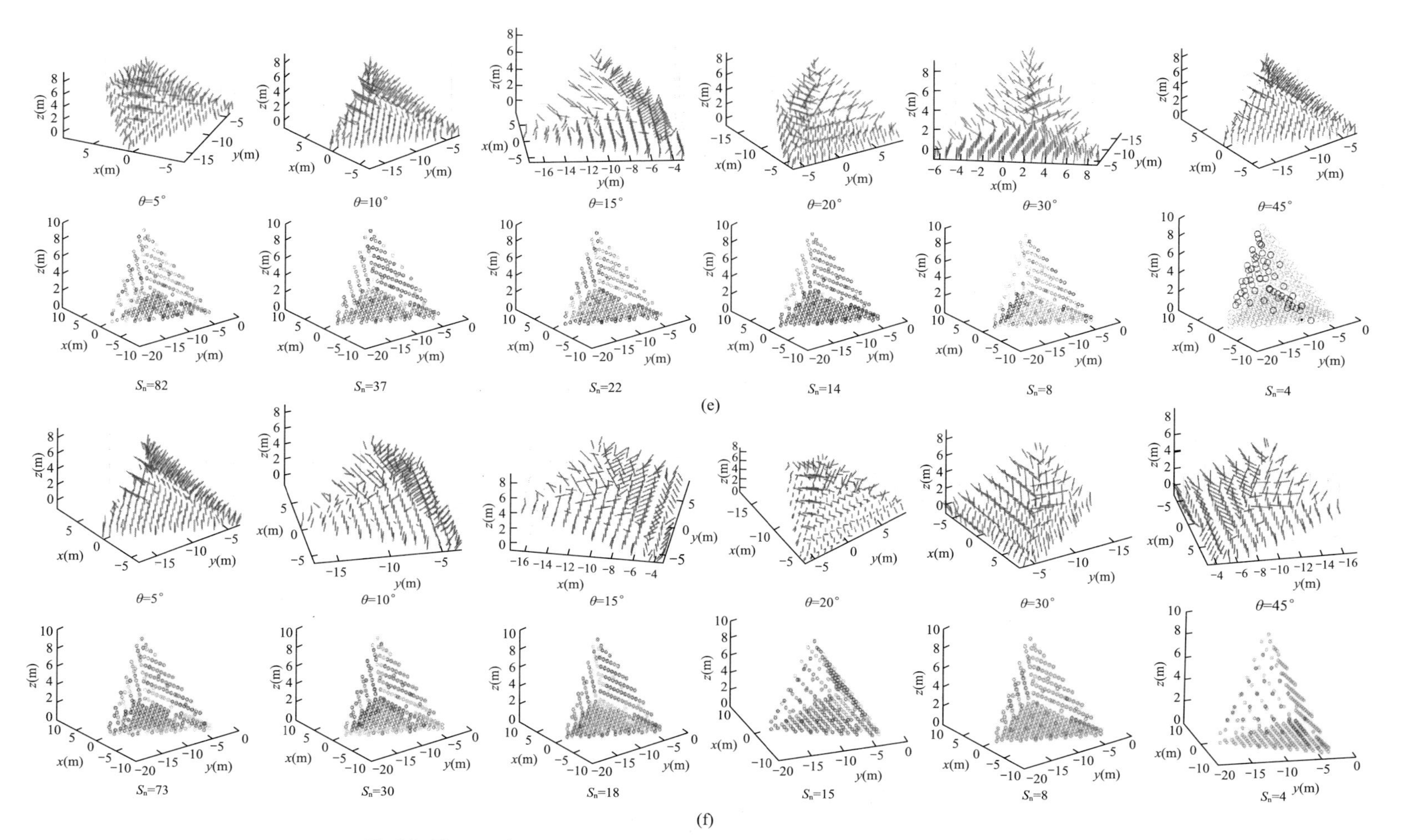

Fig. 9-6 The normal vectors and the identification of discontinuity group of the tetrahedron model (three)

(e) $K=12$; (f) $K=15$

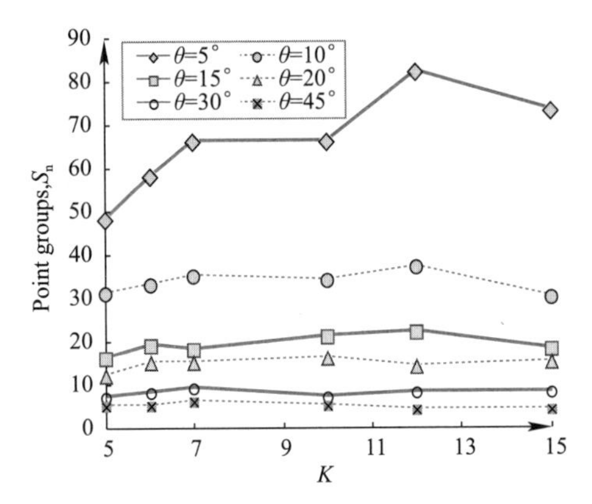

Fig. 9-7 The relationship of structure group with K values and threshold of angles

groups is 3, which is consistent with the actual result. In the case of $K=20$, $\theta=20°$, and $f=0$, the number of coplanar point groups is reduced due to the increase of the included angle threshold, which is 13 groups. With increasing the filter number f, the coplanar point group gradually decreases. According to the identification results in Table 9-3, the coplanar point group gradually changes from scattered to concentrated, which mainly affects the scattered points in the edge points. When f takes a larger value of 50, part of the edge points is effectively eliminated. The actual normal vector angle of each coplanar point of the hexahedral 3DPC is relatively high (about 90°), three sets of coplanar points could be identified in the case of threshold $\theta=45°$, even though the value of filter f is low.

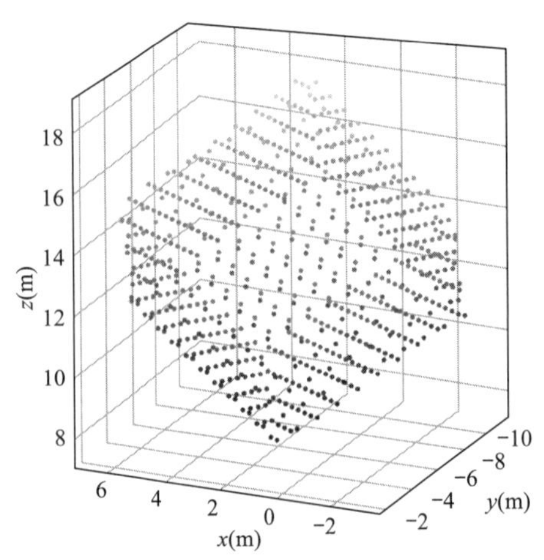

Fig. 9-8 The point cloud of hexahedron model

Fig. 9-9 shows the relationship of the number of point groups S_n identified in the hexahedral 3DPC with different K, θ, and f values. The number of effective point groups decreases with increasing the filter f values. The smaller the angle threshold is, the more significant the number of effective point groups S_n is affected by the filter value. The influence of filter value gradually decreases once the angle threshold θ is higher than 20°.

9.1.3.3 Identification of point clusters

(1) Analysis of a hexahedron 3DPC

Reasonable point groups have been acquired through the proposed algorithm by three parameters, K value, angle threshold θ, and filter f. All points belonging to a point group have similar normal vectors. However, several planes clustered by points with different intercept values may exist in the same point groups. A cluster analysis could then be performed to recognize the number of point clusters. Discontinuity information such as the density and spacing of structural surfaces of the point group could be measured through point clusters. Take the identification of the hexahedron model when $K=20$, $\theta=10°$, and $f=20$, for example, 10 point groups are examined (Fig. 9-10a). The number of groups ranges from 21 to 137. Point group 4, which contains 137 points, will be discussed by clustering analysis as shown in Fig. 9-10 (b). Two clusters are recognized as shown in Fig. 9-10 (c). The area density of the discontinuity group could then be measured according to the size of the model.

The spacing between the point clusters can be measured based on the vertical distance analysis by Eq. (9-6)—Eq. (9-8). The average values of c of each seed point in Eq. (9-2) could be obtained by Eq. (9-6).

$$e_i = \frac{\sum_{i=1}^{n_i} c_i}{n_i} \tag{9-6}$$

Where,

c_i—— the normal vector value of point i;

n_i—— the number of the points in the cluster;

e_i—— the average value of c_i.

The identification of the discontinuities groups considering different K, θ and f

Table 9-3

K value	Threshold θ	Filter f	Point group S_n	Identification result
$K=10$	$\theta=5°$	0	62	z(m) 20 15 10 5; x(m) −10 −5 0.5; y(m) 0 5 10
		10	5	z(m) 18 16 14 12 10 8 6; x(m) −10 −5 0; y(m) −4 −2 0 2 4 6 8
		20	3	z(m) 18 16 14 12 10 8; x(m) −12 −10 −8 −6 −4 −2; y(m) −5 0 5
	$\theta=10°$	0	37	z(m) 20 15 10 5 0; x(m) −2 −4 −6 −8 −10 −12; y(m) −4 −2 0 2 4 6 8
$K=20$	$\theta=5°$	0	65	z(m) 20 18 16 14 12 10 8 6 0; x(m) −5 −10; y(m) −4 −2 0 2 4 6 8
		10	11	z(m) 20 15 10; x(m) −4 −2 0 2 4 6 8; y(m) −12 −10 −8 −6 −4 −2 0
		20	3	z(m) 15 10; x(m) −2 −4 −6 −8 −10 −12; y(m) −4 −2 0 2 4 6
	$\theta=10°$	0	31	z(m) 20 15 10 5 0; x(m) −2 −4 −6 −8 −10; y(m) −5 0 5 10

Continued

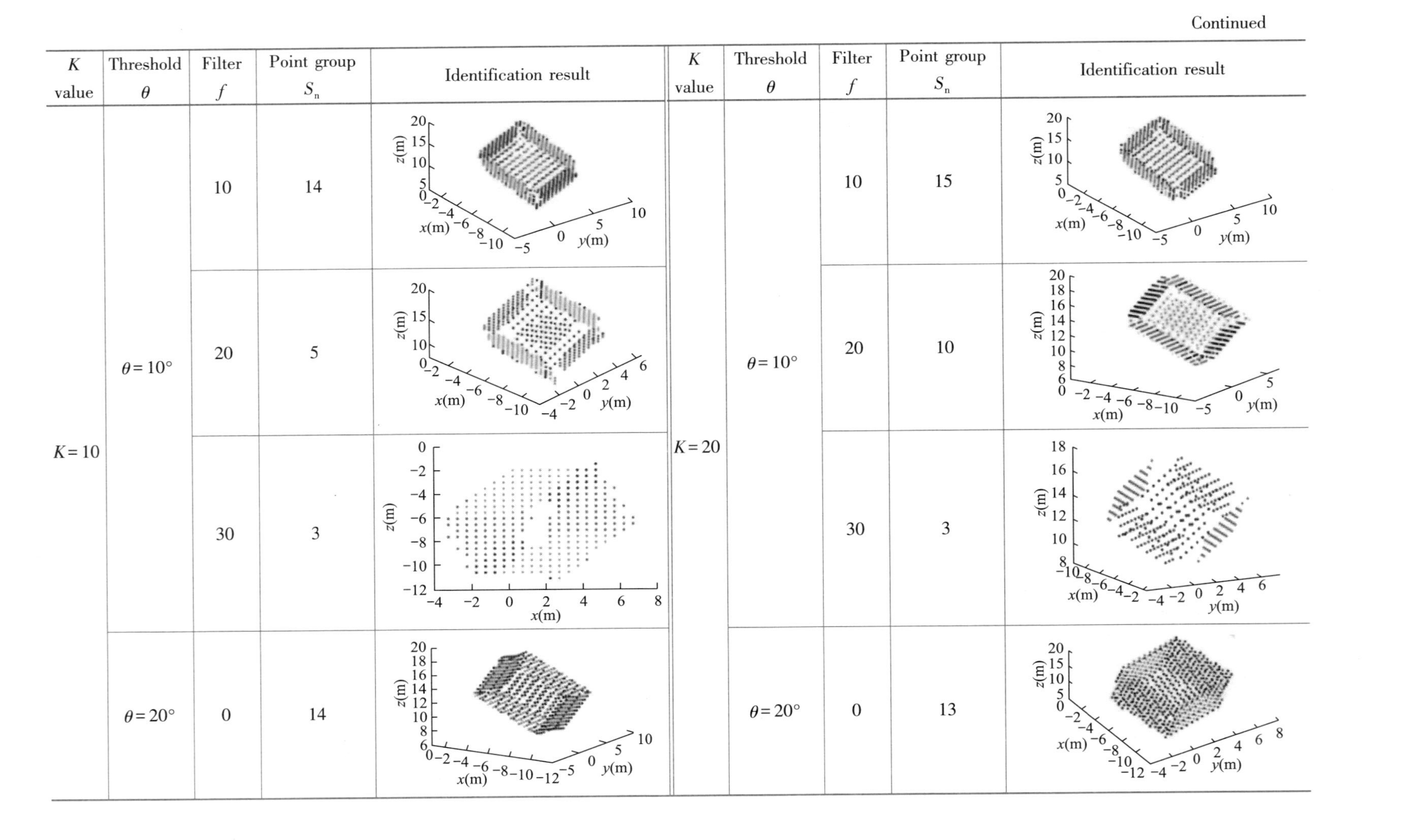

K value	Threshold θ	Filter f	Point group S_n	Identification result	K value	Threshold θ	Filter f	Point group S_n	Identification result
$K=10$	$\theta=10°$	10	14		$K=20$	$\theta=10°$	10	15	
		20	5				20	10	
		30	3				30	3	
	$\theta=20°$	0	14			$\theta=20°$	0	13	

Continued

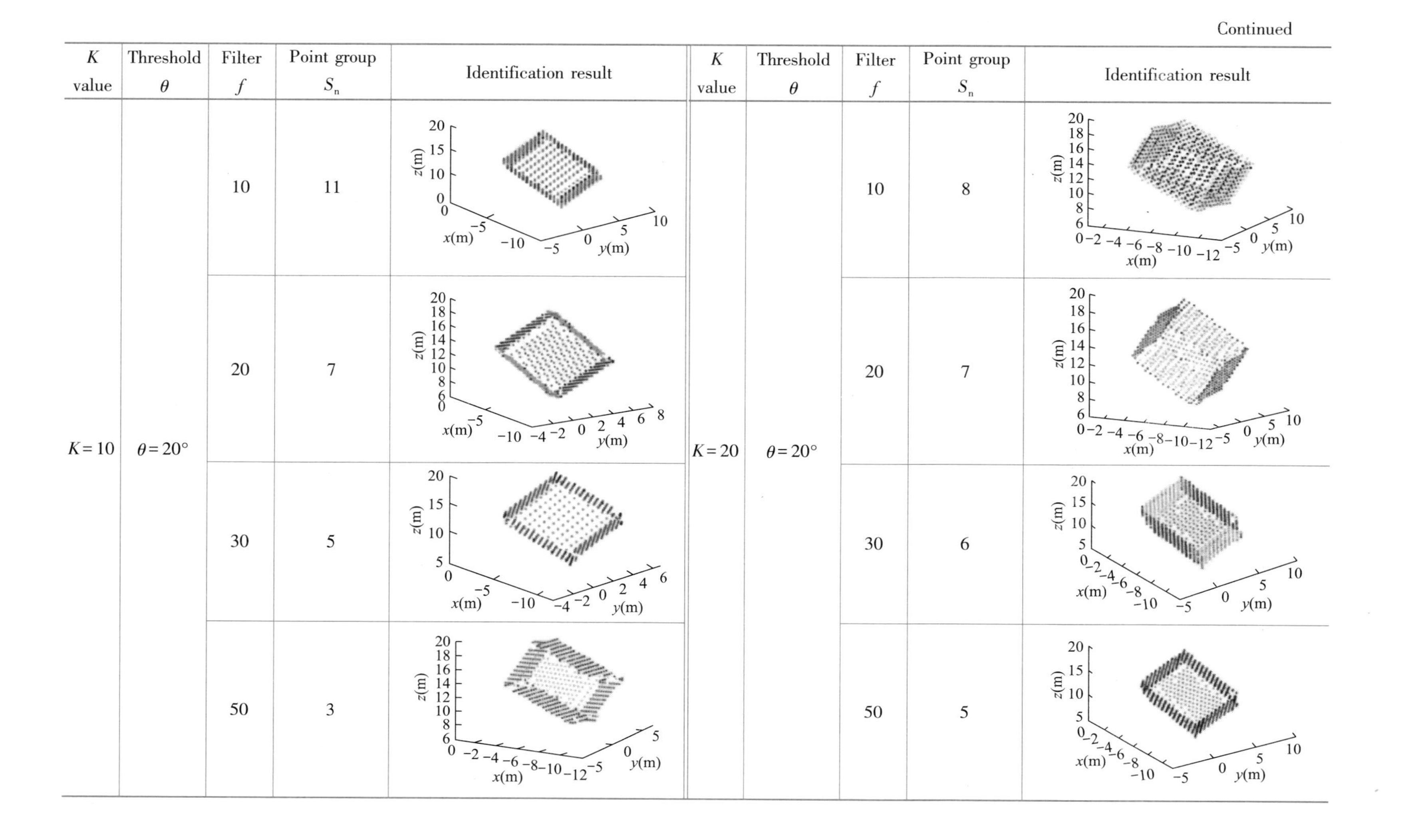

K value	Threshold θ	Filter f	Point group S_n	Identification result	K value	Threshold θ	Filter f	Point group S_n	Identification result
$K=10$	$\theta=20°$	10	11		$K=20$	$\theta=20°$	10	8	
		20	7				20	7	
		30	5				30	6	
		50	3				50	5	

Continued

K value	Threshold θ	Filter f	Point group S_n	Identification result
$K=10$	$\theta=30°$	0	6	z(m) 20 15 10 5; x(m) 0 −5 −10; y(m) −5 0 5 10
		10	6	z(m) 20 15 10 5 0; x(m) −2 −4 −6 −8 −10; y(m) −4 −2 0 2 4 6 8
		20	5	z(m) 20 15 10 5 0; x(m) −2 −4 −6 −8 −10 −12; y(m) −4 −2 0 2 4 6 8
		30	5	z(m) 20 15 10 5 0; x(m) −2 −4 −6 −8 −10 −12; y(m) −5 0 5 10
$K=20$	$\theta=30°$	0	6	z(m) 20 15 10 14 12 10 8 6 0; x(m) −5 −10; y(m) −4 −2 0 2 4 6 8
		30	4	z(m) 20 15 10 14 12 10 8 6 0; x(m) −5 −10; y(m) −4 −2 0 2 4 6 8
	$\theta=45°$	0	4	z(m) 20 15 10 5 0; x(m) −2 −4 −6 −8 −10; y(m) −5 0 5 10
		10	3	z(m) 20 15 10 14 12 10 8 6 0; x(m) −5 −10; y(m) −4 −2 0 2 4 6 8

Continued

K value	Threshold θ	Filter f	Point group S_n	Identification result	K value	Threshold θ	Filter f	Point group S_n	Identification result
$K=10$	$\theta=45°$	0	4		$K=20$	$\theta=45°$	20	3	
							30	3	

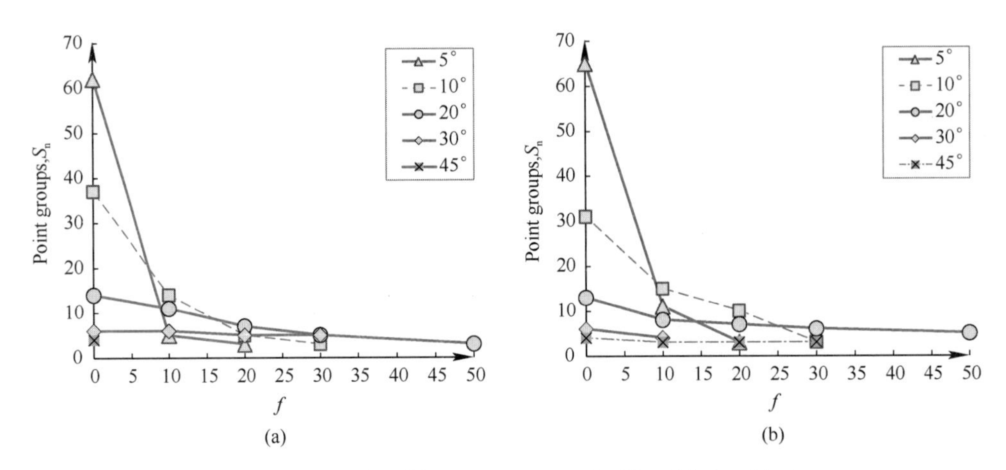

Fig. 9-9 The relationship of the structure groups with different K, θ and f values

(a) K=10; (b) K=20

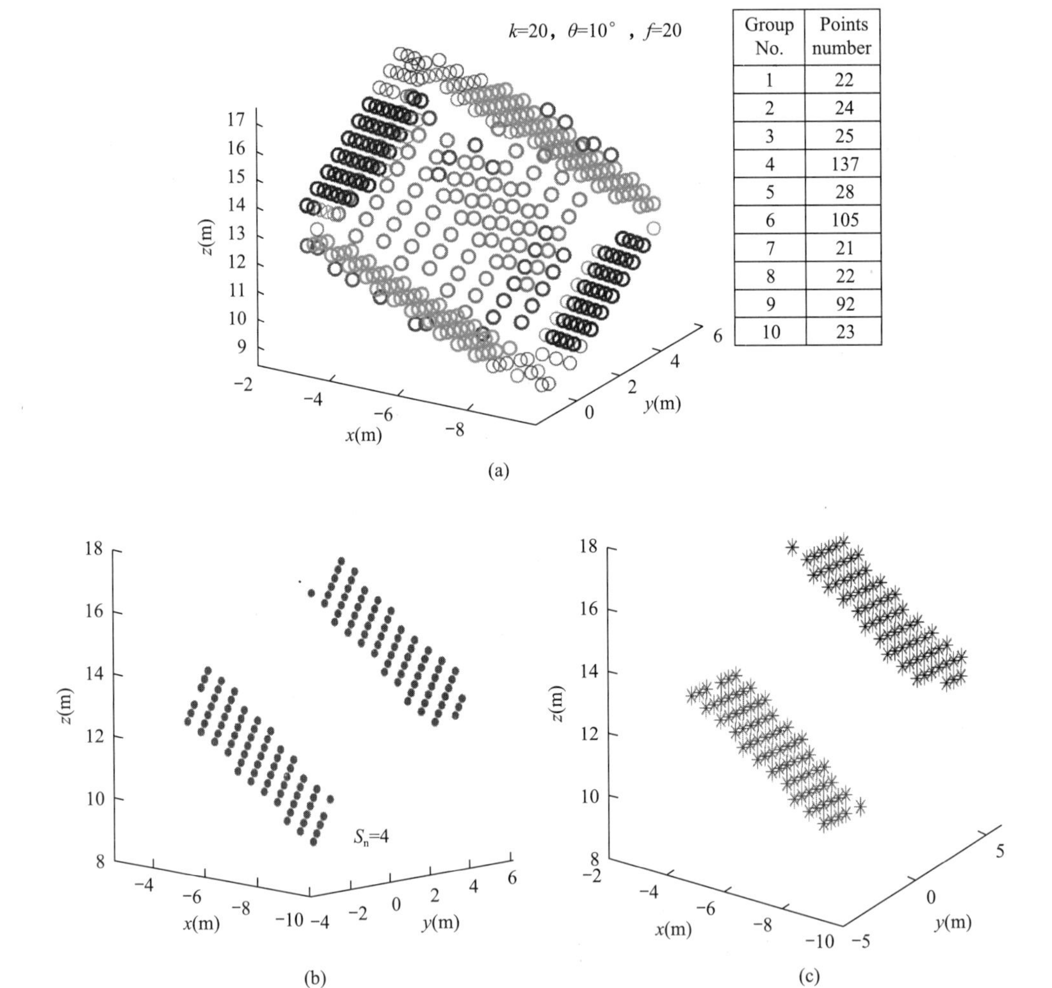

Group No.	Points number
1	22
2	24
3	25
4	137
5	28
6	105
7	21
8	22
9	92
10	23

Fig. 9-10 The structure groups and cluster analysis when K=20, θ=10°, and f=20

(a) Point groups distribution; (b) Points in Group 4; (c) Point cluster analysis

The normal vector of the point cluster is $(a, b, -1)$. Then the equation of the plane of point cluster is defined by Eq. (9-7).

$$z = ax + by + e_i \tag{9-7}$$

The spacing of the point clusters can be obtained according to Eq. (9-8).

$$d_{ij} = \frac{|e_i - e_j|}{\sqrt{a^2 + b^2 + 1}} \tag{9-8}$$

Take the identification of point cluster, for example, the normal vector of the point group is (−1.3268, 0.5142, −1). Then the equations of the two planes are $z = 23.9477 - 1.3268x + 0.5142y$ and $z = 13.0847 - 1.3268x + 0.5142y$, respectively. The value of d_{12} is 6.2460 based on Eq. (9-8). Then the spacing of the discontinuity group is 6.2460m.

(2) Case study-case I

A group of 3DPCs of rock-cut, acquired in the slope surface (Fig. 9-11a), is chosen to conduct the sensitivity analysis to verify the CGC method. The 3DPC contains 1989 points as shown in Fig. 9-11 (b). The distribution of the normal vectors is shown in Fig. 9-11 (c). The point groups recognition in the case of $K = 15$, $\theta = 20°$, and $f = 20$, is shown in Fig. 9-11 (d). Ten groups are recognized and the orientations are also analyzed. It should be addressed that in the coordinate system, the X axis, Y axis, and Z axis represent E (east), N (north), and Z (vertical), respectively. The maximum and the minimum number of points in different point groups is 799 ($S_n = 6$) and 21 ($S_n = 10$). The orientation of each point is shown in Fig. 9-12 (a). Dip direction and dip angle and the rose diagram of ten groups of the discontinuities are shown in Fig. 9-12 (b) and (c).

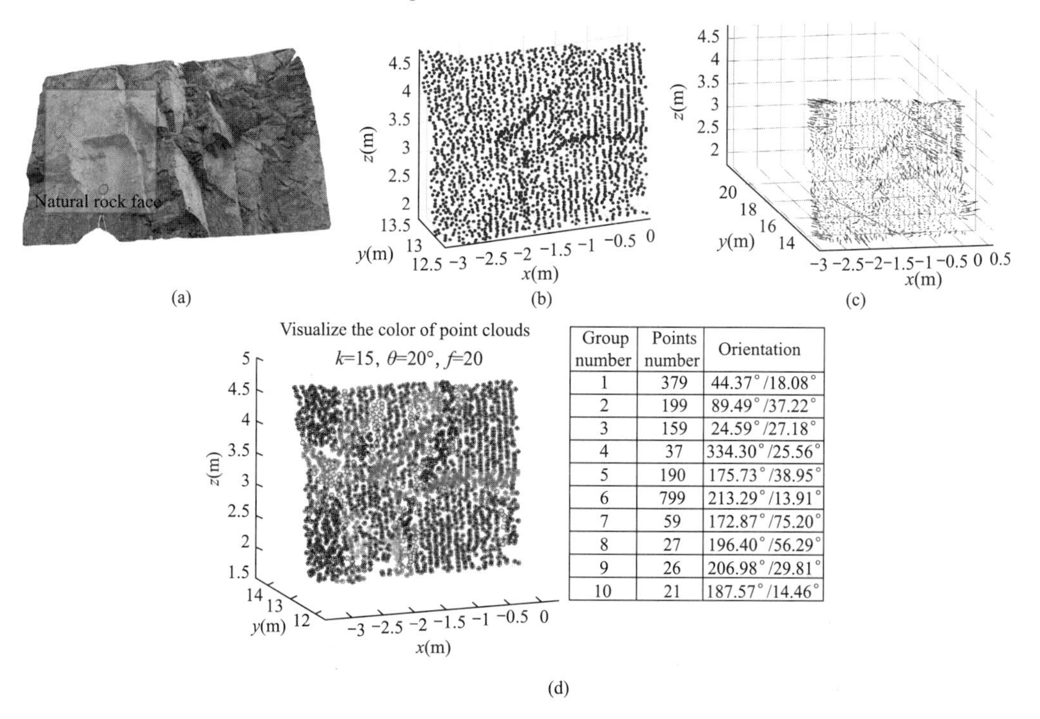

Group number	Points number	Orientation
1	379	44.37°/18.08°
2	199	89.49°/37.22°
3	159	24.59°/27.18°
4	37	334.30°/25.56°
5	190	175.73°/38.95°
6	799	213.29°/13.91°
7	59	172.87°/75.20°
8	27	196.40°/56.29°
9	26	206.98°/29.81°
10	21	187.57°/14.46°

Fig. 9-11 Point cloud and the discontinuities of different groups in Case I

(a) Natural rock face; (b) Point cloud vectors; (c) Distribution of normal; (d) Point groups recognition

The point cluster analysis is performed on Group 6, which contains 799 points, as shown in Fig. 9-13 (a). There are 4 main structural planes in this point group with several relatively discrete 3DPCs on the left side. Fig. 9-13 (b) shows the identification result of the point cluster distribution based on cluster analysis. A total of 7 point-clusters are identified and are marked with different colors. Among them, there are 6 valid point clusters. The point cluster

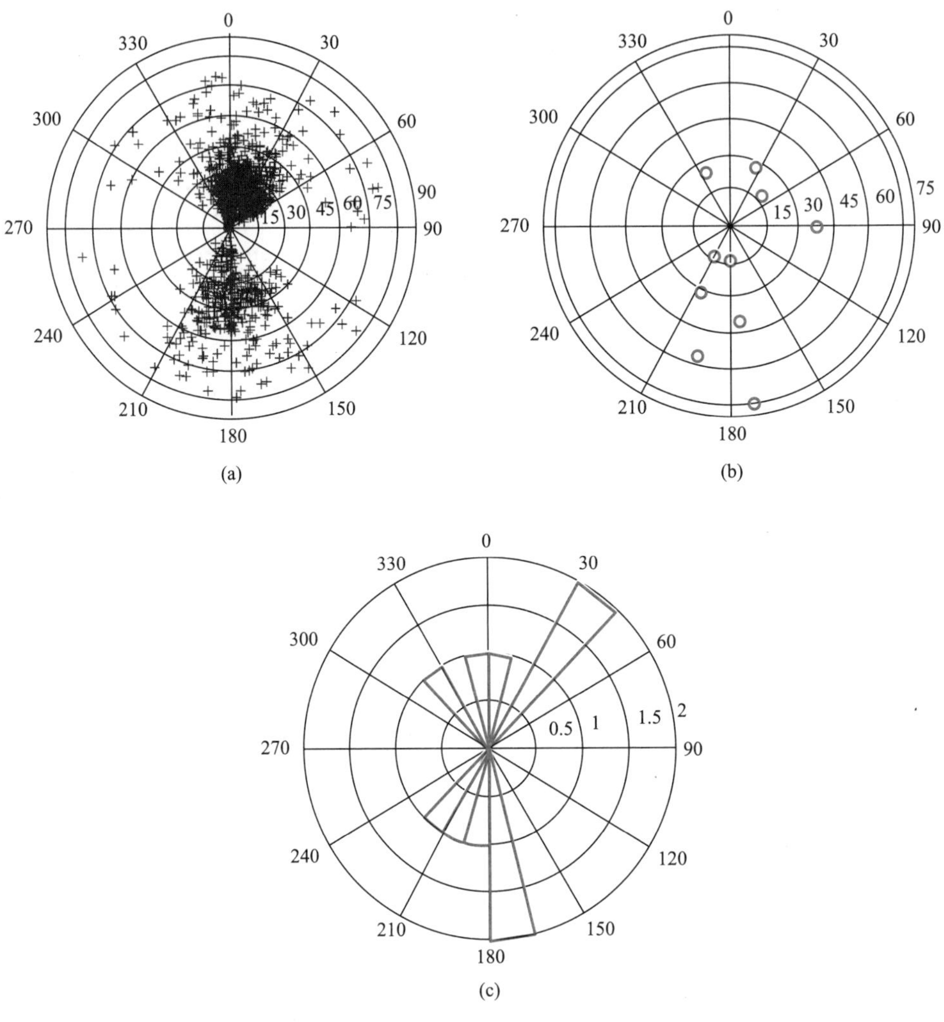

Fig. 9-12 The orientation of the discontinuities of different point groups
(a) Normal vectors of points; (b) Orientation of point groups; (c) Rose diagram of groups

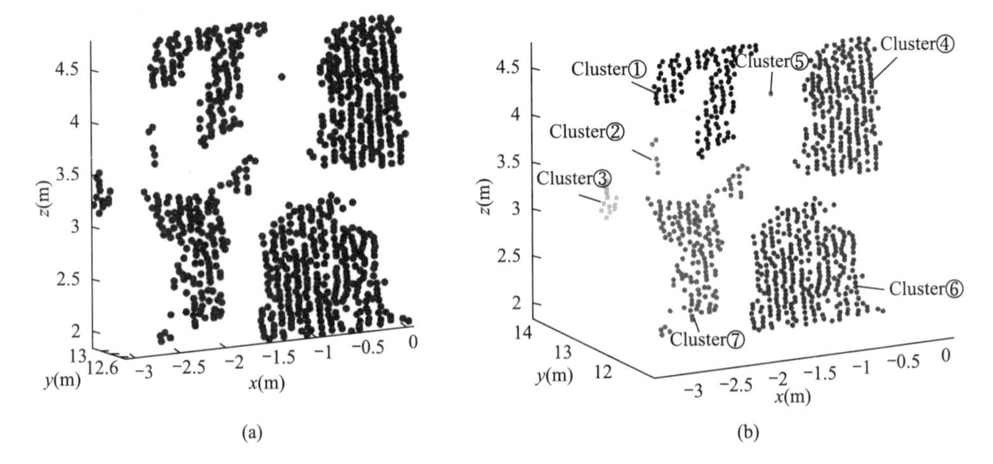

Fig. 9-13 The point cloud of Group 6 and the cluster analysis result
(a) Point group, S_n=6; (b) Point cluster analysis

⑤ only contains an independent scattered point, which can be regarded as an invalid point cluster. Considering that there may be valid points during the nearest-neighbor searching and noise points filtering procedure. The point cluster will still be considered and analyzed in this section. Fig. 9-14 shows the spatial distribution of the identified point clusters. It can be concluded that the CGC analysis method can effectively carry out the identification of different non-coplanar 3DPCs, and can provide reliable information for the extraction of structural plane spacing and density information.

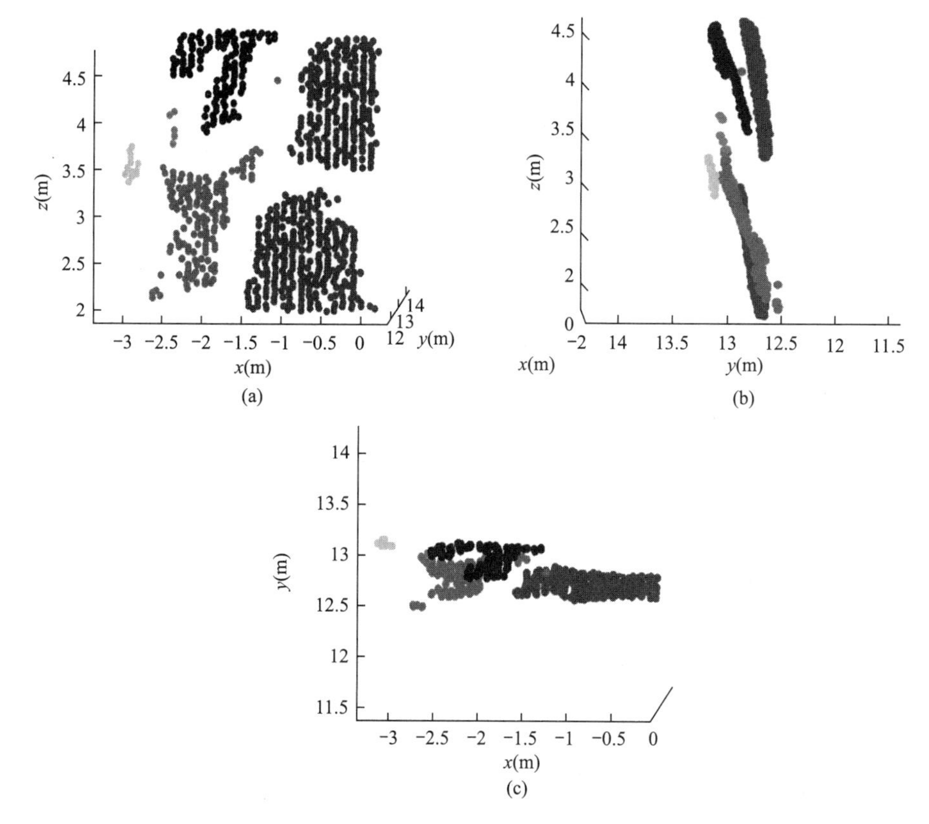

Fig. 9-14 Point clusters distribution of Group 6

(a) Front view; (b) Left view; (c) Top view

According to Eq. (9-6)—(9-8), the spacing information can be measured and listed in Table 9-4. The normal vector of the point Group 6 is (0.1978, 2.8352, −1), and the volume of the jointed rock can be determined by the hexahedron domain formed by the 3DPC. The linear density of the discontinuities can be determined based on the length of the survey line (the maximum distance in the normal direction of the point cluster). And the average spacing can be determined based on the linear density or the average spacing between the point clusters. The plane equations, volume, density information is listed in Table 9-4. According to the linear density, the structure plane spacing is 0.0847m, and the average spacing between each point cluster is 0.0878m. According to the results, the CGC method could reasonably analyze the discontinuity information of the 3DPC.

The cluster analysis of other point groups is carried out and the results are shown in Fig. 9-15. A total of 10 sets of structural planes are identified on the surface of the rock mass. There are differences in the number of point clusters identified by each group of structural planes. The least number is 2 clusters and the highest is 7 clusters. The highest linear density is in Group 6, which reaches 11.80/m, and the lowest is in Group 9 with the linear density of 1.82/m. According to the distribution of the 3DPC, the structural planes of Group 1, Group 5, and Group 6 are relatively concentrated, and the other point groups are relatively discrete.

The identification of the point clusters **Table 9-4**

Parameters	Value		
P_{max}(m)	x_{max}=0.07	y_{max}=13.16	z_{max}=4.77
P_{min}(m)	x_{min}=−3.12	y_{min}=12.51	z_{min}=1.85
Point cluster ①	$z=-32.71+0.1978x+2.8352y$		
Point cluster ②	$z=-33.11+0.1978x+2.8352y$		
Point cluster ③	$z=-33.30+0.1978x+2.8352y$		
Point cluster ④	$z=-31.89+0.1978x+2.8352y$		
Point cluster ⑤	$z=-31.78+0.1978x+2.8352y$		
Point cluster ⑥	$z=-33.57+0.1978x+2.8352y$		
Point cluster ⑦	$z=-32.05+0.1978x+2.8352y$		
Volume (m^3)	6.09		
Clusters number	7		
Normal vector of the cluster	(0.1978, 2.8352, −1)		
Length of scanning line (m)	0.5934		
Area density (/m^3)	1.15		
Linear density (/m)	11.80		
Distances of clusters (m)	0.1315, 0.0637, 0.0350, 0.0883, 0.1549, 0.0533		
Mean distance (m)	0.0878		
Spacing by linear density (m)	0.0847		

9.1.4 Application and Discussions

9.1.4.1 Underground case studies

(1) Case Ⅱ: comparison of the proposed method with the Sirovision analyzing results

We have introduced the CGC method for identifying the discontinuity information. Here, we will discuss three more case studies to verify the feasibility. The 3DPC model of Case Ⅱ is from the Aerhada underground mine, which contains 3,817,252 points, as shown in Fig. 9-16. We choose 0.1% of the 3DPC, which is 3817 points in Case Ⅱ to test the CGC method. According to the initial examination, three principal groups of discontinuities are observed from the extracted 3DPC model. Here, we will discuss the influence of filter factor f, on the point group identification. Table 9-5 lists the results in case of $K=11$, and $\theta=20°$. With the increasement of the ratio of filter points, effective point groups can be obtained. When the filter ratio is below 5%, many groups of scatters can be obtained. When increasing the filter ratio to more than 5%, the scatters can be efficiently filtered. Three point groups can be obtained when the filter ratio is about 10%. There are 1074, 903, and 457 points in each point group, respectively.

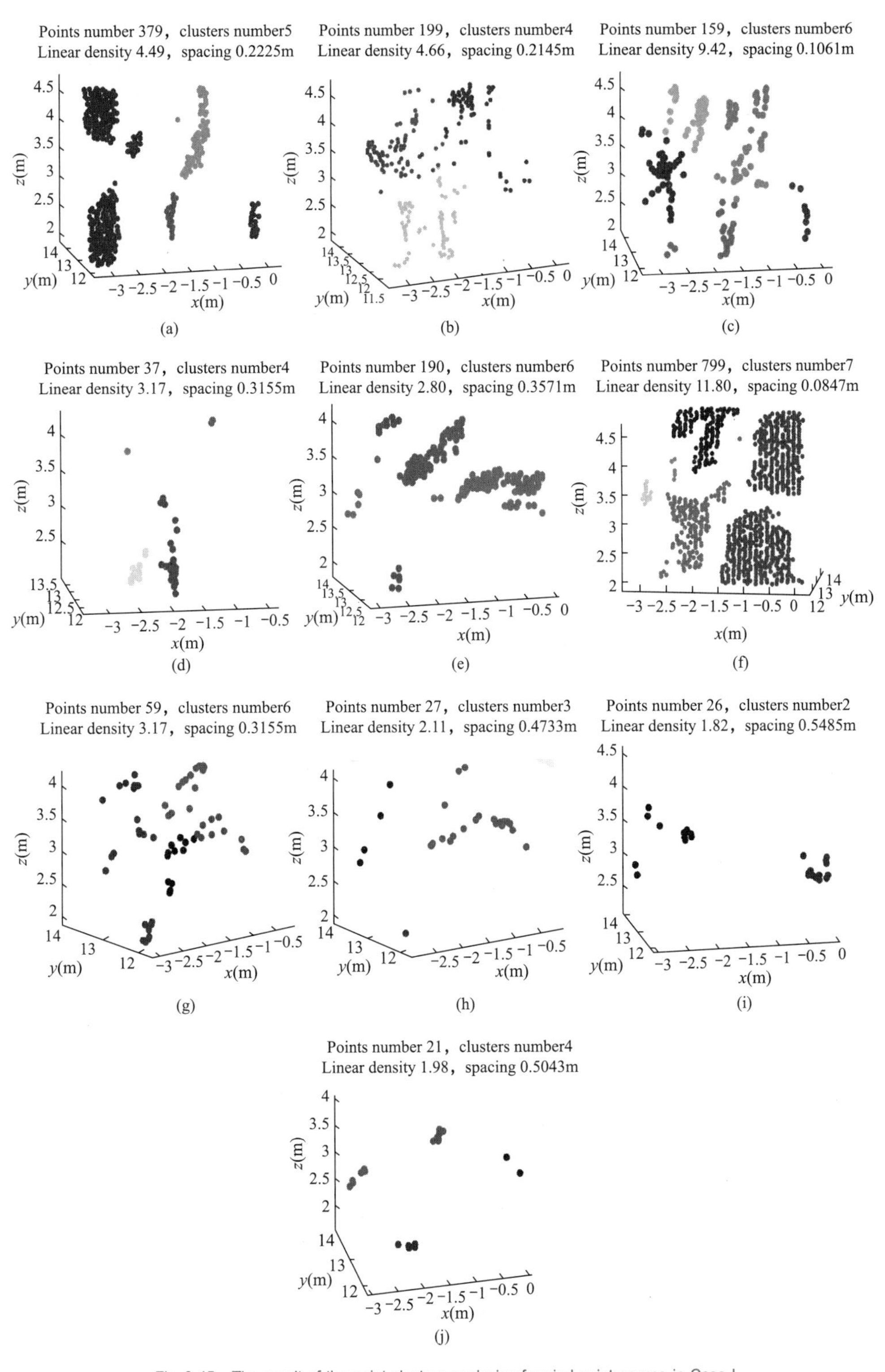

Fig. 9-15 The result of the point clusters analysis of varied point groups in Case I
(a) S_n=1; (b) S_n=2; (c) S_n=3; (d) S_n=4; (e) S_n=5; (f) S_n=6; (g) S_n=7; (h) S_n=8; (i) S_n=9; (j) S_n=10

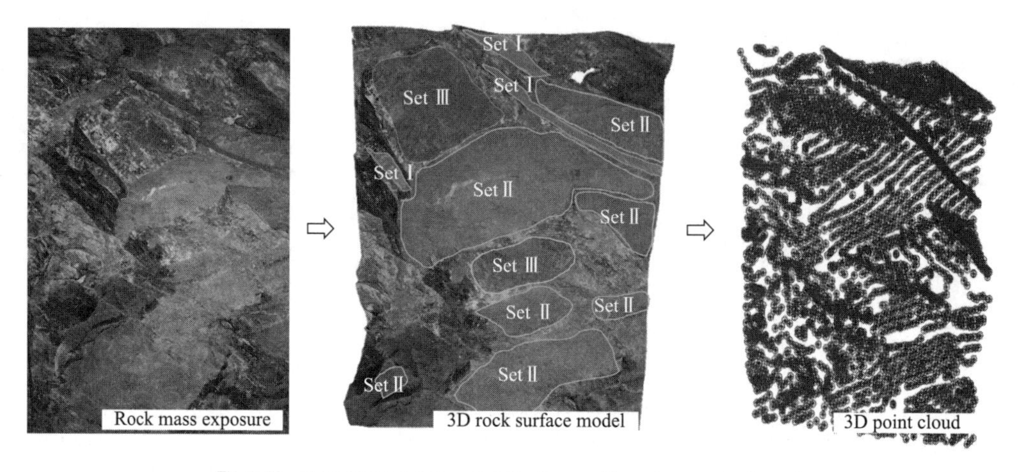

Fig. 9-16 Point cloud and the discontinuities of different groups in Case Ⅱ

Fig. 9-17 shows the relationship of the point group number with the parameters, K, θ, and f. The angle threshold and filter ratio can effectively influence the results of point groups identification. When the filter ratio is lower than 5%, the number of point groups decreases dramatically and will tend to stabilize after the value of 10%. To verify the accuracy of the CGC method, the comparison of the proposed method with the Sirovision results is conducted in Fig. 9-18 and Table 9-6. The three point groups of the CGC result ($K=10$, $\theta=20°$, $f=10\%$) as well as the projection results are shown in Fig. 9-18. According to the Sirovision test, three principal groups are obtained, the orientation of which is 323.9°/67.1°, 226.1°/63.6°, 74.8°/78°, respectively. The proposed CGC method also obtains three groups with the orientation of 322.98°/65.09°, 203.04°/56.34°, and 177.89°/49.43°, respectively. Two of the groups can accurately identify the discontinuity dip direction and dip angle. The third group of the Sirovision results is mainly the joint traces distribution. However, the rock mass exposure face is tested and identified in the CGC method. It should be the reason for the large error of the identification result. Generally, the number of rock mass exposure which is induced by excavation is always large in the 3DPC. In the future, we will improve the CGC method to consider the existence of the artificial structural surfaces.

(2) Case Ⅲ: discussion of optimum parameters of regional points in the total 3DPC model

There are generally millions of points in one certain 3DPC model of rock mass. Study on the optimum parameters using the CGC method directly on the total model is very time consuming. Here we propose an optimum parameters selection approach, which discusses the optimum parameters according the regional 3DPC. After the parameters searching from some certain segments, we can test the total 3DPC model more efficiently. Fig. 9-19 shows the 3D model of Case Ⅲ, which is also acquired from the Aerhada underground mine. There are 6168 points in the 3DPC model, which select 0.1% part of the total points (6,168,516 points in the 3D model). Four segments are selected to discuss the optimum values of K, θ and f. Part A, B, C, and D contain 1738, 1373, 1282, and 1775 scatters, respectively. In each part, the identified number of point group is 3. After numbers of trial, the optimum values of K, θ and f are 10, 15° and 10%, respectively.

To examine the feasibility of the optimum parameters, four different filter factors, which are 1%, 5%, 10% and 20%, are tested on the total 3DPC model (Fig. 9-20). When the filter ratio is 1%, 15 point groups are recognized. There are many distinct point groups in this test. When the filter ratio increases to 5%, 5 groups are recognized. The principal groups

Table 9-5 Identification results of varied filter factors in case of $K=10$, $\theta=20°$

Filter f	Filter ratio	Point group S_n	Identification result	Filter f	Filter ratio	Point group S_n	Identification result
0	0	22	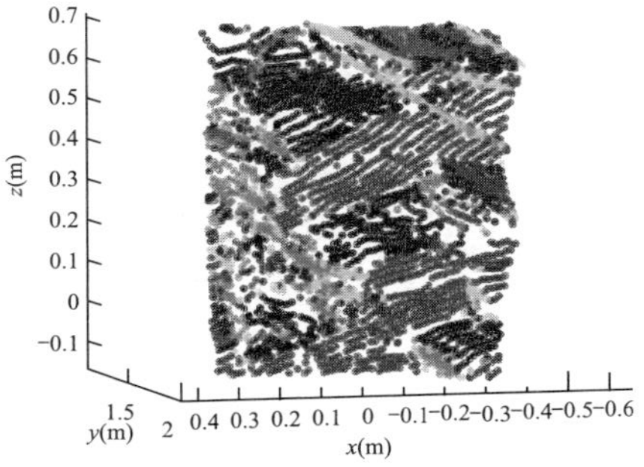 z(m): 0.7 0.6 0.5 0.4 0.3 0.2 0.1 0 −0.1 y(m): 1.5 2 x(m): 0.4 0.3 0.2 0.1 0 −0.1 −0.2 −0.3 −0.4 −0.5 −0.6	76	2%	9	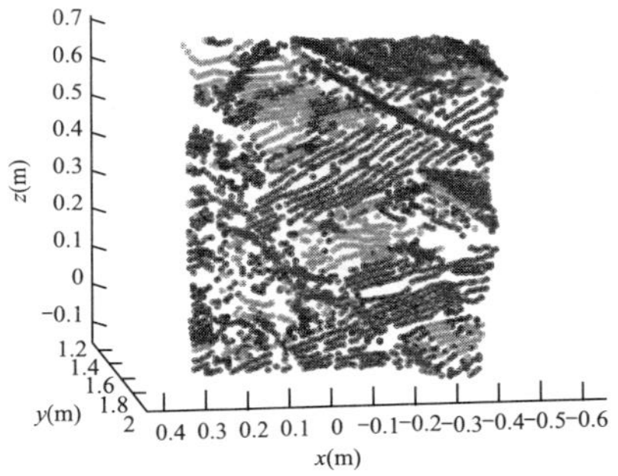 z(m): 0.7 0.6 0.5 0.4 0.3 0.2 0.1 0 −0.1 y(m): 1.2 1.4 1.6 1.8 2 x(m): 0.4 0.3 0.2 0.1 0 −0.1 −0.2 −0.3 −0.4 −0.5 −0.6
5	0.125%	17	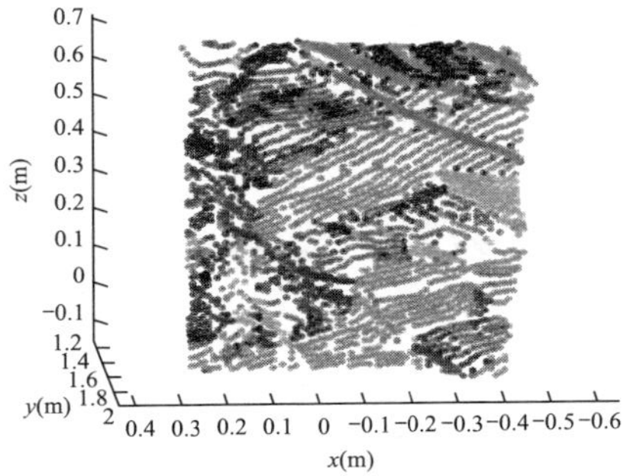 z(m): 0.7 0.6 0.5 0.4 0.3 0.2 0.1 0 −0.1 y(m): 1.2 1.4 1.6 1.8 2 x(m): 0.4 0.3 0.2 0.1 0 −0.1 −0.2 −0.3 −0.4 −0.5 −0.6	190	5%	5	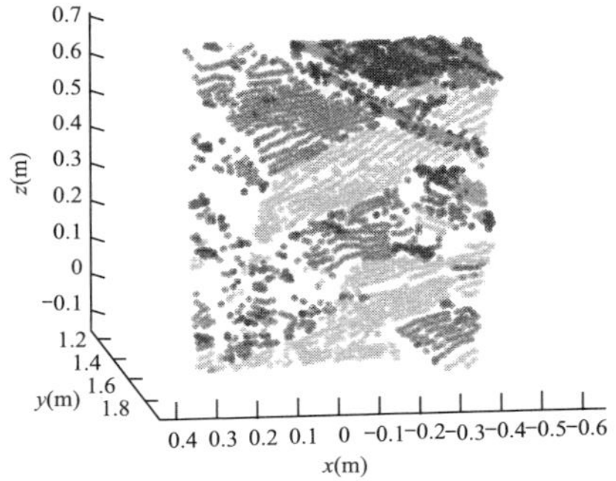 z(m): 0.7 0.6 0.5 0.4 0.3 0.2 0.1 0 −0.1 y(m): 1.2 1.4 1.6 1.8 x(m): 0.4 0.3 0.2 0.1 0 −0.1 −0.2 −0.3 −0.4 −0.5 −0.6

Continued

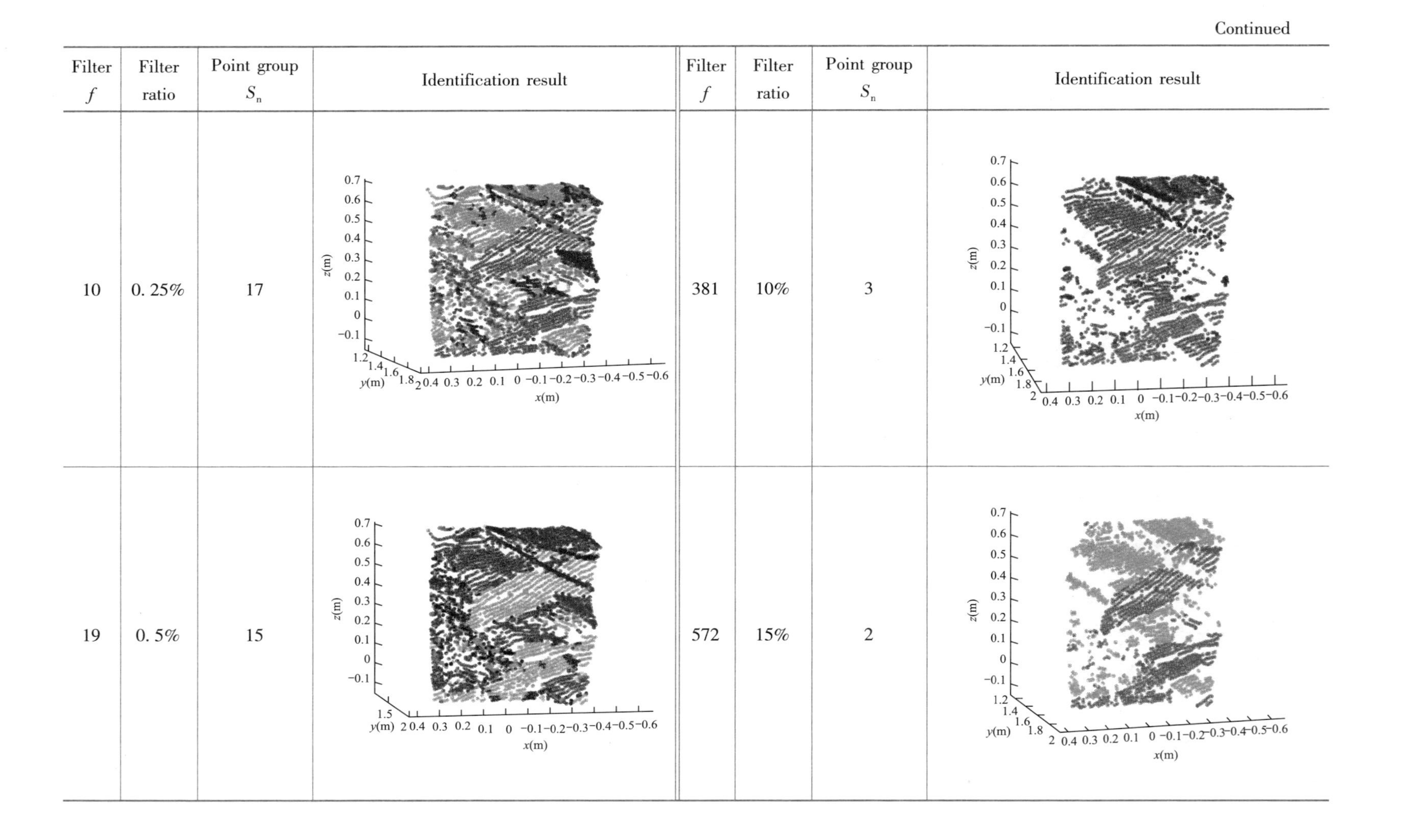

Filter f	Filter ratio	Point group S_n	Identification result	Filter f	Filter ratio	Point group S_n	Identification result
10	0. 25%	17		381	10%	3	
19	0. 5%	15		572	15%	2	

Continued

Filter f	Filter ratio	Point group S_n	Identification result	Filter f	Filter ratio	Point group S_n	Identification result
38	1%	13	z(m) 0.7 0.6 0.5 0.4 0.3 0.2 0.1 0 −0.1 y(m) 1.2 1.4 1.6 1.8 x(m) 0.4 0.3 0.2 0.1 0 −0.1 −0.2 −0.3 −0.4 −0.5 −0.6	763	20%	1	z(m) 0.6 0.5 0.4 0.3 0.2 0.1 0 −0.1 y(m) 1.2 1.3 1.4 1.5 1.6 1.7 1.8 1.9 x(m) 0.4 0.3 0.2 0.1 0 −0.1 −0.2 −0.3 −0.4 −0.5 −0.6

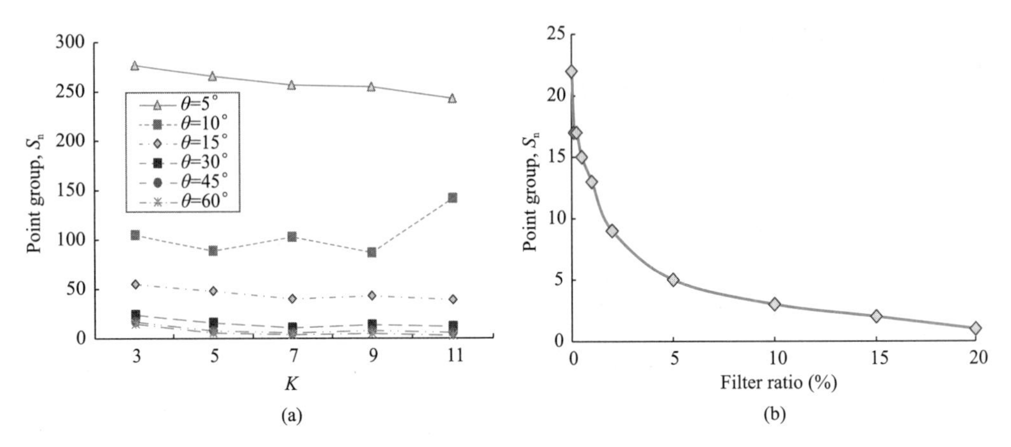

Fig. 9-17 The relationship of point group number

(a) With K values and angle threshold; (b) With filter ratio in Case Ⅱ

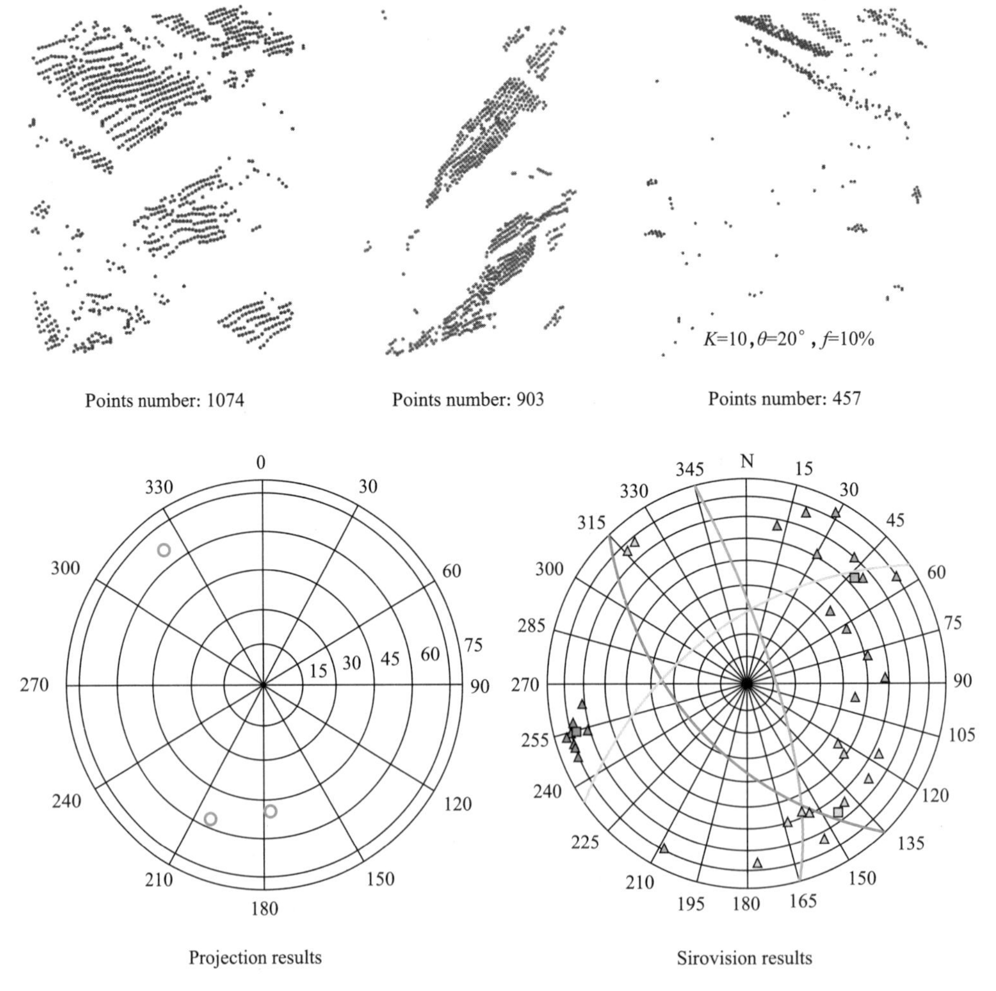

Fig. 9-18 Comparison of the identification result with the Sirovision result

Comparison of the proposed method with the Sirovision results Table 9-6

Type		Dip direction			Dip angle		
		Sirovision result (°)	CGC result (°)	Error (°)	Sirovision result (°)	CGC result (°)	Error (°)
Discontinuities groups	1	323.9°	322.98°	0.28%	67.1°	65.09°	2.99%
	2	226.1°	203.04°	10.20%	63.6°	56.34°	11.42%
Rock mass exposure	3	74.8°	177.89°	—	78.0°	49.43°	—

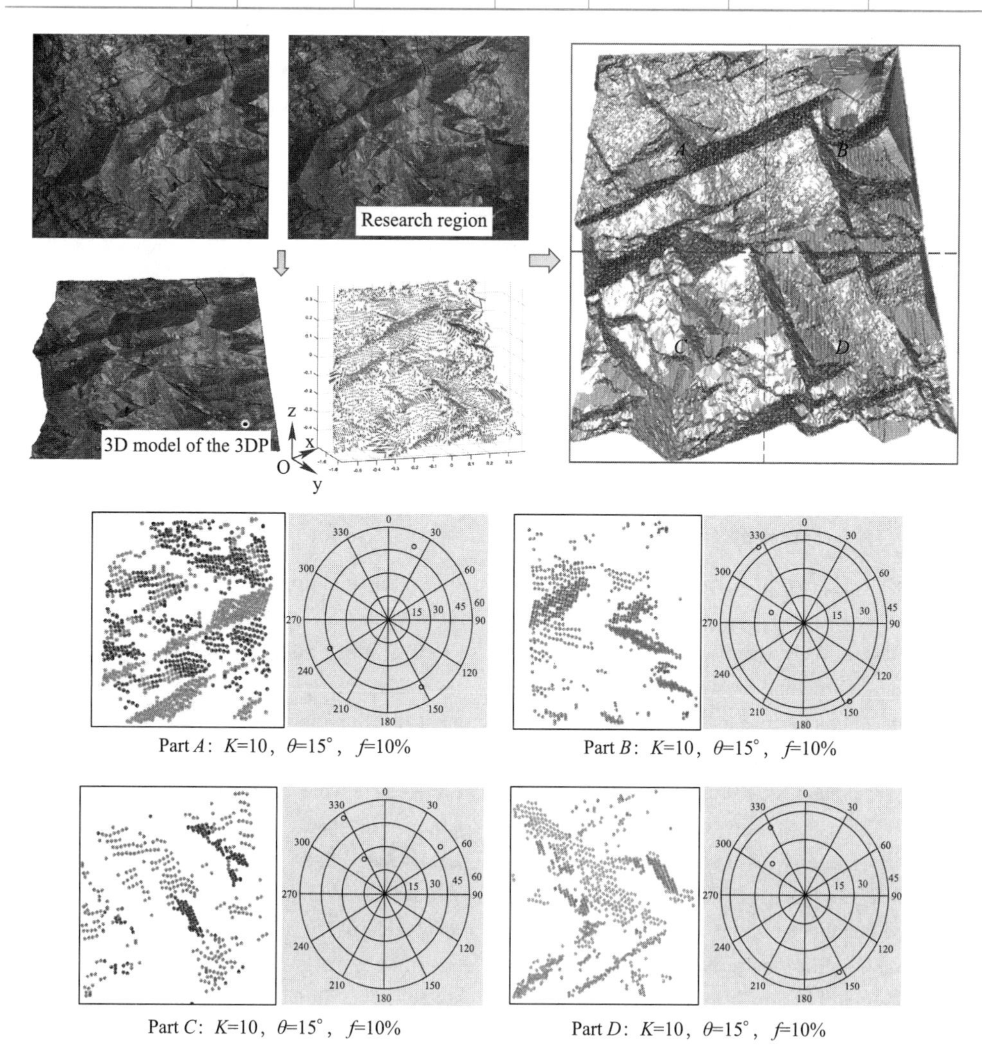

Fig. 9-19 The optimum parameters of regional 3DPC in the total model

of discontinuities can be identified with several scatters groups. This is not an optimum filter factor. In the case of f of 10%, the number of principal groups of discontinuities is 3, which coincides with the regional test results. When we continue increasing the filter ratio to 20%, only one point group, whose orientation is 270.91°/42.27°, is identified. In conclusion, the optimum parameters (K, θ, f = 10, 15°, 10%) obtained from the regional points can be applied to the total 3DPC model. This can provide an efficient approach for dealing with 3DPC models with huge number of points.

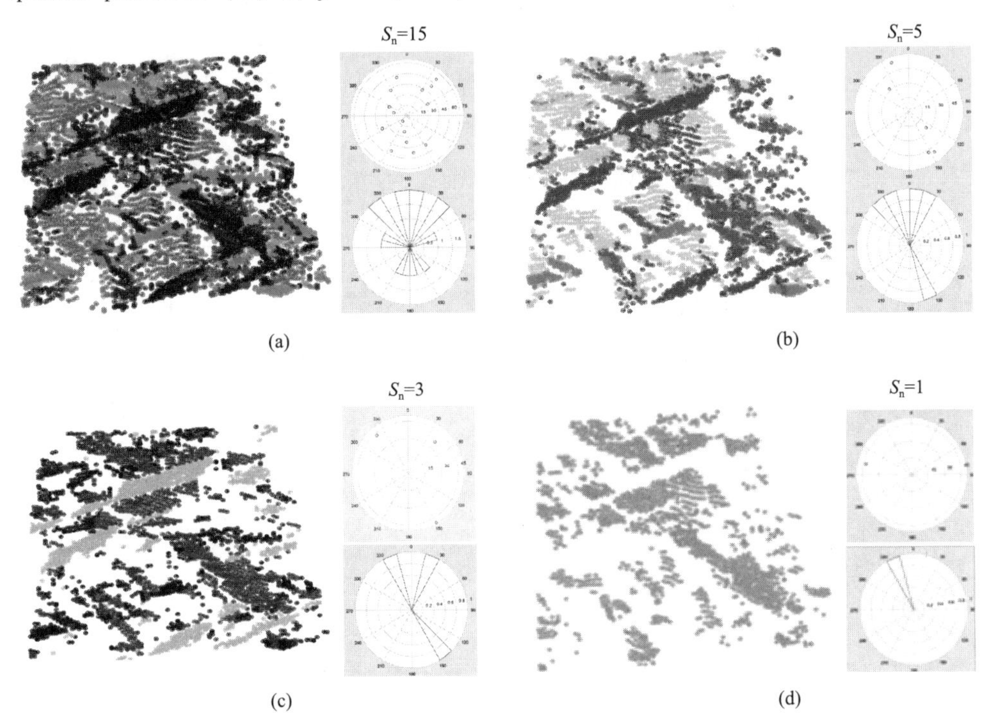

Fig. 9-20 The identification result of the 3DPC model based on the optimum parameters
(a) K=10, θ=15°, f=1%; (b) K=10, θ=15°, f=5%; (c) K=10, θ=15°, f=10%; (d) K=10, θ=15°, f=20%

9.1.4.2 Case Ⅳ: application in an open pit mine

In Case Ⅳ, we will discuss the applicability of the CGC method as well as the optimum parameters obtained in Case Ⅱ and Ⅲ. The 3DPC model in Fig. 9-21 is established from a southern rock mass exposure in Sijiaying open pit mine. The rock mass is highly fractured with at least three groups of discontinuities, which cut the rock into blocks. Meanwhile there is one set of bedding plane in the rock mass, which dominates the failure patterns of the slope. The dip direction and dip angle of the bedding plane are in the range of 90° ~ 100° and 40° ~ 60°, respectively. 22,349 points are selected from the total 22,349,163 points in the 3DPC model in Case Ⅳ. The parameters of K and θ in CGC method are 10 and 15°, respectively. Fig. 9-21 shows the identification results when the filter ratio is 1%, 5%, 10% and 20%. In the case of f= 1%, fifteen point groups are recognized with several distinct point groups. With the increasement of filter ratio, the number of point groups decreases dramatically at f= 5%, as is shown in Fig. 9-22. In the case of f= 10%, only 3 point groups are recognized. The attitude of the 3 groups is 94.81°/49.47°, 146.33°/40.05°, and 109.37°/35.57°, respectively. There are 2 point groups left when we continue to increase the filter ratio to 20%. According to the case studies, the initial optimum set of parameters, K, θ, and f, should be chosen as 10, 15°, and 10% for the varied rock mass exposures. Adjustments should be carried out considering the identification results comparing the spatial distribution of the natural discontinuities.

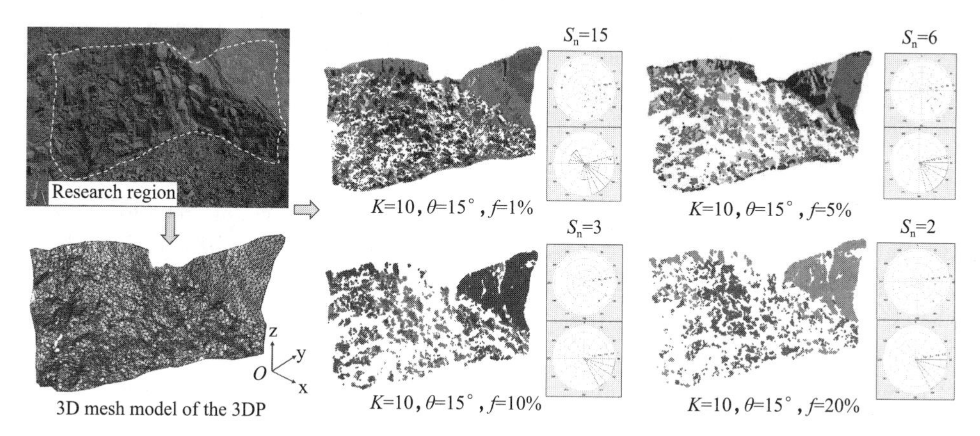

Fig. 9-21 The identification result of the 3DPC of Case Ⅳ

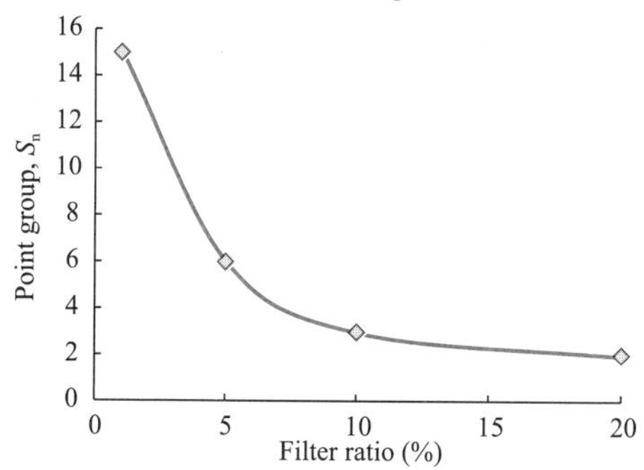

Fig. 9-22 The relationship of point group number with filter ratio in Case Ⅳ

9.1.4.3 Consuming time of varied 3DPC models

Fig. 9-23 shows the relationship of the consuming time with the number of points in varied 3DPC models. With the increasement of points number, the calculating time increases following an exponential law. For the 3DPC of the slope model, it takes nearly 1800s to deal with the 22, 349 points, which is quite time consuming. Thus, it is very important to input a series of proper parameters before executing the identification process. The study of optimum parameters according to the regional points in Case Ⅲ should be carried out before the formal CGC processing.

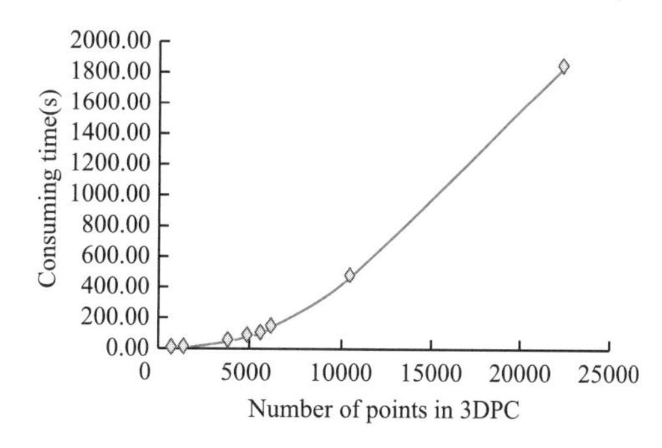

Fig. 9-23 The relationship of consuming time and the points number

According to the results of four case studies, the identification of discontinuities information based on the CGC method can be effectively applied to structural planes measurement in rock engineering. It should be addressed that the 3DPC is based on the outcrop surface of the jointed rock mass. Future efforts should still focus on validating the proposed CGC method with in-depth measurements of discontinuity of site investigation, e. g., measuring by the borehole camera, to verify the recognition method.

9.2 Rough Discrete Fracture Network Model

9.2.1 Introduction

Natural rock masses are consisted of a large number of structural planes in many different scales, which make this geological body inhomogeneous and anisotropic. The presence of structural planes weakens its mechanical properties and controls its damage mode. The structure induced deformation and failure modes of the surrounding rocks of fractured rock masses are complex under the excavation disturbance. Some on-site observations, such as Mont Terri, Mine-by, AECL underground laboratory, Jinping Hydropower Station, and Baihetan Hydropower Station, all exhibit complex deformation and asymmetric damage modes of surrounding rocks under different depths. Due to the influence of discontinuities as well as in-situ stresses, the deformation and damage patterns of the surrounding rocks in the mining area are often asymmetrically distributed under the action of mining disturbances.

The damage patterns of rock masses containing different fractures distributions differ significantly under different excavation conditions. In open pit or shallow underground mining, the damage patterns of rock masses are significantly controlled by structural planes. The damage patterns of surrounding rocks or slopes with different orientations show significant anisotropic characteristics. With the disturbance of excavation, the strength and damage mode of the surrounding rock affected by the structure surface are more complicated, showing the characteristics of asymmetric damage and zonal rupture. The structural planes play a controlling role in the deformation or damage pattern of the surrounding rock. Therefore, the geometric and mechanical properties of structural planes in rock masses are important for studying the deformation and damage patterns. Many scholars have carried out works on the influence of joint distribution on the mechanical properties of rock masses. Although many scholars point out that the deformation characteristics will gradually become stress induced failure with the increase of the surrounding pressure, the excavation opening provides space free surface for rock deformation and rupture. The structural planes still control the mechanical response of the jointed rock masses. Therefore, the existence of discontinuities within the rock mass is still one of the main factors affecting the mechanical properties and controlling the damage mode of fractured rock masses.

The natural rock mass has experienced complex tectonic movement, weathering, erosion and other geological effects, numerous joints and fractures of different sizes and directions have been made. These discontinuities cross each other to form a complex three-dimensional fractures network, which not only greatly changes the mechanical properties of the rock, but also seriously affects the seepage characteristics of the rock body. It is of great importance to accurately characterize the geometric properties of complex 3D fracture networks in the construction of many projects such as dams, nuclear waste storage, CO_2 underground storage and underground oil extraction. The complex fracture structure morphology inside the fracture network, especially the roughness characteristics of the joints, is an important factor controlling the mechanical properties of the rock mass. The characterization of the equivalent rock mass is a typical method for the numerical modeling, which constructs an equivalent rock mass model including the geometric and mechanical effects of the structural surface by establishing a DFN model (as shown in Fig. 9-24). Accurate characterization of the spatial rock structure is a prerequisite for more accurate analysis of the mechanical properties of the jointed rock masses.

Numerical simulation based the DFN models is an important method to study the mechanical properties and seepage characteristics of rock masses, which has been approved by many scholars. However, the spatial geometry of most 3D DFN models is often assumed to be planar or smooth planes (Fig. 9-24a), the profile of which is straight traces (Fig. 9-24b and c). The 3D DFN models reported by Hu, et al. (2022) and Wang, et al. (2022), as well as a 3D DFN model reported by Wang, et al. (2022) also shows the geometry and shape of the discrete fractures. The natural structural surfaces are not planar or smooth as assumed, but have certain roughness characteristics. A large number of engineering investigations found

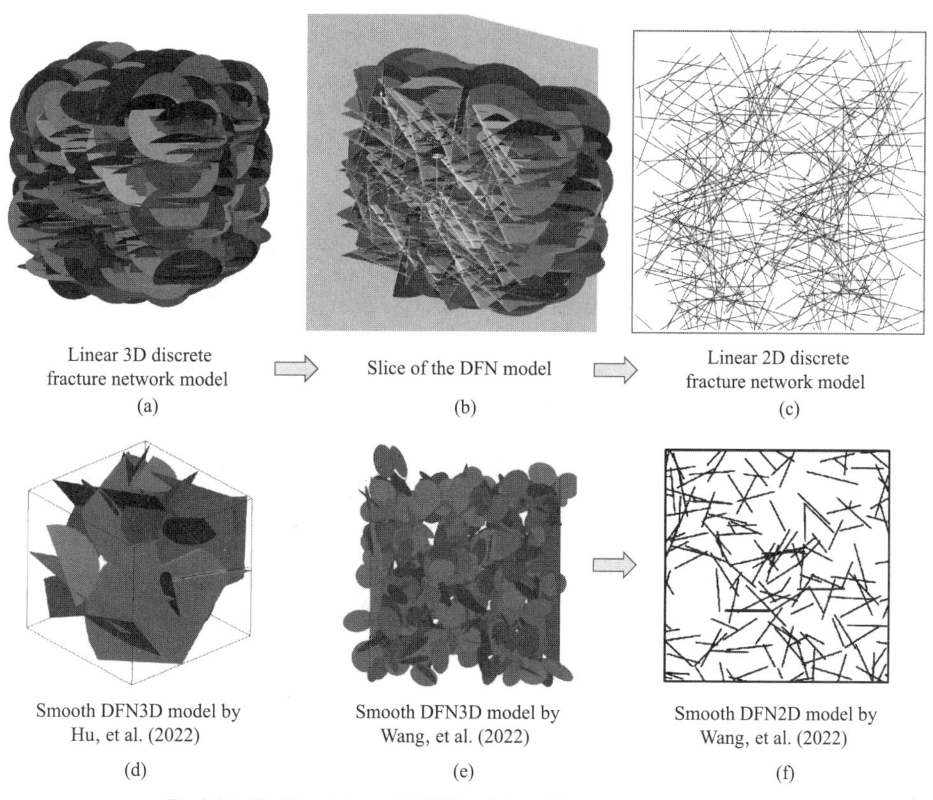

Fig. 9-24 Traditional planar 3D DFN model and the linear 2D RDFN model

that the structural faces morphologically from small scales (fractures and joints) to large scales structural faces (folds and bedding planes) show certain roughness undulations (Fig. 9-25), which will have a significant influence on the mechanical properties of rock masses.

Rough discontinuities planes existed in natural rock masses have a great impact on the mechanical behavior and more and more researches have been focused on the aspects. The joint roughness coefficient (JRC) defined by Barton and Choubey is widely used to characterize the roughness pattern of joints. The JRC relies on the subjective judgment of JRC on the roughness of structural surfaces, which is highly subjective and empirical. Numerous scholars have conducted a lot of research on characterizing the geometric parameters of rocks in quantitativeways. The geometry describing roughness developed by Mandelbrot has been recognized by

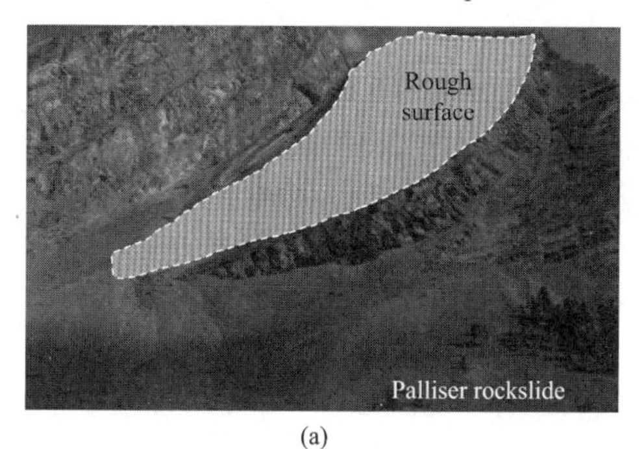

(a)

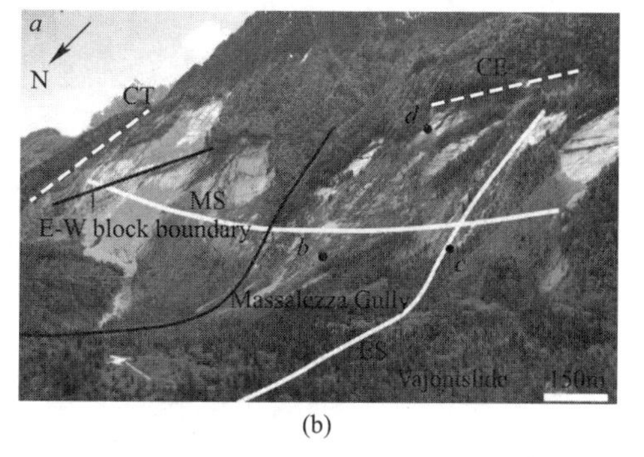

(b)

Fig. 9-25 The natural rough structures (one)

(a), (b) Modified from D Stead and Wolter A (2015)

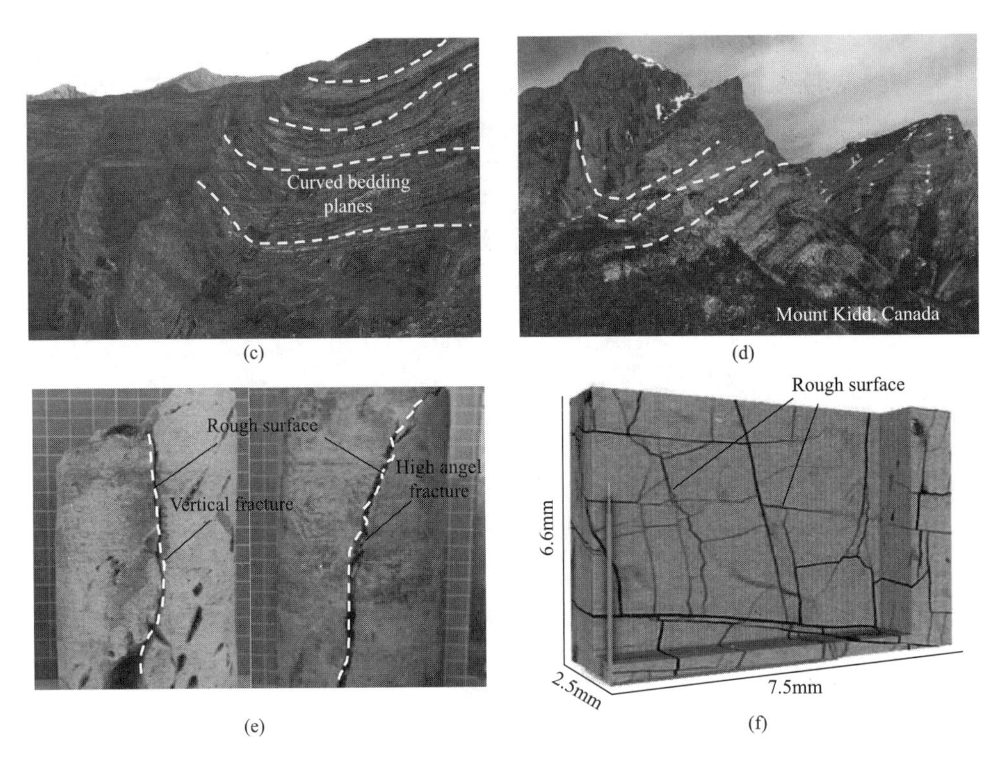

Fig. 9-25 The natural rough structures (two)
(c) Modified from D Stead and Wolter A (2015); (d) Modified from Fleuty M J (1964);
(e) Modified from Zhang et al. (2022); (f) A 3D representation of the micro-CT image of a coal sample by Ramandi, et al. (2022)

many scholars. Considering the roughness of single fracture in the discrete fracture network, Wang, et al. proposed a two-dimentional rough discrete fractures network (RDFN2D) model which fully considers the rough characteristics of the fracture trace. The RDFN2D model is established including the sinusoidal traces, triangular traces, stepped traces and fractal-based traces (Fig. 9-26), which provides a new model to simulate the spatial distribution and rough geometric

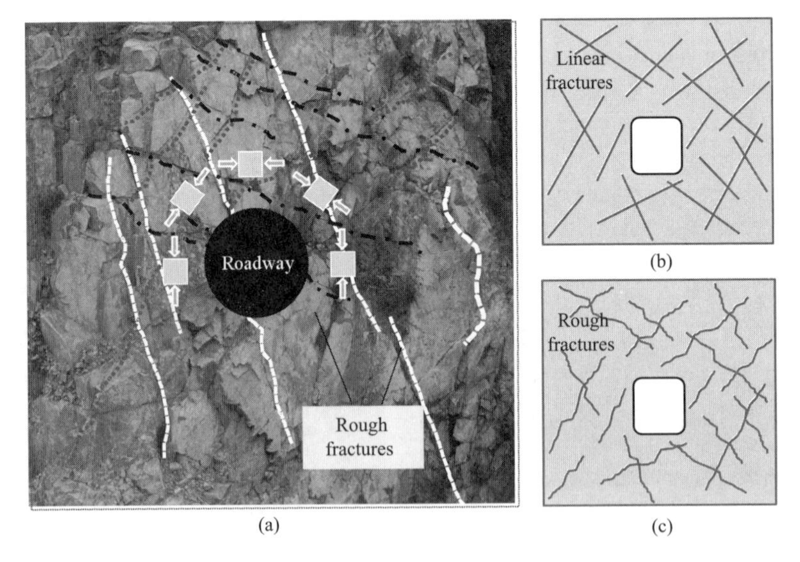

Fig. 9-26 Comparison of rough DFN model with linear DFN model (modified from Wang et al., 2022)

characteristics. The effect of the rough fracture network on the fracture mechanism of the jointed rock mass is studied. The dominating of the joint roughness on the compressive and shear fractures is verified. They also suggest that the influence of the joint roughness should not be neglected in the stability analysis of rock engineering.

However, it should be noted that although the DFN modeling method can efficiently represent the rock structure, the full potential of the DFN model to capture stochastic variations can only be realized if the generated DFN geometry is statistically the same or similar to the real fracture network. It should be especially noted that although scholars have paid attention to the rough properties of 3D discontinuities, they have not conducted studies for rough 3D structural surface networks, but still for smooth types. The main reasons are (1) the 3D surface characterization is relatively complex, and it is difficult to establish the connection with the actual structural surface; (2) the 3D smooth structural surface network is relatively easy to model through the 3D modeling software, but the 3D rough structural surface network modeling mechanism is complex and difficult to model; (3) in engineering, not enough attention is paid to the influence of rough structural surface network models.

Here, we creat a new three-dimensional RDFN model considering the roughness and waviness of structure planes, which improves the characterization methods of discrete discontinuities. The single 3D structure is characterized based on the W-M fractal surface. The influence of the dominating parameters on the surface waviness is systematically discussed. Then the rough discrete fracture network model in three-dimension is established considering the orientation of varied sets of discontinuities. The differences between RDFN3D and DFN3D models generated by the same approach are compared and discussed. The export from RDFN3D model to the commercial codes, e. g., COMSOL Multiphysics, AUTOCAD and the PFC3D code, are realized, which will broaden the applicability in the stability analysis in rock engineering.

9.2.2 The Characterization of Rough Surfaces

9.2.2.1 W-M geometry in three-dimension

Yan and Komvopoulos (1998) proposed a modified Weierstrass-Mandelbrot (W-M) function as expressed by Eq. (9-9) and Eq. (9-10).

$$z(x,y)=L\left(\frac{G}{L}\right)^{D-2}\left(\frac{\ln\gamma}{M}\right)^{\frac{1}{2}}\sum_{m=1}^{M}\sum_{n=0}^{n_{\max}}\gamma^{n(D-3)}\cdot\left\{\cos\varphi_{m,n}-\cos\left\{\frac{2\pi\gamma^{n}(x^{2}+y^{2})^{\frac{1}{2}}}{L}\cdot\cos\left[\tan^{-1}\left(\frac{y}{x}-\frac{\pi m}{M}\right)\right]+\varphi_{m,n}\right\}\right\} \tag{9-9}$$

$$n_{\max}=\operatorname{int}\left[\frac{\log\left(\frac{L}{L_s}\right)}{\log\gamma}\right] \tag{9-10}$$

Where,

z —— the height of the asperity of the rough surface;

x and y —— the surface coordinates;

L —— the sample length;

G —— the fractal roughness of the surface ($G>0$);

D —— the fractal dimension ($2<D<3$);

γ —— the parameter that controls the frequency distribution ($\gamma>1$, γ takes the value of 1.5 according to Komvopoulos and Ye);

M —— the number of superimposed rough peaks on the surface of the rough structure and takes the value of 10;

n —— the frequency;

$n_{\max}$ —— the upper limit, which is given by Eq. (9-10);

L_s —— the cutoff length that limits the highest roughness frequency up to $1/L_s$;

$\varphi_{m,n}$—— a random quadrant distributed within $[0, 2\pi]$.

The 3D W-M surfaces are built based on the MATLAB code. Given a certain size in the xy plane, a point cloud grid can be obtained, and the z-directional position is obtained according to the W-M function. To efficiently build the 3D surfaces, we introduce another parameter, the surface precision P_r of the 3D surface, which represents precision of the mesh grid. The fractal dimension D can affect the frequency of the roughness of the surface. A higher fractal dimension will decrease the waviness of the rough surface. The fractal roughness G can affect the amplitude of the rough surfaces. The maximum value of the rough surface will increase when increasing the value of G. The surface precision P_r determines the number of the grids which determines the complexity of the rough surfaces. A smooth plane will be obtained when the value of P_r is zero. The more of the surface precision, the more complex the rough surface will behave. In the 3D W-M function, the fractal dimension of D, the height magnitude of G, and the accuracy P_r play a dominant role in defining the function surface. To illustrate the effect of the three parameters on the surface, some typical W-M surfaces are generated to demonstrate the role of fractal dimension D, height amplitude G, and precision P_r. To discuss the influence of the three parameters, the sample surface size is set to be 10cm × 10cm. The minimum resolution of the simulated sample is 0.1cm, i.e., L = 10cm, L_s = 0.1cm, and n_{max} = 11. As shown in Fig. 9-27, at a fixed height amplitude $G = 10^{-5}$ and $P_r = 50$, the absolute height of the function surface decreases sharply by increasing the size of the fractal dimension D. The larger the value of D, the smoother the surface.

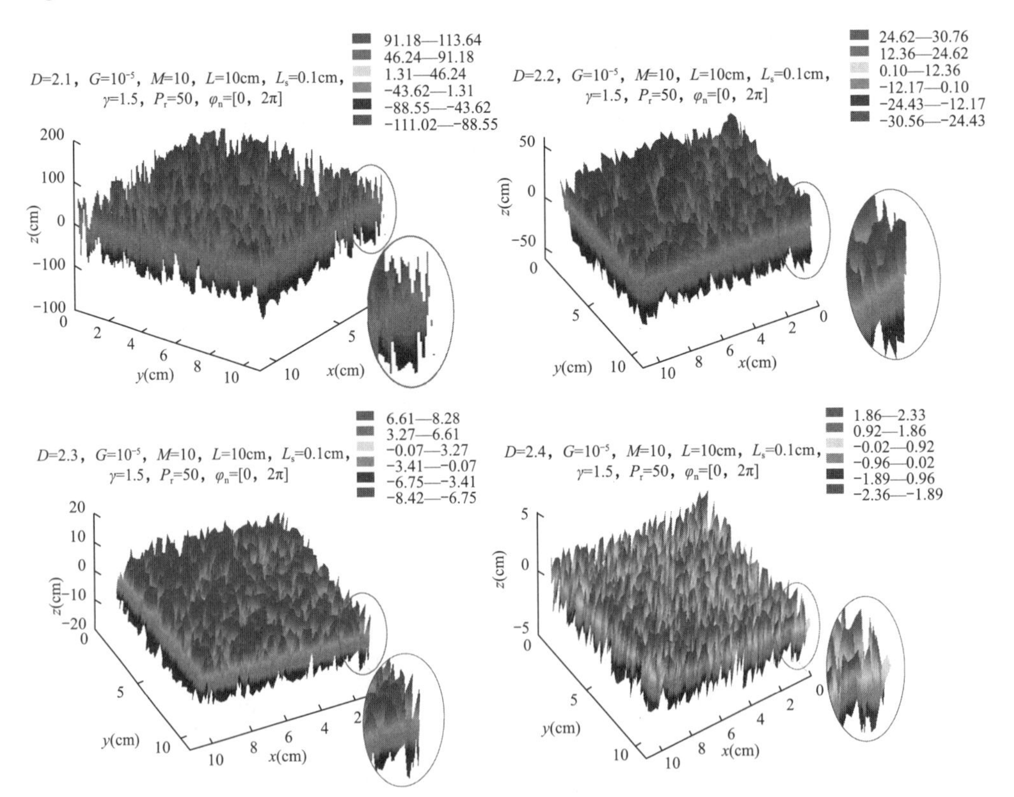

Fig. 9-27 The 3D W-M surfaces of varied values of the parameter of D (D=2.1, 2.2, 2.3, 2.4, 2.5, 2.6, 2.7, 2.8, and 2.9) (one)

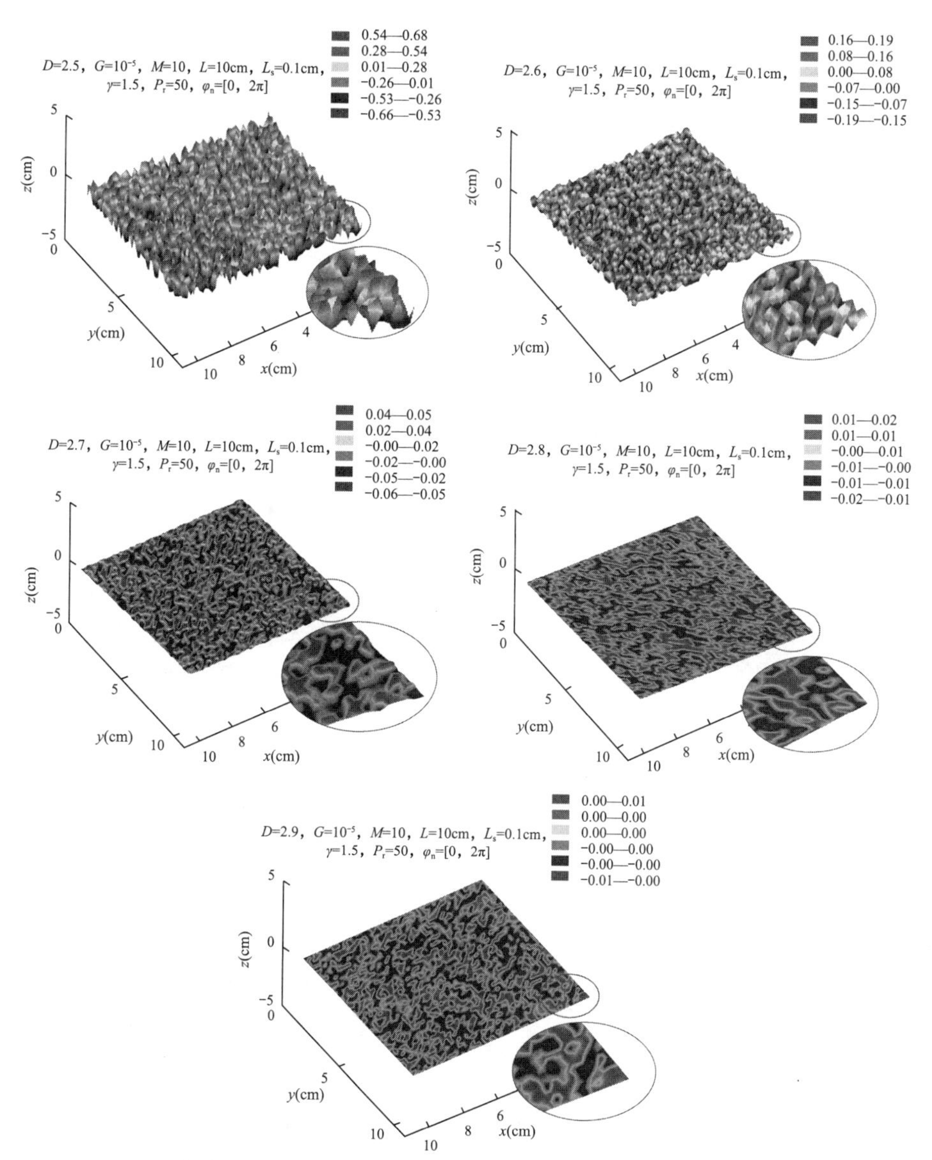

Fig. 9-27 The 3D W-M surfaces of varied values of the parameter of D (D=2.1, 2.2, 2.3, 2.4, 2.5, 2.6, 2.7, 2.8, and 2.9) (two)

As shown in Fig. 9-28, at a fixed fractal dimension of $D=2.5$ and $P_r=50$, the absolute height of the function surface increases gradually by increasing the value of G. The number of peaks spread in the rough surface is not affected by the changing of G value. The larger the value of G, the rough of the surface.

As shown in Fig. 9-29, the influence of the precision of P_r is compared. At a fixed height amplitude of G and fractal dimension of D, the number of the peaks spread in the rough surface deceases when decreasing the value of P_r. According to the 2D profile from each 3D surface, the waviness and roughness are affected by the precision. The change of P_r does not change the height of the rough surface; however, the number of peaks decreases when decreasing the precision of the surface. It should be noted that the 3D W-M surface will become a smooth plane when P_r equals to 1. This could enable the generation of both rough surfaces and smooth surfaces.

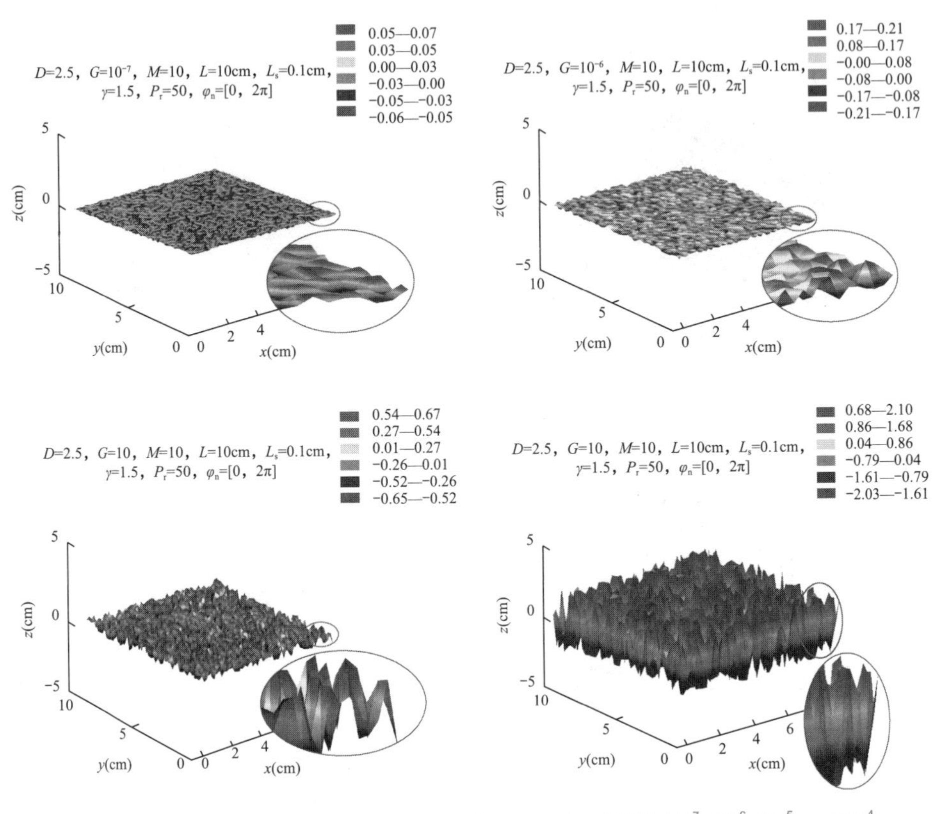

Fig. 9-28 The 3D W-M surfaces of varied values of the parameter of G ($G=10^{-7}$, 10^{-6}, 10^{-5}, and 10^{-4})

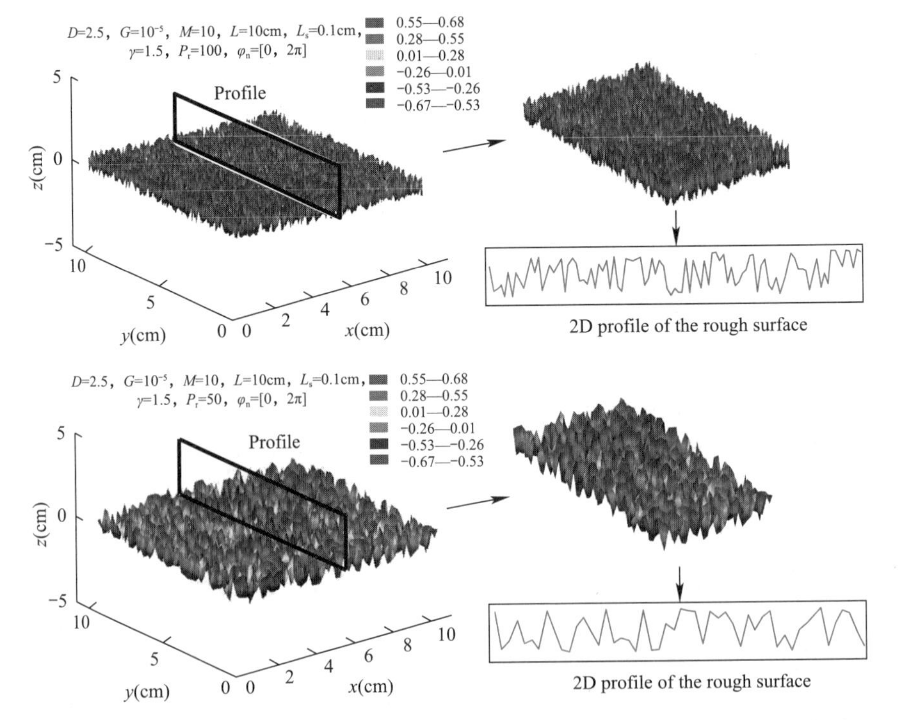

Fig. 9-29 The 3D W-M surfaces of varied values of the parameter of P_r ($P_r=100$, 50, 20, and 1) (one)

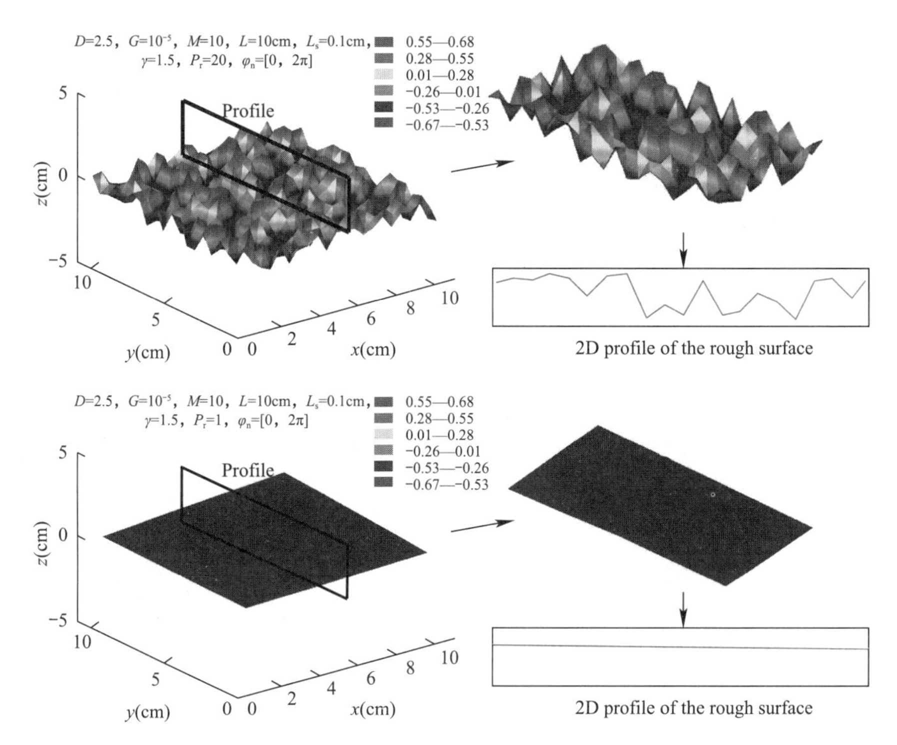

Fig. 9-29 The 3D W-M surfaces of varied values of the parameter of P_r(P_r = 100, 50, 20, and 1) (two)

9. 2. 2. 2 Relationship of the 3D rough surfaces with the JRC curves

It is important to establish the relationship of the W-M surface with the natural discontinuities. Once the relationship is obtained, the mathematical W-M surface can have its physical meaning in the application of rock engineering. We discuss the relationship of the 3D rough surfaces with the JRC curves by Barton and Choubey (1977). Table 9-7 lists the comparison of the proposed 3D meshes with JRC curves. Two key parameters G and D are modified to match the 2D JRC curves when other parameters are set: $P_r = 20$, $M = 10$, $L = 100$, $L_s = 0.1$, $\gamma = 1.5$, $\varphi = [0, 2\pi]$. The combination of G and D can affect the shape and roughness of the surfaces. Take the JRC curve of 6—8 for example, the 2D profile at $x = 50$mm along the W-M surface matches well when the lg$G$ and D are -5.6 and 2.50, respectively. This result can provide a reference for the characterizing of the natural discontinuities in jointed rock masses. Further relationship of the 3D W-M meshes with the natural rough surface needs working out in the future researches.

9. 2. 3 Construction of the RDFN3D Model

9. 2. 3. 1 Principle of the construction of RDFN3D model

Modelling the discrete fractures network is a key for the characterization of the equivalent rock mass. A 3D rough discrete fracture network model (RDFN3D) is established according to the 3D W-M surfaces based on the Monte Carlo method. The characterization of the geometric properties of the rough structural surface is realized. The modeling principle of the fractional dimensional RDFN3D model is shown in Fig. 9-30. First, we should determine the rough surfaces based on the W-M function according to the shapes of natural structures. Here, the rough surface is mainly determined by the parameters including the fractal dimension of D, the amplitude of G, and the precision of P_r. Then, the determination of the information of

Comparison of the 3D rough W-M meshes with the JRC profiles

Table 9-7

No.	JRC	The JRC profiles (100mm)	The 3D surfaces by the W-M function			
			3D surfaces (100mm×100mm)	2D Profiles at $x=50$mm	Parameters	
					lg$G$	D
1	0—2		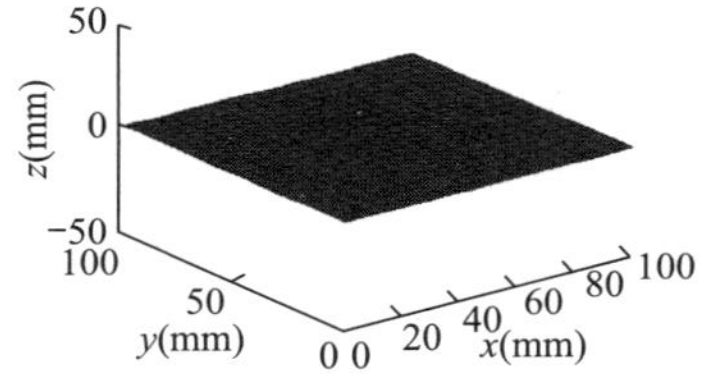	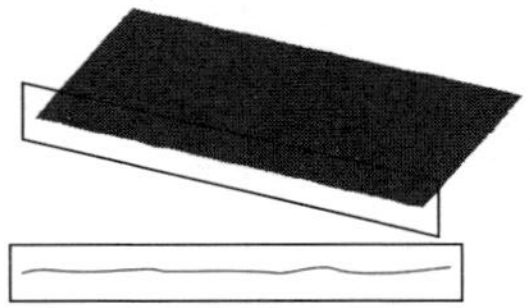	-6.0	2.54
2	2—4		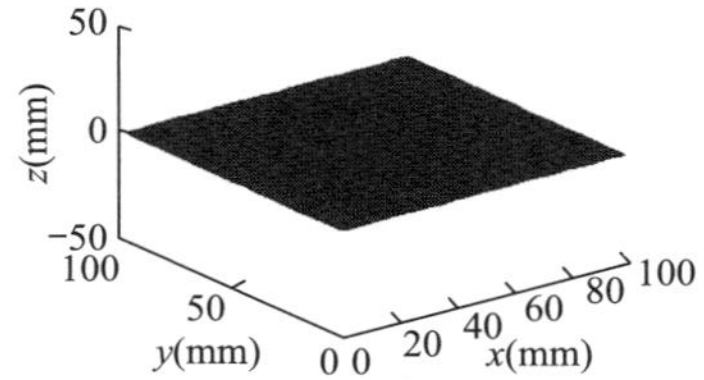	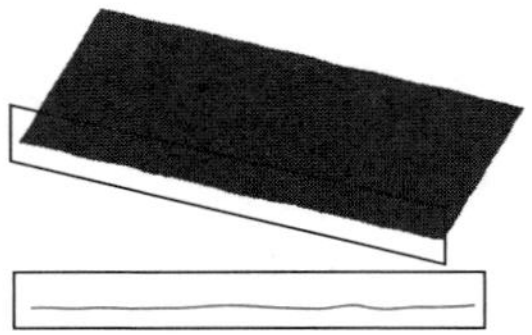	-6.1	2.52
3	4—6		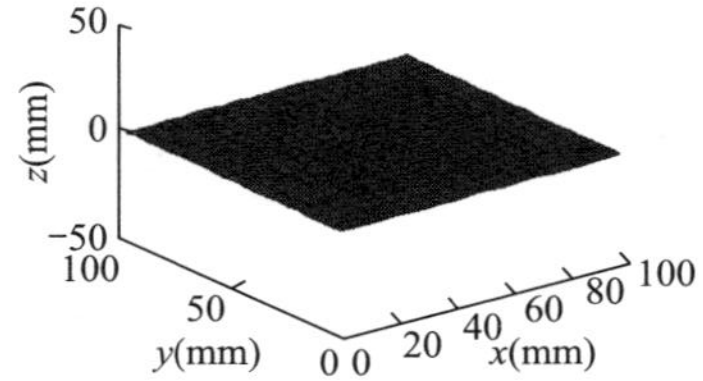	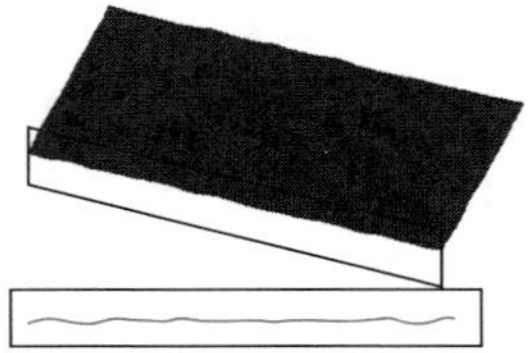	-5.7	2.53

Continued

No.	JRC	The JRC profiles (100mm)	The 3D surfaces by the W-M function			
			3D surfaces (100mm×100mm)	2D Profiles at $x=50$mm	Parameters	
					lg$G$	D
4	6—8		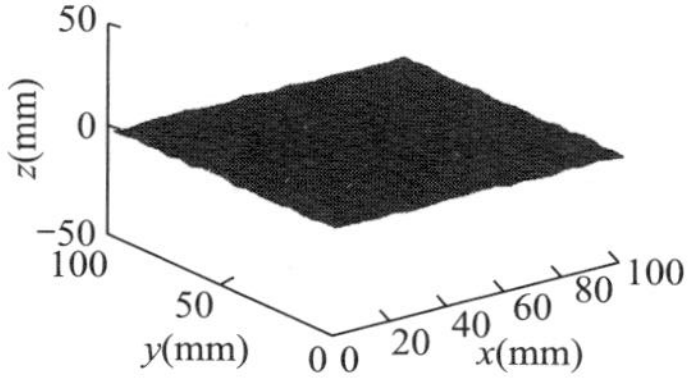	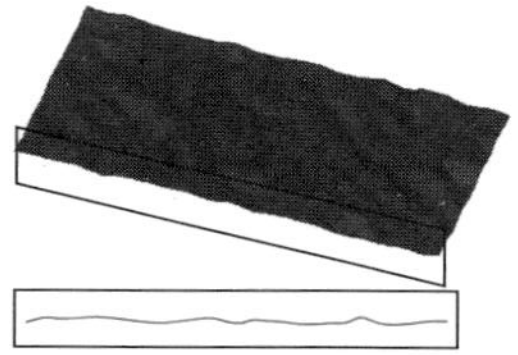	-5. 6	2. 50
5	8—10		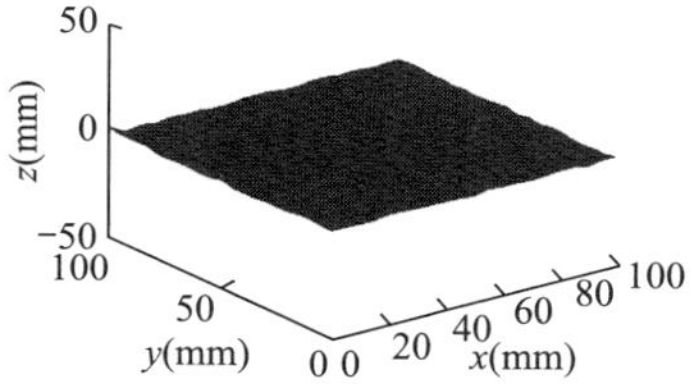	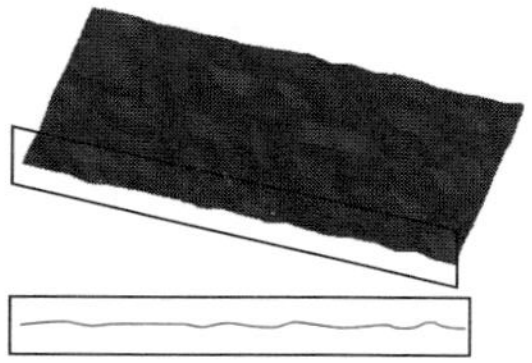	-5. 8	2. 49
6	10—12		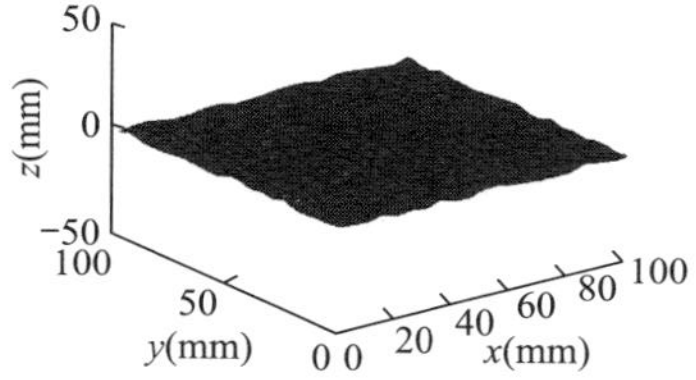	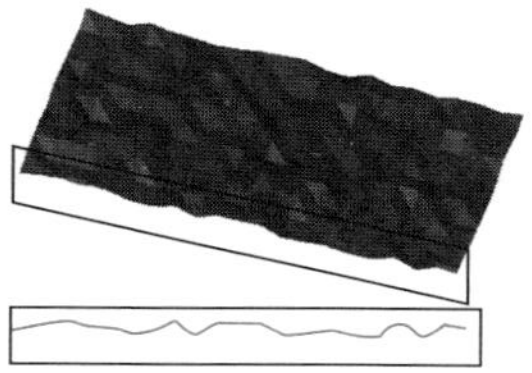	-5. 6	2. 47

Continued

No.	JRC	The JRC profiles (100mm)	The 3D surfaces by the W-M function			
			3D surfaces (100mm×100mm)	2D Profiles at $x=50$mm	Parameters	
					lg$G$	D
7	12—14		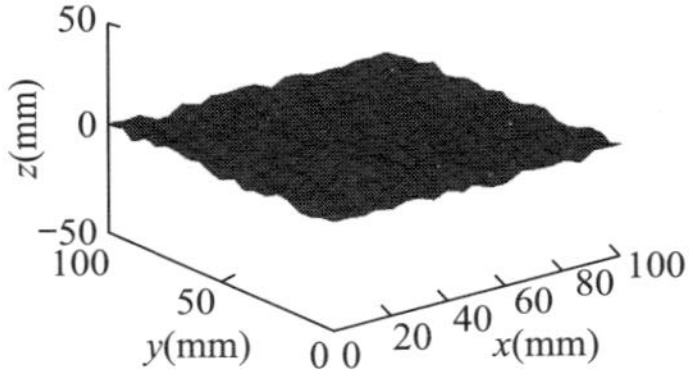	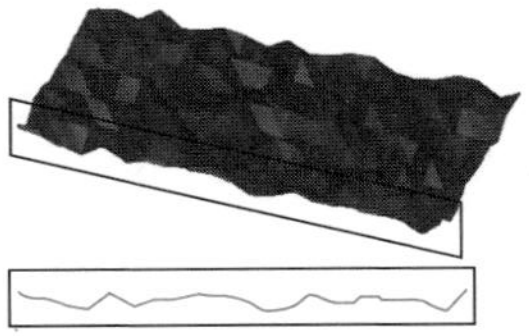	−5. 7	2. 43
8	14—16		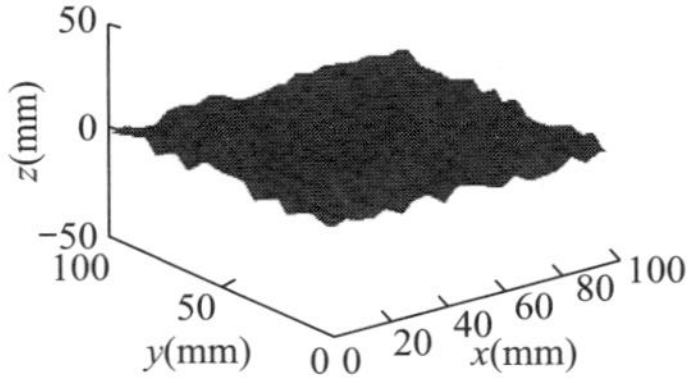	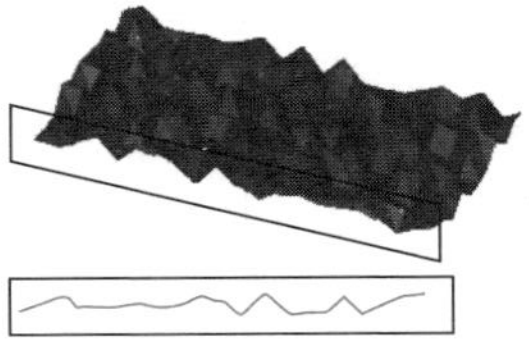	−5. 2	2. 44
9	16—18		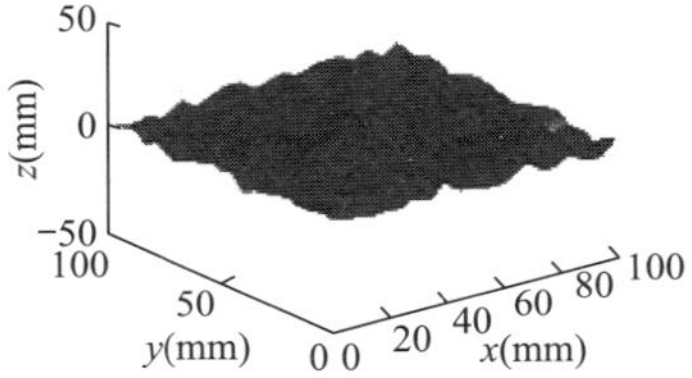	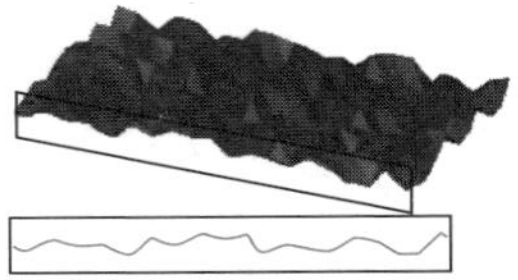	−5. 4	2. 42

Continued

No.	JRC	The JRC profiles (100mm)	The 3D surfaces by the W-M function			
			3D surfaces (100mm×100mm)	2D Profiles at $x=50$mm	Parameters	
					lg$G$	D
10	18—20		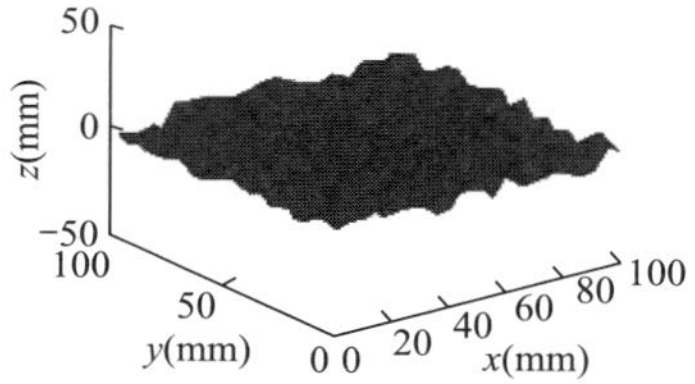	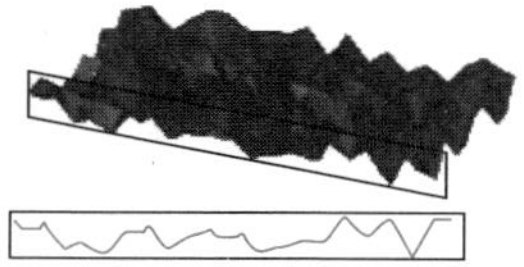	−5. 5	2. 39

$P_r=20$, $M=10$, $L=100$, $L_s=0.1$, $\gamma=1.5$, $\varphi=[0, 2\pi]$

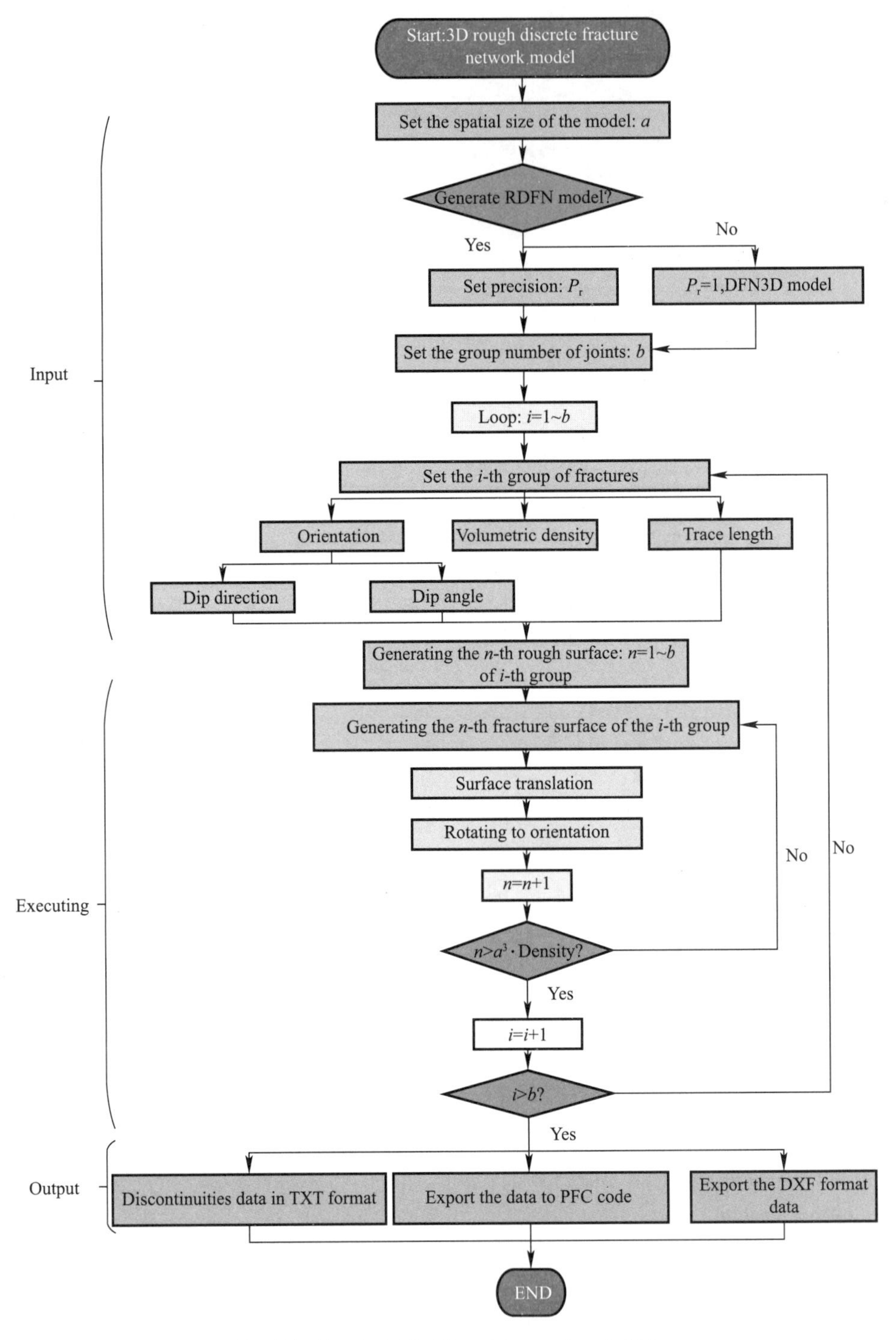

Fig. 9-30 Flowchart of RDFN model generation algorithm

the 3D structural planes includes the orientation, spatial size, spacing. The statistical model of the 3D discontinuities should be established according to the on-site survey. Then the RDFN3D model can be established based on the Monte Carlo method considering the positions, areas and directions of W-M surfaces. Since all the rough surfaces are determined using the mathematical model in the MATLAB code, we can export the information of all the nodes composing the rough surfaces. Many format files can be exported, such as the *. DXF files, and the *. TXT. Here we export the RDFN3D model to AUTOCAD, COMSOL Multiphysics, and the particle flow code in three-dimension (PFC3D), which enable the computational analysis of the mechanical behavior of the RDFN3D model.

9.2.3.2 Influence of dominating parameters

In this section, three case studies including both the RDFN3D and DFN3D models are established and the influence of the parameters of D, G and P_r is discussed. The RDFN3D model including only one set of fractures is introduced and the information is listed in Table 9-8. The average values and deviation of the dip direction, dip angle as well as the trace length are set uniformly distributed. Fig. 9-31 shows the varied rough fractures network considering different values of D, G and P_r. Increasing the fractal dimension D from Fig. 9-31 (a) to Fig. 9-31 (b) results in a significant decrease in the absolute height of the fracture surfaces and a decrease in the degree of undulation. A smoother appearance of the fracture surfaces is obtained. As compared with those in Fig. 9-31 (b) to Fig. 9-31 (c), with the increase of height amplitude G, the absolute height of each fracture surface also increases along the profile.

The geometrical parameters of the rough fractures **Table 9-8**

Group	Color	Dip direction (°)		Dip angle (°)		Trace length (mm)		Density (traces/cm^3)
		Uniform distribution		Uniform distribution		Uniform distribution		
		Avg.	Std.	Avg.	Std.	Avg.	Std.	
1	Blue	60	10	-30	10	60	20	20

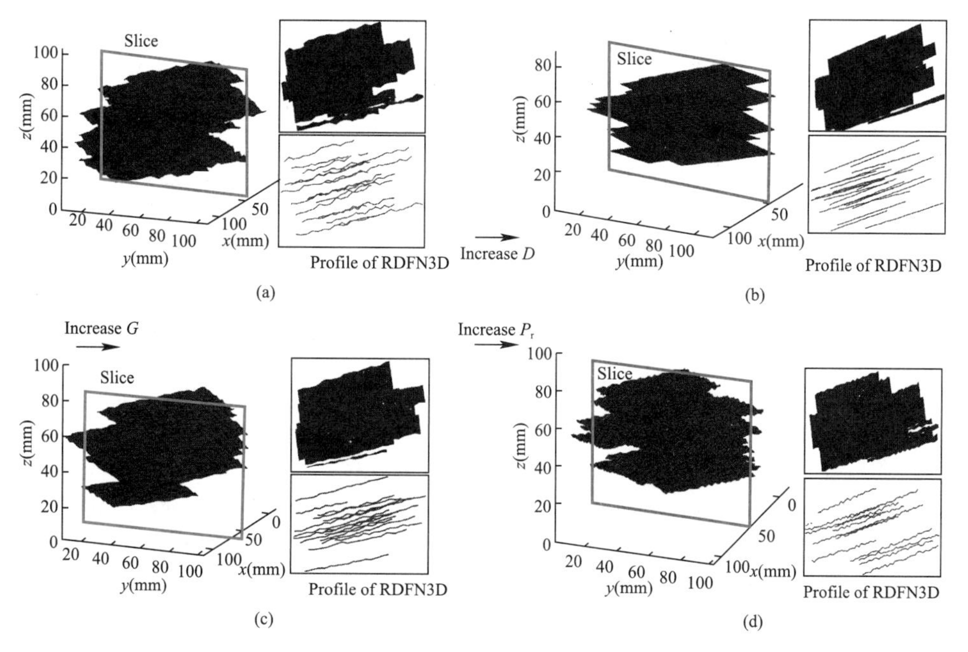

Fig. 9-31 The profiles of RDFN3D models of varied values of the parameter of D, G and P_r(at x=50mm)

(a) D=2.5, $G=10^{-5}$, M=10, L=100mm, L_s=1mm, γ=1.5, P_r=15, φ_n=[0, 2π]; (b) D=2.6, $G=10^{-5}$, M=10, L=100mm, L_s=1mm, γ=1.5, P_r=15, φ_n=[0, 2π]; (c) D=2.6, $G=10^{-4}$, M=10, L=100mm, L_s=1mm, γ=1.5, P_r=15, φ_n=[0, 2π]; (d) D=2.6, $G=10^{-4}$, M=10, L=100mm, L_s=1mm, γ=1.5, P_r=15, φ_n=[0, 2π]

The undulation frequency does not change and the roughness of fracture surface is increased. Comparing Fig. 9-31 (c) and (d), the absolute height of each fracture surface does not change when increasing the precision P_r. However, the degree of waviness increases significantly. The roughness of Fig. 9-31 (d) and (a) show a similar pattern. The parameters of G and P_r can significantly increase the roughness of the 3D surfaces. However, the fractal dimension of D decreases the surface roughness and waviness.

(1) Case study Ⅰ: RDFN3D model with one set of fractures with high persistence

The RDFN3D model of the case I study is shown in Fig. 9-32 and the corresponding parameters are listed in Table 9-9. In case I, there is only one set of discrete fractures with relatively high persistence. The scale size of the model is 100mm×100mm×100mm. The dip direction, dip angle and the trace length are uniformly distributed. Fig. 9-32 (a) shows the spatial distribution of the fractures and Fig. 9-32 (b) shows the profiles of varied sections.

(2) Case study Ⅱ: RDFN3D model with three sets of fractures with high persistence

Both the RDFN3D model and DFN3D model are shown in Fig. 9-33 in case Ⅱ study and the corresponding parameters are listed in Table 9-10. As is shown in case Ⅱ, there are three sets of discrete fractures with high persistence. The scale size of the model is also 100mm×100mm×100mm. Fig. 9-33 (a) shows the spatial distribution of the fractures and the profiles of varied sections of RDFN3D model and Fig. 9-33 (b) shows the spatial distribution of the fractures and the profiles of varied sections of DFN3D model. The roughness of each joint trace can be clearly obtained.

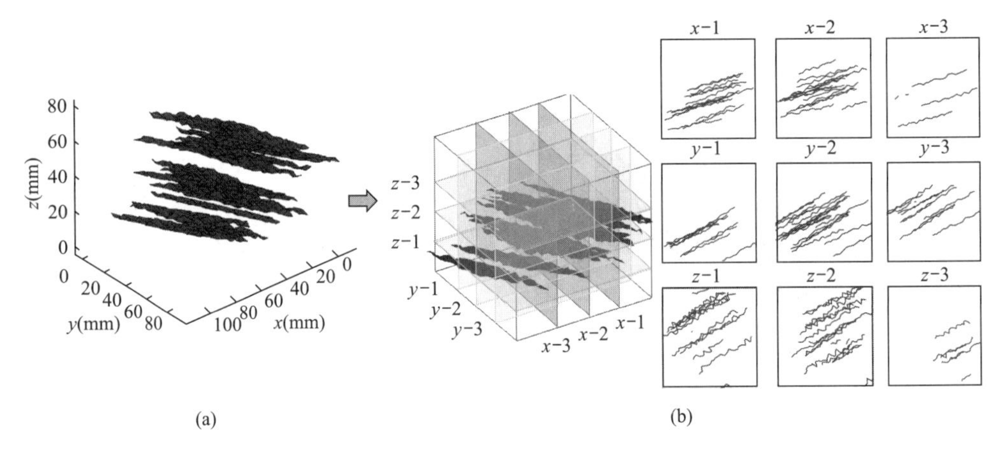

Fig. 9-32 The distribution of RDFN3D model with one joint set (case I, with high persistence)

(a) RDFN3D model with one set joints; (b) Profiles of varied sections in the RDFN3D model

(3) Case study Ⅲ: RDFN3D model with three sets of fractures with low persistence

In natural structures, the fractures are generally distributed with low persistence. In case study Ⅲ, the RDFN3D model as well as DFN3D model with lower persistence are shown in Fig. 9-34. Table 9-11 lists the corresponding information of the three sets of fractures. The scale size of the model is also 100mm×100mm×100mm. The corresponding JRC curve is 10—14.

The fractal parameters of the rough fractures in case I **Table 9-9**

Group	Color	Dip direction (°)		Dip angle (°)		Trace length (mm)		Density (traces/cm^3)
		Avg.	Std.	Avg.	Std.	Avg.	Std.	
1	Blue	60	10	−30	10	60	20	20
$D=2.5$, $G=10^{-7}$, $M=10$, $L=100$mm, $L_s=1$mm, $\gamma=1.5$, $P_r=15$								

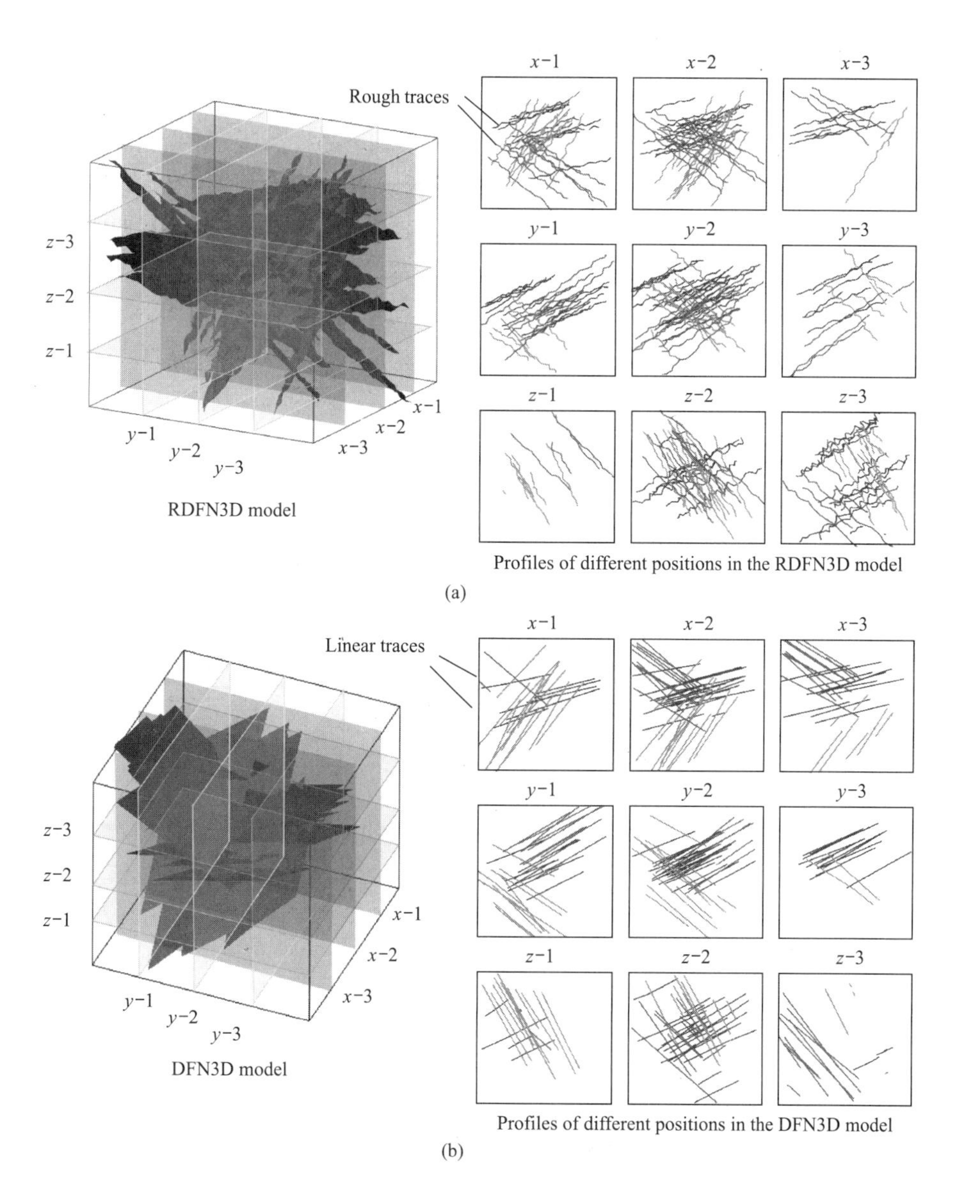

Fig. 9-33 The distribution of RDFN3D and DFN3D models with threes joint sets (case Ⅱ, with high persistence)

The fractal parameters of the rough fractures in case Ⅱ **Table 9-10**

Group	Color	Dip direction (°)		Dip angle (°)		Trace length (mm)		Density
		Avg.	Std.	Avg.	Std.	Avg.	Std.	($traces/cm^3$)
1	Blue	60	10	-30	10	60	20	20
2	Green	150	10	60	20	70	20	15
3	Red	135	20	-45	10	70	30	10

$D=2.5$, $G=10^{-5}$, $M=10$, $L=100mm$, $L_s=1mm$, $\lambda=1.5$, $P_r=15$ (for RDFN3D model), $P_r=1$ (for DFN3D model)

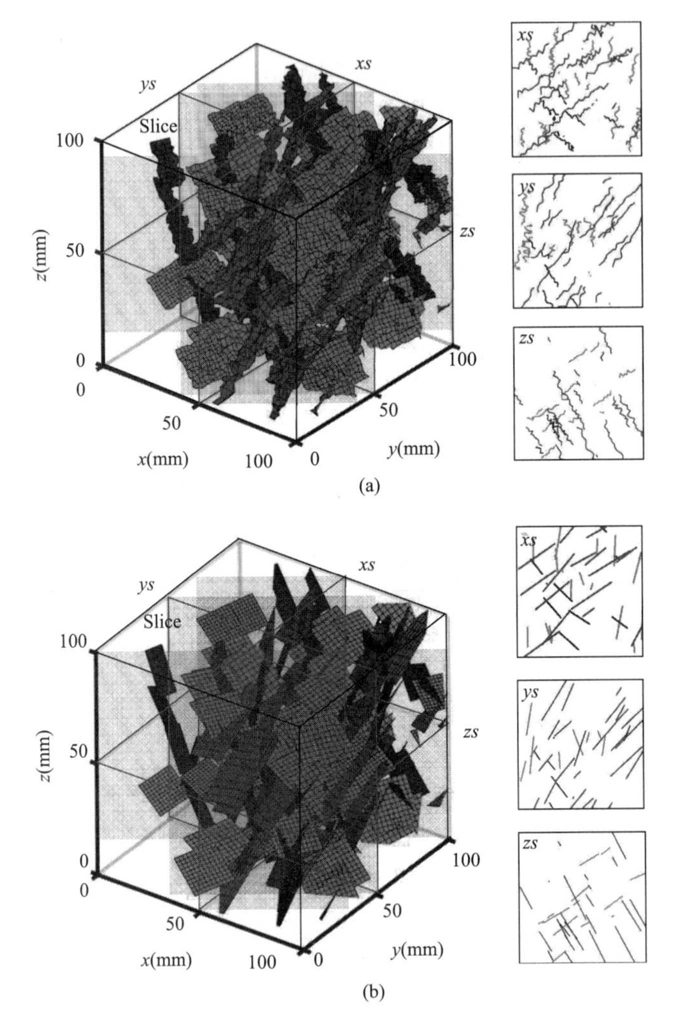

Fig. 9-34 The distribution of RDFN3D and DFN3D models with threes joint sets (case Ⅲ, with low persistence)

Distribution of RDFN3D model with three sets of fractures Table 9-11

Group	Color	Normal distribution				Exponential distribution		Density (traces/cm^3)
		Dip direction (°)		Dip angle (°)		Trace length (mm)		
		Avg.	Std.	Avg.	Std.	Avg.	Std.	
1	Blue	60	10	30	20	20	0	30
2	Green	90	30	60	20	20	10	60
3	Red	120	30	60	10	30	20	60

$D=2.5$, $G=10^{-5}$, $M=10$, $L=100$mm, $L_s=1$mm, $\lambda=1.5$, $P_r=15$ (for RDFN3D model), $P_r=1$ (for DFN3D model)

9.2.4 Analysis and Discussion

9.2.4.1 Application in PFC3D modeling

The establishment of numerical synthetic rock mass (SRM) is proposed based on the particle flow code (PFC) to study the mechanical behavior of jointed rock masses. The SRM model has been widely applied in the PFC model to study themechanical properties of rock masses. It is necessary to establish a more detailed geomechanical model of the rock matrix to conduct the numerical simulation. The RDFN3D modeling method has been introduced based on the MATLAB code. Here, the export with DXF format and PFC3D commanding code are also realized, which provides a computational model for carrying out numerical simulation, such as in the finite element method and the distinct element method. The application of the RDFN3D model will be presented and discussed below with PFC3D and COMSOL models.

As shown in Fig. 9-35, each rough surface of the 3D rough network model can be regarded as composed of multiple nodes. The conventional structural surface can be regarded as a projection of each discrete point of the plane, and the rough surface can also be determined through the projection with discrete particles showing certain undulations. The point cloud data can be imported into a TXT text file. And a series of point cloud coordinates are obtained by importing the text file into PFC3D. Fig. 9-36 shows a RDFN3D model in the PFC3D numerical model. Fig 9-36 (a) shows the RDFN3D mathematical model, in which three sets of fractures are distributed. Fig. 9-36 (b) shows the corresponding PFC3D model. The fracture opening can be achieved by defining the particle diameter. Fig. 9-37 shows the numerical PFC3D model based on the digital RDFN3D model. Fig. 9-37 (a) shows the original 3D rough surface. Fig. 9-37 (b) shows the established PFC3D model. The contact bond distribution and the contact force chains are shown in Fig. 9-37 (c) and (d). The numerical model can provide a new SRM model based on the PFC3D code, for performing the analysis of mechanical properties such as uniaxial compression, triaxial compression, splitting tensile and shear mechanical properties. Further studies will be carried out to discuss the influence of joint roughness on the mechanical behavior based on the numerical RDFN3D model.

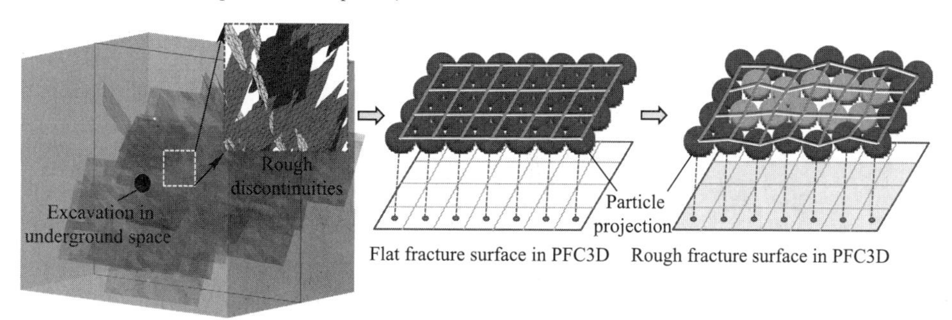

Fig. 9-35 The generation of rough structure planes in PFC3D

9.2.4.2 Application in COMSOL modeling

Fig. 9-38 (a) shows the comparison of the established RDFN3D and DFN3D models. Fig. 9-38 (b) shows the exported DXF file in AUTOCAD software. And Fig. 9-38 (c) shows the 2D profiles of both the RDFN3D and DFN3D model. It should also be noted that the model can be exported to STL format for the future 3D printing. Here, the RDFN3D model can be exported as a DXF format file and then imported to the FEM model. Fig. 9-39 (a) shows

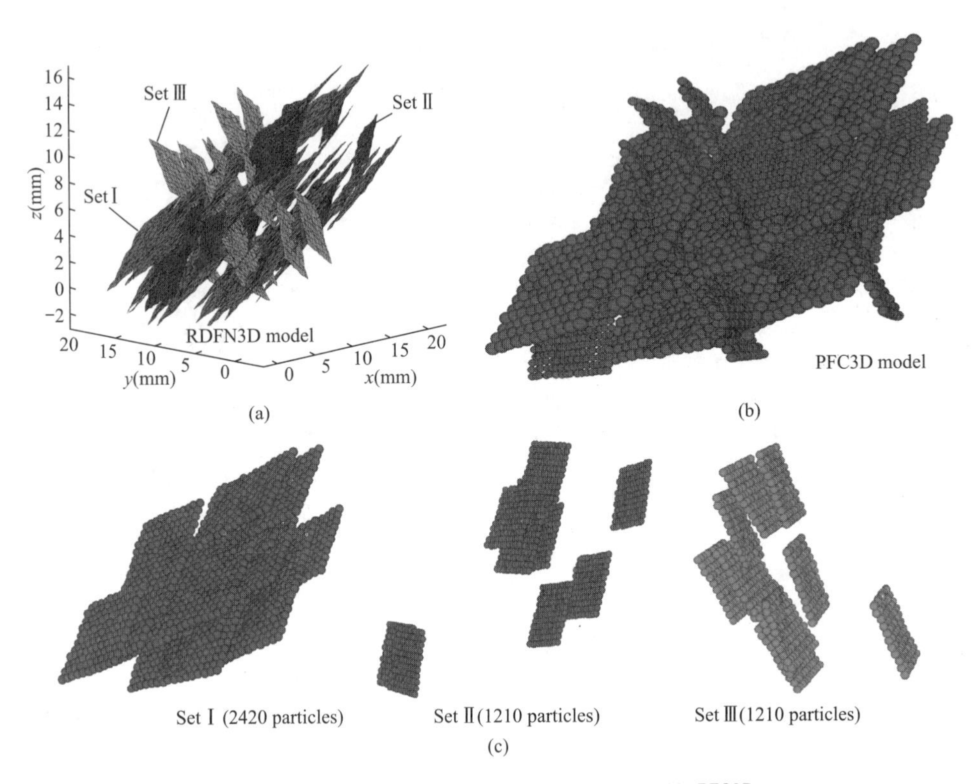

Fig. 9-36 The 3D modeling of the discontinuities model in PFC3D

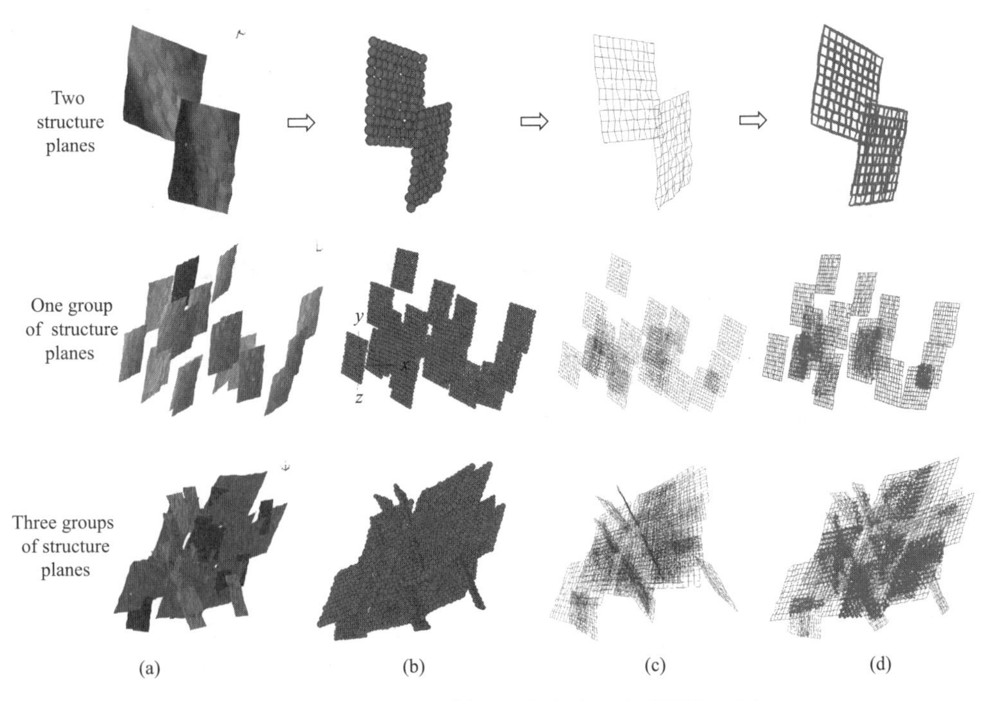

Fig. 9-37 The comparison of the rough structures in PFC3D models

(a) The mesh of rough planes; (b) PFC fractures particles model; (c) Contact bonds; (d) Contact forcechains

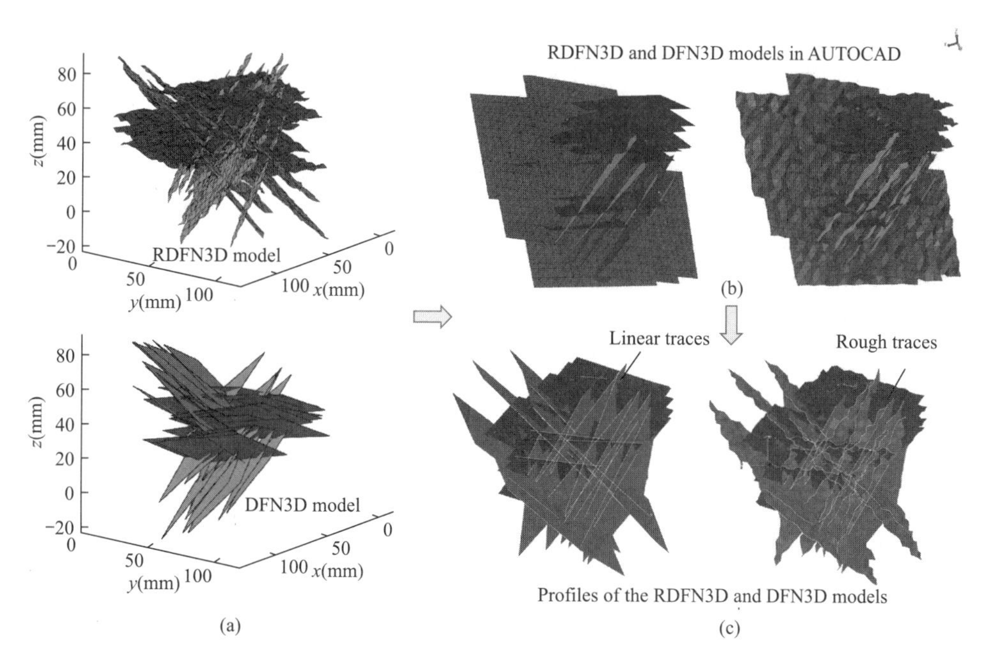

Fig. 9-38 The comparison of RDFN3D and DFN3D models

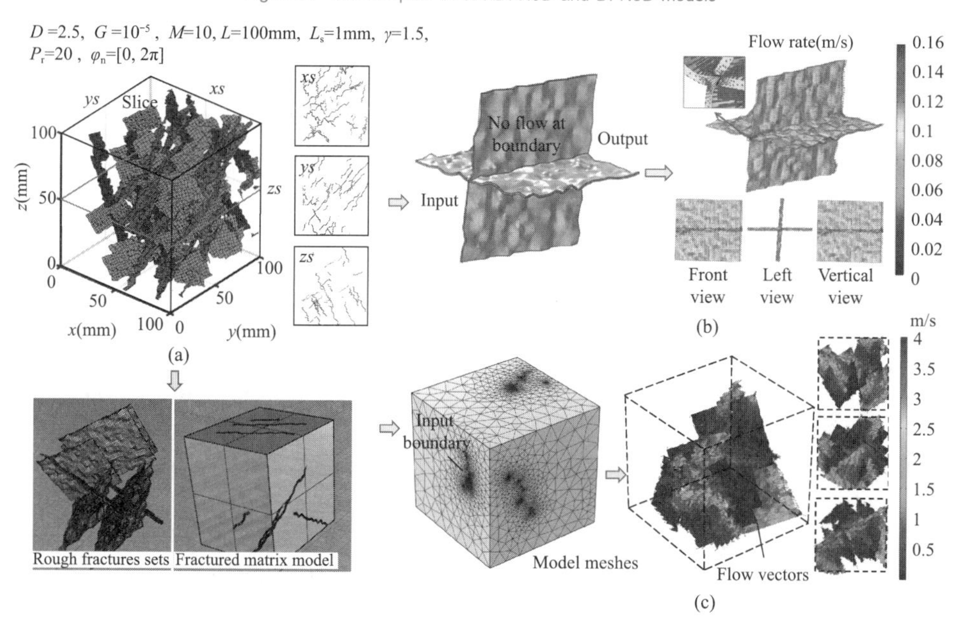

Fig. 9-39 The water flow simulation in rough structures

(a) Profiles of different positions in the RDFN3D model; (b) Flow in rough orthogonal surfaces; (c) Flow in rough fracture network model

the FEM models in the COMSOL Multiphysics code. The characteristics of hydraulic flow of two RDFN3D models are discussed based on Darcy's flow relationship. Fig. 9-39 (b) shows two crossing fractures and the flow vectors along the rough surfaces. The water pressure in the input and output is 1kPa and 0kPa, respectively. Other boundaries are set to impermeable. The density and the dynamic viscosity of the fluid are $1000 kg/m^3$ and $1.001 \times 10^{-6} Pa \cdot s$, respectively. The permeability coefficient is $1 \times 10^{-8} m^2$. The movement of fluid within the rough surface flows forward around the concave and convex regions, forming a high flow velocity region at a relatively smooth location. While in the concave and convex body, a distinct low flow velocity region is formed in the fluid in the rough fissure channel. It means a resistance will be formed on the concave and convex area. The fluid flow in the rough crossing fractures shows the same pattern as in the single surface. Fig. 9-39 (c) shows the seepage flow in a RD-FN3D model. The aperture of the fractures is 1.0mm. The surface roughness has a key influence on the flow patterns along the connected planes. Meanwhile, the orientation of the flow vectors is affected by the waviness, which means the roughness of fractures should not be neglected when dealing with the fluid flow or solute transport in the fracture-matrix systems.

9.3 3D Printing in Geotechnical Engineering

9.3.1 Introduction

Jointed rock masses are discontinuous and anisotropic due to the existence of distributed complex fractures. These inherent weak planes can cause low shear strength and contribute significantly to the failure of rock masses. Therefore, understanding the influence of geometry and distribution of pre-existing interfaces on the mechanical behavior of rock masses is crucial for design and stability assessments in rock engineering. In rock engineering, especially in the slope engineering, the effect of fractures network on the shear behavior of rock mass is important and the slide surface is greatly influenced by the joint roughness and geometry. The shear behavior of intact and jointed rock masses has been the focus of many studies and considerable progress has been made in this field.

Discrete fracture network (DFN) models, which consider joint geometrical parameters (i. e., the joint-orientation angle, spacing, and persistence), have proven to be an effective approach for studying rock mass structures. Until recently, the fractures (or joints) distributed in the DFN models neglect the roughness, although the effects of the roughness of joint planes on shear strength and failure patterns have been proven to be important. We propose a rough discrete fracture network (RDFN) model, which considers both the joint geometry and random distribution, and tested the mechanical behaviors using PFC code. The roughness in this model indicates that the fracture orientation and shape is treated as complex geometry compared to linear shape. And the joint roughness and geometry is found to have a large influence on the mechanical behaviors of the rock mass. The shear strength of a rock mass is underestimated if the joint roughness is ignored. Thus, laboratory experiments have been widely used to study the shear behavior of jointed rocks and remarkable achievements have been made to understand the mechanical response. However, the failure characteristic of jointed rock is still not clearly revealed because the existence of natural roughness makes it difficult to perform duplicate experiments.

Three-dimensional (3D) printing, which is also called rapid prototyping or additive manufacturing, is an innovative technology that enables the printing of complex structures through the accumulation of successive layers. It has advantages over traditional methods since it allows for precise fabrication, flexible preparation, and high repeatability. 3D printing technology has attracted attention inthe field of geoengineering. With the development of printing technology and materials, 3D printing can provide an innovative approach for the study of rock masses. In addition, considerable attention has been given to 3D printing regarding rock mechanics in recent years. Here, 3D printing technology is used to study the mechanical behavior of a synthetic rock mass based on fused deposition modeling

(FDM) with polylactic acid (PLA) material. DFN models and RDFN models considering joint roughness are prepared using 3D printing technology. A sulphoaluminate cement, which is rapid hardening, provide the intact rock in the fracture network model. Direct shear tests are then carried out on the jointed rock specimens with different types of DFN models at various sizes. The shear strength and fracture patterns are compared with the DFN models by considering joint roughness and the size effect of the shear behavior is examined. The findings provide a promising way to replicate and visualize joints and understand their influence on the mechanical behavior of rock masses, which will facilitate a better understanding of the failure mechanisms.

9.3.2 Modeling of the Discrete Fracture Network

(1) Experiment Models

Rock masses containa large number of fractures of different sizes. These fractures play a dominant role in the mechanical response and make the material discontinuous and anisotropic. It is important to characterize the fracture geometry and analyze the relevant properties of the rock mass. The geometrical distribution of natural fractures is statistically distributed. To investigate the mechanical behavior of a jointed rock mass, Wang, et al. (2018) have reported the results of uniaxial compression on 3D-printed jointed rock specimens. RDFN model, which considers both the spatial distribution of discrete fractures, and the complex geometry of single joint, is an improved DFN model. The roughness in this model means the orientation and shape is treated as rough geometry compared to linear shape. The corresponding digital DFN and RDFN models by Wang, et al. (2020) are shown in Fig. 9-40. The fracture geometries (trace length, dip angle, and spatial connectivity) of both the DFN and RDFN models have the same probability distribution. Here, the sinusoidal function is used to represent the arc-shape joints. The roughness of the rock joints is determined by a series of sinusoidal curves, as shown in Eq. (9-11).

$$y=A\sin(\omega x+\varphi) \tag{9-11}$$

Where,

A —— the asperity of the joint trace;

ω —— the frequency that affects the joint roughness;

φ —— the phase position that affects the joint.

The curve is modified to match the surface of a natural rock joint by changing the relevant parameters as shown in Fig. 9-40. The joint dip angle θ, is characterized by rotating the curve within the coordinate system. The joint trace length L, is quantified by the linear length of the rough joints. The density, dip direction, dip angle and trace length of three different types of joints are listed in Table 9-12. The RDFN model is then established using MATLAB according to Wang, et al. (2020), as shown in Fig. 9-40.

Two uniformly distributed sets of rough joints are established to study the influence of roughness on the jointed rock mass. The asperity intensity A, ω, φ, dip angle θ, and trace length of each set of joints are as follows: $A_1=0.5$ and $A_2=0.8$, $\omega_1=\omega_2=1.0$, $\varphi_1=\varphi_2=0$,

Geometrical parameters and distribution of the discontinuities Table 9-12

RDFN model	Joint set	Density (/m)	Dip angle (°)			Trace length (m)			Asperity A(m)	ω	φ	Length of the asperity
			Type	Mean value	Deviation	Type	Mean value	Deviation				
RDFN	#1(in red)	0.3	Uniform distribution	0	15	Uniform distribution	6	1	0.3 · Rand	1	0	0

Continued

RDFN model	Joint set	Density (/m)	Dip angle (°)			Trace length (m)			Asperity A(m)	ω	φ	Length of the asperity
			Type	Mean value	Deviation	Type	Mean value	Deviation				
RDFN	#2(in green)	0.3	Uniform distribution	45	15	Uniform distribution	3	1	0.15 · Rand	1	0	0
	#3(in blue)	0.3	Uniform distribution	135	15	Uniform distribution	2	1	0.10 · Rand	1	0	0

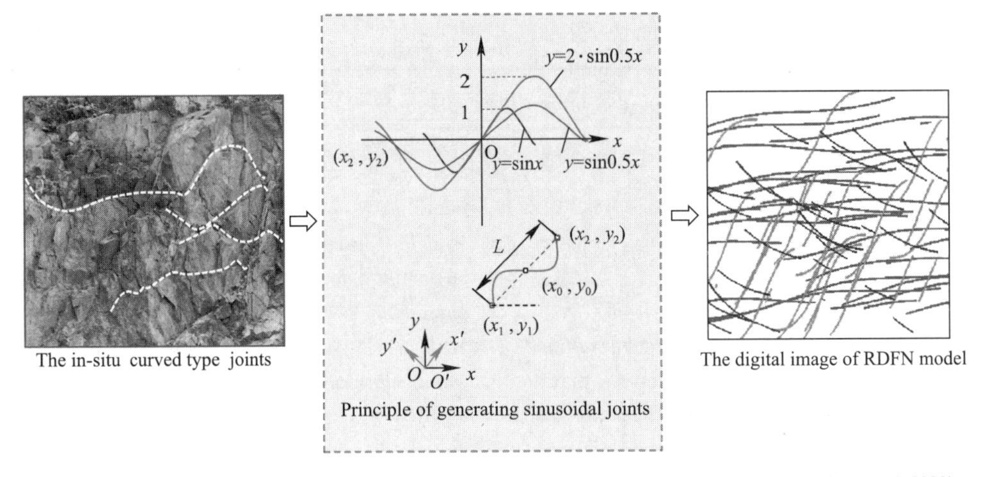

Fig. 9-40 The principle of generating sinusoidal type RDFN model based on natural rough joints (Wang et al. 2020)

$\theta_1 = 45°$ and $\theta_2 = 105°$, 3m and 2m. The principal improvement in the RDFN model is accomplished with the consideration of joint geometry and orientation (Wang, et al. 2020). Wang, et al. (2020) introduced the principle of the RDFN model and conduced numerical direct shear test on the jointed rock models. In this work, the corresponding rock models will be experimentally printed using 3D printing technique. As shown in Fig. 9-41, specimens of varied sizes (2m×2m, 3m×3m, 4m×4m, 5m×5m, and 6m×6m) can be generated by increasing the geometric size of the fracture network. The corresponding sizes of the experimental specimens are downsized with the scale of 1 : 100, which are 20mm×20mm, 30mm×30mm, 40mm× 40mm, 50mm × 50mm, and 60mm × 60mm, respectively. Then, the digital fracture models can be printed using the 3D printing system. The process for establishing the fracture network model is shown in Fig. 9-42.

Detailed process is listed below.

(a) First, a digital image of the fracture network model is generated using MATLAB code according to Wang, et al. (2020), as shown in Fig. 9-42 (a). Then the image of the joint traces is imported into AUTOCAD software and then widened in the xy plane. It should be noted that the width of each trace should be as thin as possible. However, the resolution of the 3D printing system should also be considered.

(b) After the trace is widened and stretched in the z direction, the 3D DFN model is established (Fig. 9-42a). The 3D digital model will be exported as *. STL format file and can be then imported into XYZ ware Pro software, where the 3D model can be sliced into many thin layers.

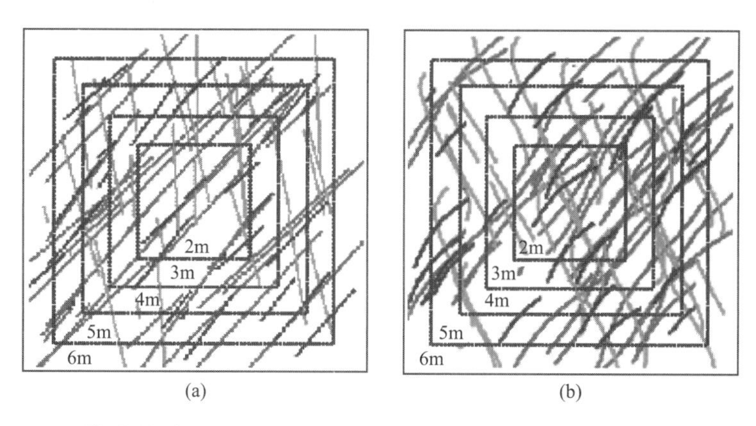

Fig. 9-41 Schematic of size effect analysis of shear test on jointed models

(a) DFN model; (b) RDFN model

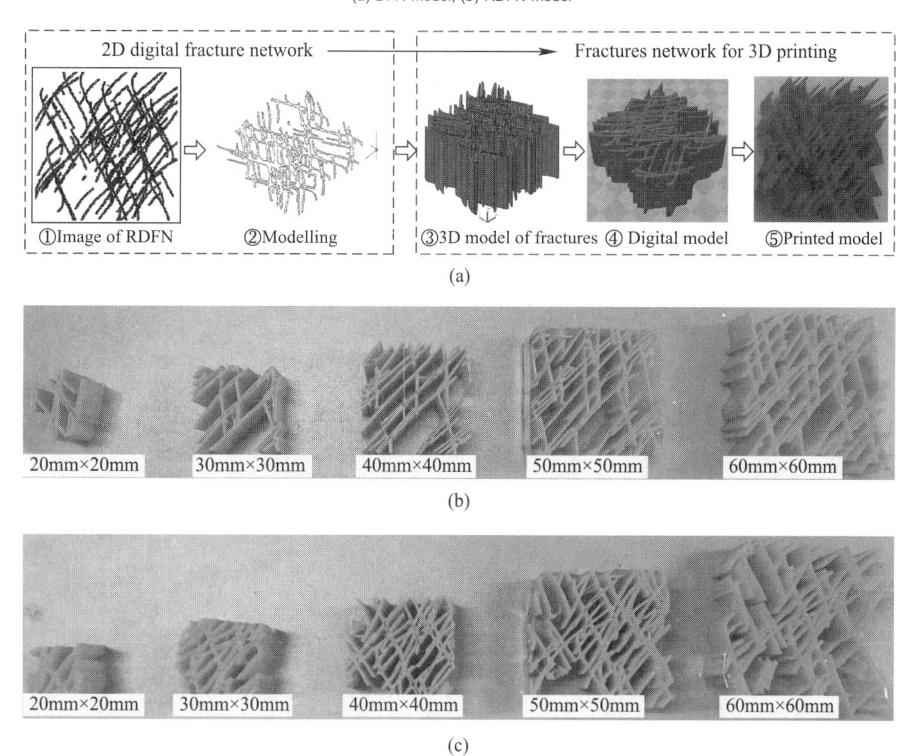

Fig. 9-42 Technique of establishing the fracture network models using 3D printing

(a) The preparation method; (b) DFN models; (c) RDFN models

(c) After the digital 3D model is sliced, the 3D-printed fracture model can then be realized, as shown in Fig. 9-42 (a).

It should be noted that one limitation of the technique is that isolated fractures or rock mass with low fracture density will be difficult to print using the 3D printing system. Several supporting pillars will be needed to print the isolated parts, which may influence the mechanical behaviors of the printed specimens.

(2) Printing Equipment and Materials

The stereolithography (STL) format files are inputted into the XYZ Printing DaVinci 3.0 printer. The thickness of the printed model is controlled by the resolution of the nozzle of the 3D printing machine. The x-y-z resolution of the

3D printing machine is 0.1mm, which is controlled by the stepping motor. Although the thickness of the minimum 0.1mm can be chosen, it is difficult to successfully print the fractures model. In this test, the effective layer thickness is set to be 0.3mm after many times of debugging. The printing method of the equipment is based on fused deposition modeling (FDM). Polylactic acid (PLA), which is an environmentally friendly thermoplastic aliphatic polyester derived from corn starch, is used as the printing material. The PLA material is heated to its melting point, pushed through a nozzle into a thin stream, and subsequently accumulated in layers as the platform descends. The average elastic modulus, Poisson's ratio and uniaxial yield stress of the specimens with 50% fill density is 1.391GPa, 0.203 and 31MPa, respectively. The distance between the nozzle and the platform is set to be 1.68mm. The heating temperatures of the nozzle and platform are 200℃ and 50℃, respectively. For the FDM method, the temperature will influence the stability of the first layer printed on the platform. If the temperature is too low or too high, the PLA material will solidify soon or still in melting status after being fused from the hot nozzle, which may cause unstable and even fail the printing process. Another important printing parameter, the filling ratio, controls the fill density of the printed specimen. When the filling ratio is 0%, the printed specimen is an empty box filled with no PLA material. If the filling ratio is 100%, the printed specimen will be fully filled with PLA material. For the XYZ Printing DaVinci 3.0 system, there are two filling patterns, i.e., mixed filling and honeycomb filling, respectively. Considering the linear or arcs shape of printed fractures, the mixed filling pattern, which will print the specimen faster, is chosen in this study. And the filling ratio is set to be 0% to assure no PLA material is filled inside the model to replicate the low resistance of joint fractures during shear loading.

9.3.3 Direct Shear Tests

9.3.3.1 Experiments on Jointed Specimens

The DFN and RDFN models, with sizes of 20mm×20mm, 30mm×30mm, 40mm×40mm, 50mm×50mm, and 60mm×60mm, are shown in Fig. 9-42 (b) and Fig. 9-42 (c). The thickness of all specimens is 20mm. It should be noted that the fill ratio (or fill density) of the joints, which is set to be 0% to replicate the low resistance of joints, still resulted in an interface with some thickness and tensile strength. Further, a printed PLA fracture network model is placed in a 3D-printed square box and filled with sulphoaluminate cement. The density of the sulphoaluminate cement is about 2.82g/cm^3. The specific surface area of the sulphoaluminate cement is more than 350m^2/kg. The initial setting time and the final setting time is about 15min and 30min, respectively. The size of the square box is 1.2 times the size of the fracture network area, i.e. 24mm × 24mm, 36mm × 36mm, 48mm × 48mm, 60mm × 60mm, and 72mm×72mm. The purpose of enlarging the dimension of the box is to make sure the cement is properly filled and around the fractures. If the size of the box is too large, more cement around the fracture model will be brought, which may influence the comparison of the RDFN and DFN model. The water-cement ratio by weight is 55%. The synthetic specimens are water-cured for seven days at a constant temperature of 20℃. According to Wang, et al. (2018), the uniaxial compression strength (UCS) of standard specimen is 50 to 70MPa. The UCS of the rock samples with sizes of 20mm, 30mm, 40mm, 50mm, and 60mm is 8.62MPa, 6.76MPa, 5.54MPa, 11.16MPa, 5.90MPa, respectively. In each scenario, three specimens are prepared and the total number is 30; 15 intact specimens (with no fracture network) are also produced. Direct shear tests are performed with the BC-100D testing system. The constant normal load (CNL) condition is set and the normal load is set to be 0kN in this test. The shear loading is initiated through shear displacement control at a rate of 0.3mm/min until failure; both shear load (pressure sensor, ±0.001kN) and displacement (displacement sensor, ±0.001mm) are recorded by TestMaster software during the shear test.

9.3.3.2 Results and Analysis

(1) Direct shear behavior

The experimentsmainly focus on the influence of structural changes to reveal the importance of joint distribution and roughness. The direct shear test on jointed specimens is conducted as shown in Fig. 9-43. Half of the specimen is fixed inside the fixing box (Fig. 9-43) to ensure that the specimen is shear failure, rather than bending shear failure. The typical shear stress-shear displacement curves for three types of rock specimens with varying sizes are shown in Fig. 9-44. Considering the low ultimate shear load, the initial shear load (about 0.5kN) is not reset to be 0. With the increase in specimen sizes, the peak shear stress gradually decreased. After reaching a maximum magnitude, the shear stress of the synthetic specimens decreases. For the jointed rock specimens, the shear stress decreases gradually and reaches a relatively stable stress level. As expected, the peak shear stress of the intact specimens is higher than that of the jointed rock models. The shear strength exhibited by the RDFN model is higher than that of the DFN model. During the post-peak stage, the intact specimens show a brittle-type failure response. The jointed rock models show a more ductile failure mode. The values of the residual shear strength in most RDFN models are higher than those in the DFN models. It should also be noted that the ductile fracture material may contribute to the ductile post-peak response of rock mass.

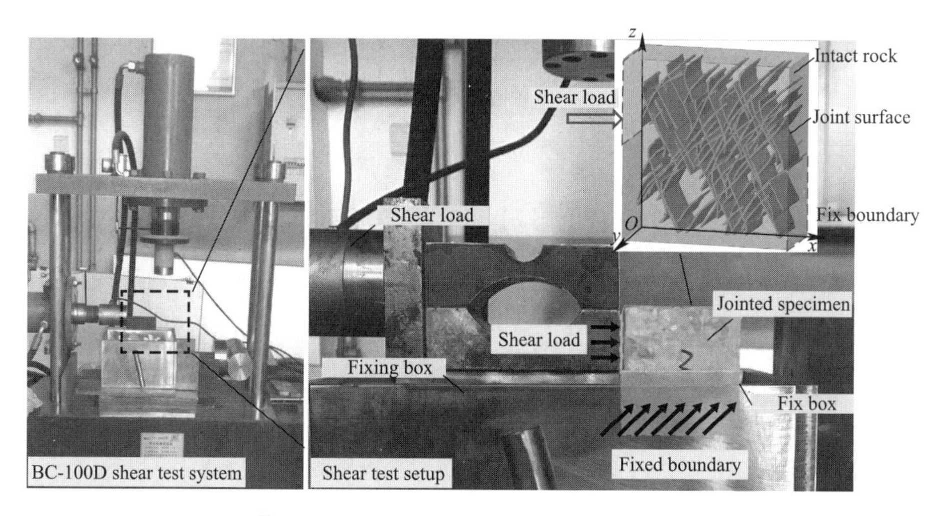

Fig. 9-43 Direct shear test and the jointed rock model

(2) Size effect of 3D-printed specimens

In general, for a given rock mass, as the volume of the rock mass increases, the mechanical parameters of the rock mass will change until the volume reaches a critical size, after which the parameters will remain stable. This critical size is called the representative elementary volume (REV). The REV is usually regarded as the volume of a heterogeneous material that is sufficiently large to be statistically representative of the composite (Kanit, et al. 2003). Fig. 9-45 shows the variations in the peak shear stress as a function of specimen size and an obvious size effect is observed from the experimental results. As shown in Fig. 9-45 (a), the value of the peak shear stress decreased gradually with increasing size. The quasi-brittle nature of the cement in the intact specimens is a

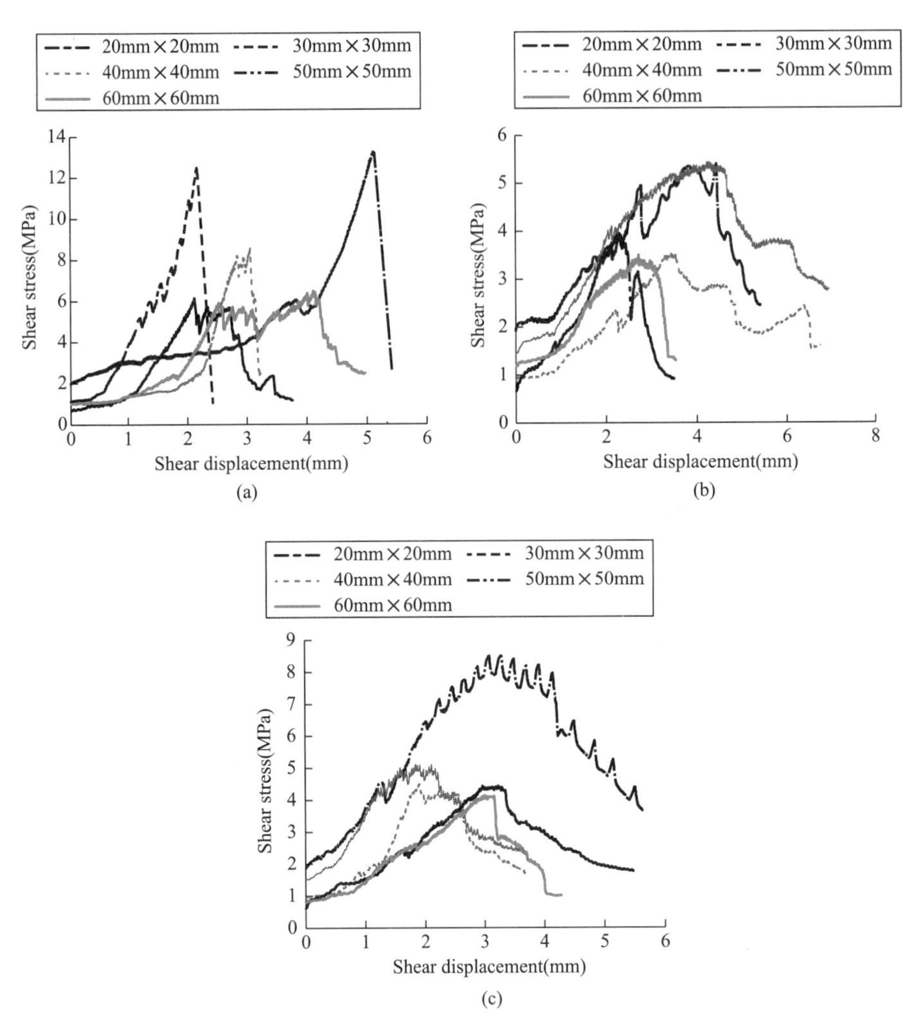

Fig. 9-44 Shear stress-shear displacement curves of different rock mass models

(a) Intact model(with no fractures); (b) DFN model; (c) RDFN model

key factor influencing the size effect. For the jointed rock models, the values of the peak shear stress all decreased gradually with increasing size. In the RDFN case in Fig. 9-45 (b), the peak shear stress decreases when the sample size increases from 30 to 40mm and it does not change with further increase in size. However, for the DFN case shown in Fig. 9-45 (c), the peak shear stress does not decrease when the sample size increases from 30 to 40mm, but it does decrease with further increase in size. The peak shear stress values in the RDFN model are higher than those in the DFN model, which is consistent with the numerical results in Wang, et al. (2020).

(3) Fracture patterns of specimens

The failure patterns of the three types of rock specimens at varying sizes are shown in Fig. 9-46. For intact specimens, the fracture planes exhibited relatively regular patterns and typical brittle failure is observed, as shown in Fig. 9-46 (a). The failure patterns in the jointed rock specimens are relatively complex. As shown in Fig. 9-46 (b) and (c), one or two joint traces play important roles in the failure pattern. As the size increases, the influence of a single joint on the observed failure patterns of the RDFN or DFN models becomes less significant. The failure patterns of the RDFN models and DFN models for various sizes are different,

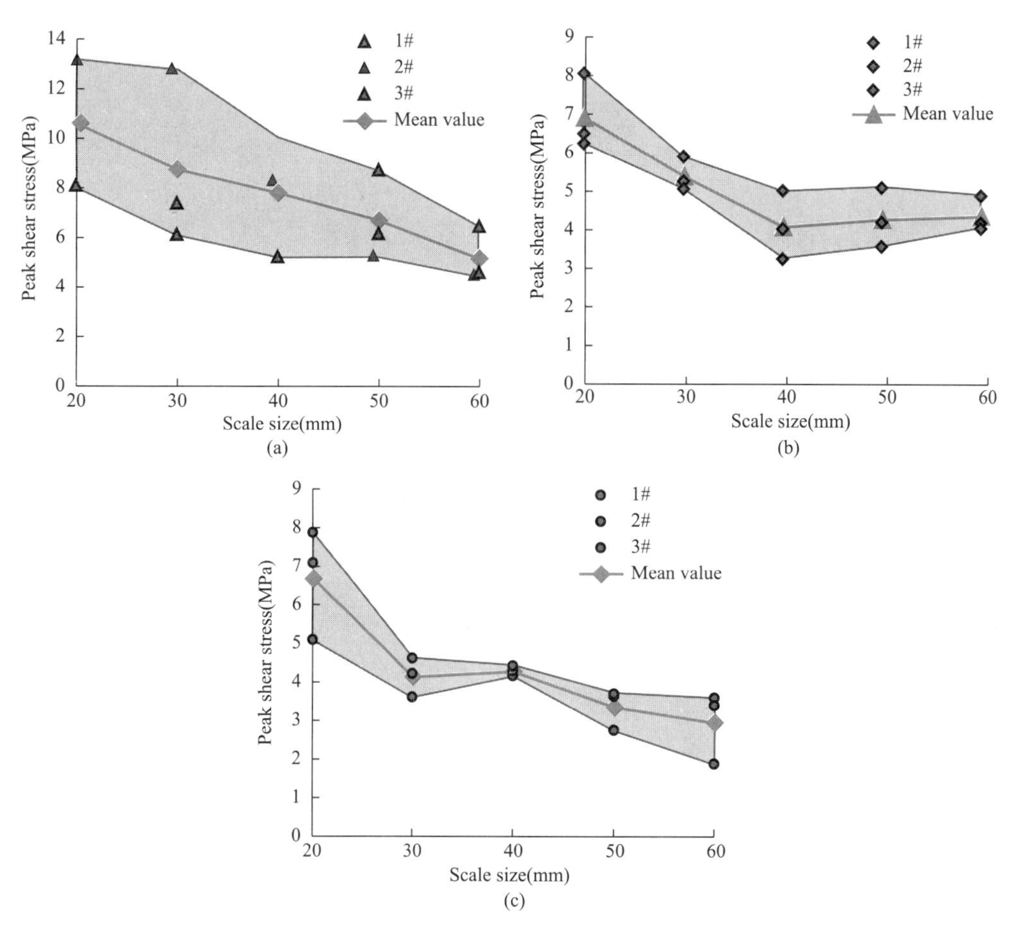

Fig. 9-45 Peak shear stress of three types of rock models with increase of specimen size

(a) Intact model(with no fractures); (b) RDFN model; (c) DFN model

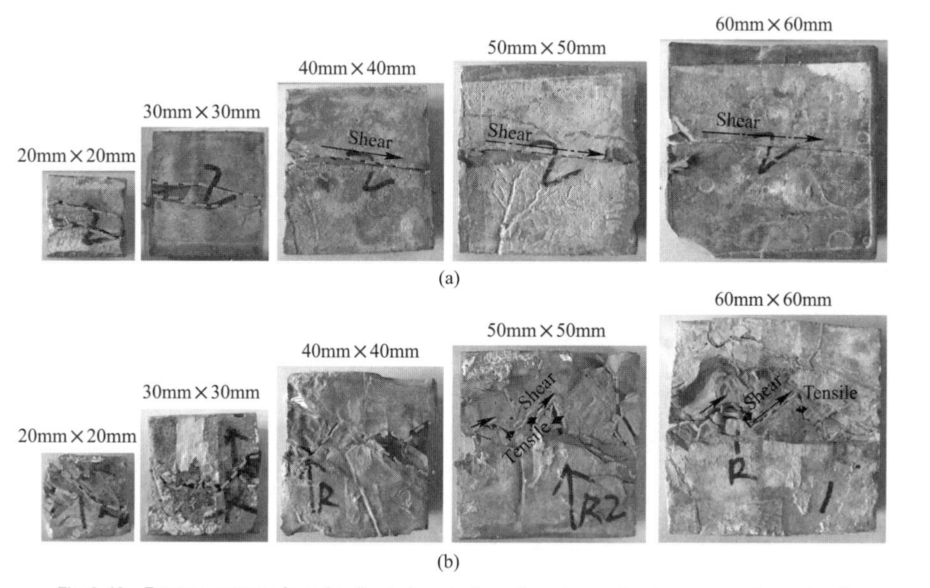

Fig. 9-46 Fracture patterns from the direct shear tests on three types of rock mass specimens (one)

(a) Intact models (with no fractures); (b) RDFN models

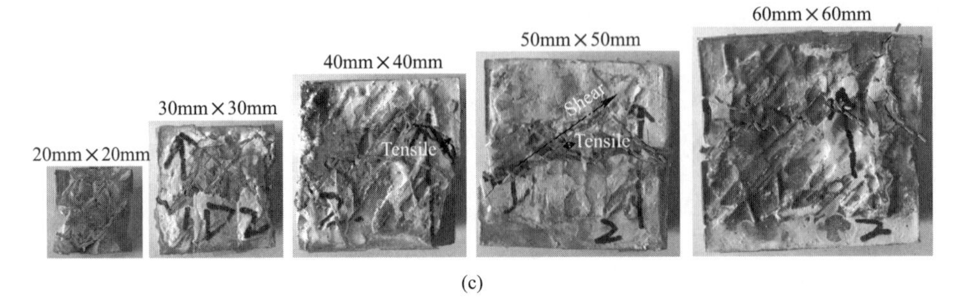

(c)

Fig. 9-46 Fracture patterns from the direct shear tests on three types of rock mass specimens (two)
(c) DFN models

and no apparent convergence is ob-served. Meanwhile, differences are observed between RDFN and DFN models of the same size, as shown in Fig. 9-46. This means that joint roughness can influence the failure patterns, which is consistent with the numerical results by Wang, et al. (2020). The joint roughness is found experimentally important for the shear behaviors and can not be neglected when modeling the fractrued rock mass.

9.3.4 Discussions

The 3D printer can create identical fractures specimens. However, the various microstructures or pores in the jointed rocks may cause the heterogeneity of specimens. An examination of the shear behaviors, i. e., fracture pattern and shear stress-shear displacement curves, is carried out to discuss the feasibility of 3D printing technology in the study of mechanical behaviors of jointed rock masses. Fig. 9-47 is the fracture patterns of DFN and RDFN models with specimen size of 50mm×50mm. It is indicated that the DFN models exhibited an approximate fracture trend along the joints traces. Some of the joint traces are the key influencers that determined the macro shear failure. However, the fractures of one of RDFN models along shear direction show different patterns. According to the numerical shearing result in Fig. 9-48 by Wang, et al. (2020), the shear fractures are dominated by the distribution of fracture networks. Joint roughness is also one of the key influencers which affect the shearing cracks. For the DFN model, where the joints are linearly distributed, the shearing fracture may initiate and propagate easily. However, the propagation in RDFN model may be difficult due to the influence of joint roughness. The peak shear strength exhibits higher than the DFN model due to the effect of rock teeth in RDFN model (Fig. 9-47b).

The shear stress-shear displacement curves are shown in Fig. 9-49. The pre-peak region of the shear stress-shear displacement curves of DFN models exhibites an approximate tendency. The curves of 2# and 3# behave nearly the same. The peak shear stress of the DFN specimens is 2.75MPa, 3.64MPa, and 3.70MPa, respectively. The average value is 3.36MPa with the value of standard deviation of 0.43MPa. The shear stress-shear displacement curves of RDFN models show a relatively large deviation. The peak shear stress of the DFN specimens is 3.57MPa, 4.20MPa, and 5.10MPa, respectively. The average value of RDFN models is 4.29MPa with the value of standard deviation of 0.63MPa. The residual shear strength of DFN models is 0.92MPa, 0.97MPa, and 0.82MPa, respectively. For the RDFN models, the residual shear strength is 0.90MPa, 1.16MPa, and 1.28MPa, respectively. It indicates that the DFN models exhibit a higher convergence than the RDFN models. The main reason should be the heterogeneity of cement in and around the PLA fractures.

According to the size effect analysis, the shear strength decrease when increasing the sample sizes. The number of joint elements in

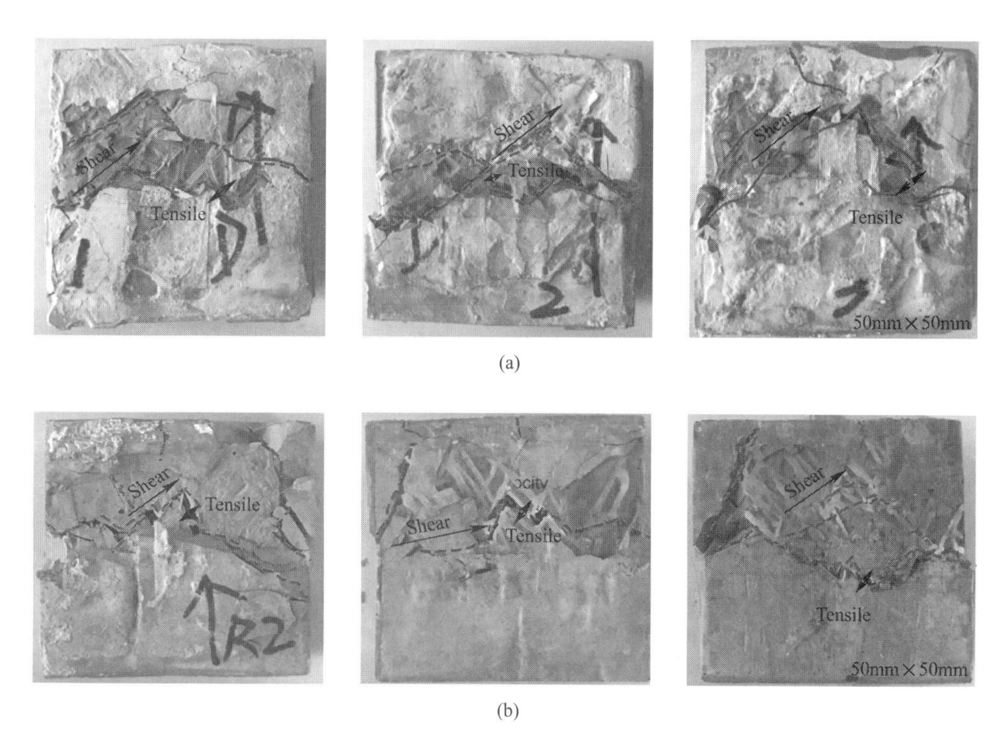

Fig. 9-47 Failure patterns of jointed specimens with size of 50mm×50mm

(a) DFN models; (b) RDFN models

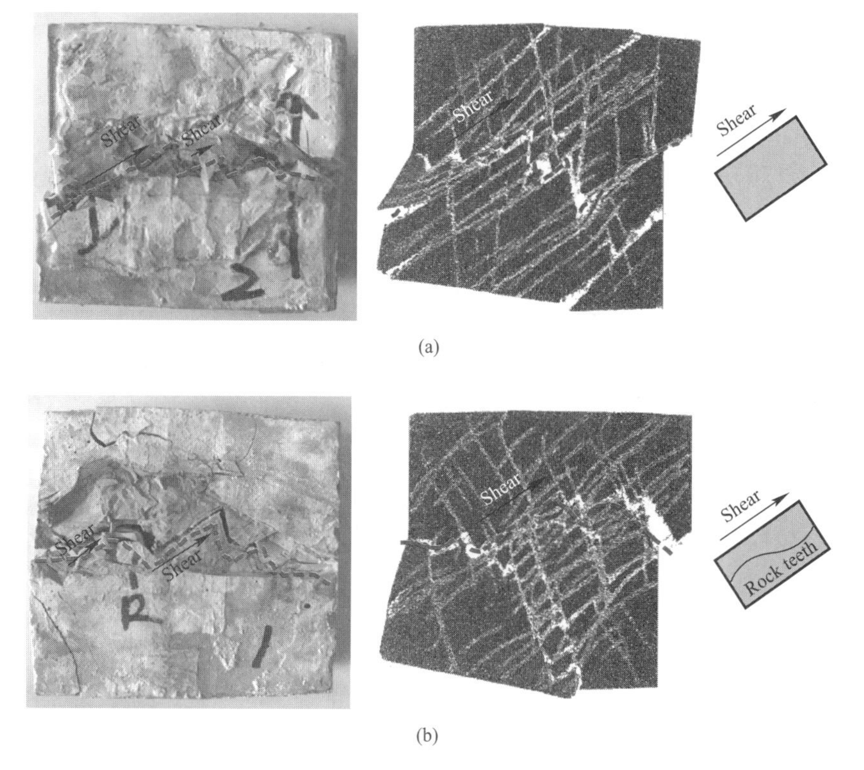

Fig. 9-48 Comparison of failure patterns between experimental results and PFC test (Wang, et al. 2020)

(a) The fracture patterns of DFN models; (b) The fracture patterns of RDFN models

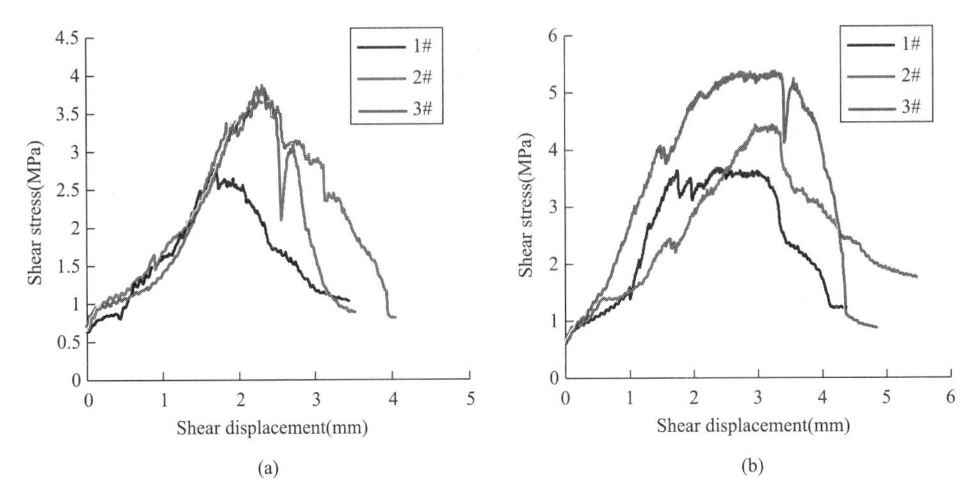

Fig. 9-49 Shear stress-shear displacement curves of jointed specimens of 50mm×50mm
(a) DFN models; (b) RDFN models

the jointed models increases when increasing the sample size. The ratio of joint elements to the total number of elements in both the RDFN and DFN models are discussed in Table 9-13. Fig. 9-50 shows the trends of both RDFN models and DFN models. The ratio in RDFN and DFN model both gradually decreases when increasing the sample sizes. It is also shown that the ratio in RDFN models is higher than the DFN models. However, the values of the peak shear strength of RDFN models are higher than the DFN models. It means that the roughness of joints network plays a more important role in the shear behaviors of jointed rock mass.

By considering the spatial distribution and roughness of the joints, the 3D-printed jointed rock specimen can model the mechanical behavior and fracture pattern of rock masses. It should also be noted that the joint sets in the models are composed of the PLA material and are made with the same strikes. The width of each joint is relatively large due to the limitation of the 3D printing system. Then a true 3D rock-like specimens in larger scale can be examined to study the shear behaviors of jointed rock masses.

Comparison of the RDFN and DFN models **Table 9-13**

Size (mm×mm)	Joints elements number		Total elements number in the model		Ratio of joint elements to total elements	
	RDFN	DFN	RDFN	DFN	RDFN	DFN
20×20	10,703	9193	31,152	31,150	34.36%	29.51%
30×30	19,132	16,010	55,460	54,747	34.50%	29.24%
40×40	27,758	25,814	86,730	87,912	32.01%	29.36%
50×50	37,017	31,039	124,608	122,830	29.71%	25.27%
60×60	42,300	35,113	170,569	171,396	24.80%	20.49%

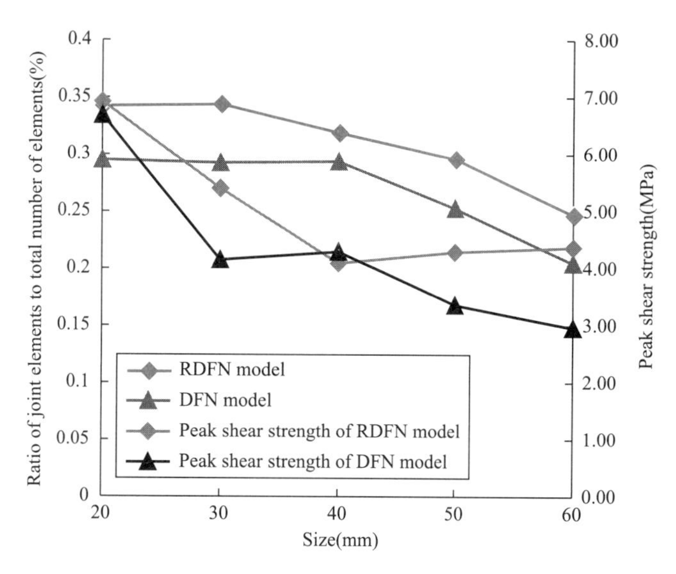

Fig. 9-50 Comparison of ratio of joints elements in the area of the models

9.4 Anisotropic Properties of Jointed Rock Mass

9.4.1 Introduction

Underground mining has been considered a high risk activity worldwide. Violent roof failure or rock burst induced by mining have always been a serious threat to the safety and efficiency of mines in China. Accurate and detailed characterization for rock masses can control stable excavation spans, support requirements, cavability and subsidence characteristics and thus influence the design of mining layouts and safety of mines. Rock mass is a geologic body composing by the discontinuities which have a critical influence on deformational behavior of blocky rock systems. The mechanical behavior of this material depends principally on the state of intact rock whose mechanical properties can be determined by laboratory tests and existing discontinuities containing bedding planes, faults, joints and other structural features. The distributions and strength of these discontinuities are both the key influencing factors for characterizing the discontinuous and anisotropic materials. The key work is to study the principal direction of elasticity or permeability and then assume the jointed rock mass as transversely isotropic geomaterial.

In rock engineering like mining and tunnel engineering, interactions between in-situ stress and seepage pressure of groundwater have an important role. Groundwater under pressure in the joints defining rock blocks reduces the normal effective stress between the rock surfaces, and therefore reduces the potential shear resistance which can be mobilized by friction. Since rock behavior may be determined by its geohydrological environment, it may be essential in some cases to maintain close control of groundwater conditions in the mine area. Therefore, accurate description of joints is an important topic for estimating and evaluating the deformability and seepage properties of rock masses.

9.4.2 Constitutive Relation of Anisotropic Rock Mass

Due to the existence of joints and cracks, the mechanical properties (Young's modulus, Poisson's ratio, strength, et al.) of rock masses are generally heterogeneous and anisotropic. Three preconditions should be confirmed in this section: (1) Anisotropy of rock mass is mainly caused by Ⅳ or Ⅴ-class structure or rock masses containing a large number of discontinuities; (2) rock masses according with (1) can be treated as homogeneous and anisotropic elastic material; (3) seepage tensor and damage tensor of rock mass with multi-set of joints can be captured by the scale of represent element volume (REV).

(1) Stress analysis

Elasticity represents the most common constitutive behavior of engineering materials, including many rocks, and it forms a useful basis for the description of more complex behavior. The most general statement of linear elastic constitutive behavior is a generalized form of Hooke's Law, in which any strain component is a linear function of all the stress components, i. e.:

$$\varepsilon_{ij}=[S]\sigma_{ij} \tag{9-12}$$

Where,

$[S]$ —— the flexibility matrix;

ε_{ij} —— the strain;

σ_{ij} —— the stress.

Many underground excavation design analyse involving openings where the length to cross section dimension ratio is high are facilitated considerably by the relative simplicity of the excavation geometry, as shown in Fig. 9-51. In this section, the roadway is uniform cross section along the length and can be properly analyzed by assuming that the stress distribution is identical in all planes perpendicular to the long axis of the excavation. Thus, this problem can be analyzed in terms of plane geometry.

The state of stress at any point can be defined in terms of the plane components of stress σ_{11}, σ_{22}, σ_{12} and the components σ_{33}, σ_{23}, σ_{31}.

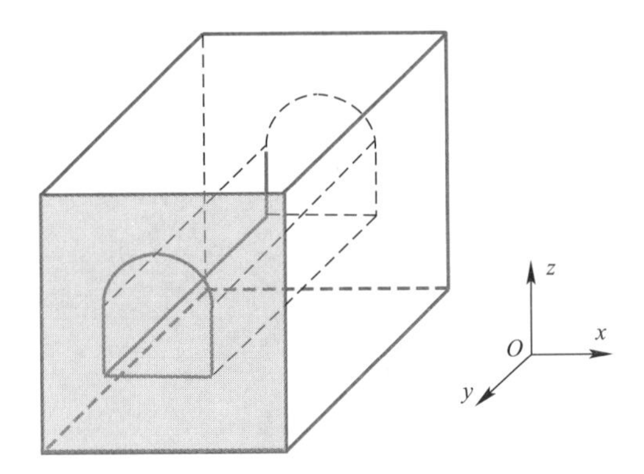

Fig. 9-51 A diagrammatic sketch for underground excavation

In this research, the z direction is assumed to be a principal axis and the antiplane shear stress components will vanish. The plane geometric problem can then be analyzed in terms of the plane components of stress, since the σ_{33} component is frequently neglected. Eq. (9-12), in this case, may be recast in the form of Eq. (9-13).

$$\begin{bmatrix}\varepsilon_{11}\\ \varepsilon_{22}\\ \varepsilon_{12}\end{bmatrix}=[S][\sigma]=\begin{bmatrix}\dfrac{1}{E_1}-\dfrac{v_{31}^2}{E_3} & -\left(\dfrac{v_{12}}{E_1}+\dfrac{v_{31}^2}{E_3}\right) & 0\\ -\left(\dfrac{v_{12}}{E_1}+\dfrac{v_{31}^2}{E_3}\right) & \dfrac{1}{E_1}-\dfrac{v_{31}^2}{E_3} & 0\\ 0 & 0 & \dfrac{2(1+v_{12})}{E_1}\end{bmatrix}\begin{bmatrix}\sigma_{11}\\ \sigma_{22}\\ \sigma_{12}\end{bmatrix} \tag{9-13}$$

Where,

$[S]$ —— the flexibility matrix of the material under plane strain conditions.

The inverse matrix $[S]^{-1}$ (or $[E]$) can be expressed in the form of Eq. (9-14).

$$\begin{bmatrix}\sigma_{11}\\ \sigma_{22}\\ \sigma_{12}\end{bmatrix}=[E][\varepsilon]=\begin{bmatrix}\dfrac{E_1^2(E_3-E_1v_{31}^2)}{\Delta} & \dfrac{-E_1^2(E_3v_{21}+E_1v_{31}^2)}{\Delta} & 0\\ \dfrac{-E_1^2(E_3v_{21}+E_1v_{31}^2)}{\Delta} & \dfrac{E_1^2(E_1v_{31}^2-E_3)}{\Delta} & 0\\ 0 & 0 & \dfrac{E_1}{2(1+v_{12})}\end{bmatrix}\begin{bmatrix}\varepsilon_{11}\\ \varepsilon_{22}\\ \varepsilon_{12}\end{bmatrix} \tag{9-14}$$

Where Δ is expressed as follows:

$$\Delta=-E_1E_3+E_1E_3v_{21}^2+E_1^2v_{31}^2+E_1^2v_{31}^2+2E_1^2v_{21}v_{31}^2 \quad (9\text{-}15)$$

The moduli E_1 and E_3 and Poisson's ratio v_{31} and v_{12} can be provided by uniaxial strength compression or tension in the 1(or 2) and 3 direction.

The mechanical tests including laboratory and in-situ tests for rock masses in large scale can hardly capture the elastic properties directly. Hoek-Brown criterion can calculate the mechanical properties of weak rock masses by introducing the Geological Strength Index (GSI). Nevertheless, the anisotropic properties can not be captured using this criterion and thus the method for analyzing anisotropy of jointed rock mass needs a further study. In this paper, the original joint damage in rock mass is considered as macro damage field. In elastic damage mechanics, the elastic modulus of the jointed material may degrade and the Young's modulus of the damaged element is defined as follows:

$$E=E_0(1-D) \quad (9\text{-}16)$$

Where,

D — the damage variable;

E and E_0— the elastic moduli of the damaged and intact rock sample, respectively.

In this equation, all parameters are scalar.

With the geometric information of the fracture sample, the damage tensor can be defined as Eq. (9-17).

$$D_{ij}=\frac{l}{V}\sum_{k=1}^{N}\alpha^{(k)}[n^{(k)}\otimes n^{(k)}](i,j=1,2,3) \quad (9\text{-}17)$$

Where,

N — the number of joints;

l — the minimum spacing between joints;

V — the volume of rock mass;

$n^{(k)}$ — the normal vector of the k-th joint;

$a^{(k)}$ — the trace length of the k-th joint (for 2 dimensions). According to the principal of energy equivalence, the flexibility matrix for the jointed rock sample can be obtained as shown by Eq. (9-18).

$$S'_{ij}=(1-D_i)^{-1}S_{ij}(1-D_j)^{-1} \quad (9\text{-}18)$$

Where,

S_{ij}— the flexibility matrix for intact rock;

D_i and D_j— the principal damage value in i and j direction, respectively.

For plane strain geometric problem, the constitutive relation, where the coordinates and principal damage have the same direction, can be expressed as follows:

$$\begin{bmatrix}\sigma_{11}\\ \sigma_{22}\\ \sigma_{12}\end{bmatrix}=\begin{bmatrix}\dfrac{E_0(v_0-1)(D_1-1)^2}{2v_0^2+v_0-1} & \dfrac{-E_0v_0(D_1-1)(D_2-1)}{2v_0^2+v_0-1} & 0\\ \dfrac{-E_0v_0(D_1-1)(D_2-1)}{2v_0^2+v_0-1} & \dfrac{E_0(v_0-1)(D_2-1)^2}{2v_0^2+v_0-1} & 0\\ 0 & 0 & \dfrac{E_0(D_1-1)(D_2-1)}{2(1+v_0)}\end{bmatrix}\begin{bmatrix}\varepsilon_{11}\\ \varepsilon_{22}\\ \varepsilon_{12}\end{bmatrix} \quad (9\text{-}19)$$

Where,

E_0— the Young's modulus of intact rock;

v_0— Possion's ratio for intact rock. The parameters can be easily captured by laboratory tests. Eq. (9-18) gives the principal damage values D_1 and D_2. Based on geometrical damage mechanics, all elements in this matrix can be obtained by the method mentioned above and the anisotropic constitutive relation of jointed rock sample can finally be confirmed.

(2) Seepage analysis

The seepage parameters of rock mass are quantized form of permeability, and also are the basis for solving seepage field of equivalent continuous medium. Based on the attribute of fractured rock mass and by taking engineering design into account, fractured rock mass is often considered to be anisotropic continuous medium. In the fracture network shown in Fig. 9-52, a total number of N cross points or water heads and M line elements are contained. Parameters can be obtained with the model such as water head, the related line elements, equivalent mechanical fissure width, seepage coefficient, et al. The corresponding coordinates of each point can be acquired. For a fluid flow analysis based on the law of mass conservation, the fluid equations on a certain water head.

$$\left(\sum_{j=1}^{N'} q_j\right)_i + Q_i = 0 \quad (i = 1, 2, \cdots, N) \tag{9-20}$$

Where,

q_j — the quantity of flow from line element j to water head i;

N' — the total number of line elements intersect at i;

Q_i — the fluid source term.

In the joint network, each line element will be assigned a length l_j and fissure width b_j to investigate the permeability.

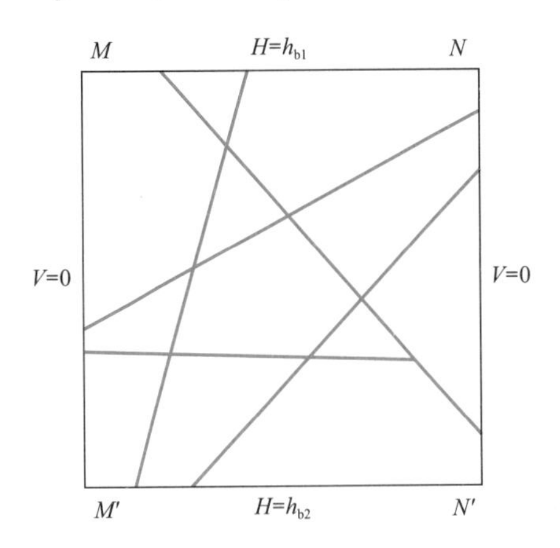

Fig. 9-52 Fracture network schematic diagram in seepage area

According to the hydraulic theory, for a single joint seepage, the flow quantity q_j of line element j can be expressed as Eq. (9-21).

$$q_j = \frac{\rho b_j^3}{12\mu} \cdot \frac{\Delta h_j}{l_j} \tag{9-21}$$

Where,

Δh_j — the hydraulic gradient;

μ — the coefficient of flow viscosity;

ρ — the density of water;

b_j — the fissure width of joint.

On the basis of Eq. (9-20) and Eq. (9-21), the governing equations can be represented as Eq. (9-22) for seepage in fracture network.

$$\left(\sum_{j=1}^{N'} \frac{\rho b_j^3}{12\mu} \cdot \frac{\Delta h_j}{l_j}\right)_i + Q_i = 0 \tag{9-22}$$

Based on the discrete fracture network method, an equivalent continuum model for seepage has been established. The hydraulic conductivity can be acquired based on Darcy's Law on the basis of water quantity in the network.

Boundary conditions in the model shown in Fig. 6 are as follows.

(a) MN and $M'N'$ as the constant head boundary;

(b) MM' and NN' as the impervious boundary with flow V of 0.

Then the hydraulic conductivity can be defined as:

$$K = \frac{\Delta q \cdot M'M}{\Delta H \cdot MN} \tag{9-23}$$

Where,

Δq — the total quantity of water in the model region (m^2/s);

ΔH — the water pressure difference between inflow and outflow boundaries (m);

K — the equivalent hydraulic conductivity coefficient in the MM' direction (m/s);

MN and $M'M$ — the side length of the region (m).

Based on Biot equations, the steady flow model is given by:

$$K_{ij} \nabla^2 p = 0 \tag{9-24}$$

Where,

K_{ij} — the hydraulic conductivity;

p — the hydraulic pressure.

For plane problems, the dominating equation of seepage flow is as follows in Eq. (9-25). The direction of joint planes is considered to be the principal direction of hydraulic conductivity.

$$K_{ij} = \begin{bmatrix} K_{11} & K_{12} \\ K_{21} & K_{22} \end{bmatrix} \tag{9-25}$$

9.4.3 Coupling Mechanism of Seepage and Stress

(1) Seepage inducted by stress

The coupling action between seepage and stress makes the failure mechanism of rock complex. The investigations on this problem have pervasive theoretical meaning and practical value. The principal directions associated with the symmetric crack tensor are coaxial with those of the permeability tensor. The first invariant of the crack tensor is proportional to the mean permeability, while the deviatoric part is related to the anisotropic permeability (Oda, 1985). Generally, the change of stress which is perpendicular to the joint plane is the main factor leading to the increase or decrease of the ground water permeability. In the numerical model, seepage is coupled to stress describing the permeability change induced by the change of the stress field. The coupling function can be described as Eq. (9-26) as given by Louis (1974).

$$K_f = K_0 e^{-\beta\sigma} \tag{9-26}$$

Where,

K_f—— the current groundwater hydraulic conductivity;

K_0—— the initial hydraulic conductivity;

σ —— the stress perpendicular to the joint plane;

β —— the coupling parameter (stress sensitive factor to be measured by experiment) that reflects the influence of stress; β is, the greater the range of stress induced the permeability. the larger

(2) Stress inducted by seepage

On the basis of generalized Terzaghi's effective stress principle, the stress equilibrium equation can be expressed as Eq. (9-27) for the water-bearing jointed specimen.

$$\sigma_{ij} = E_{ijkl}\varepsilon_{kl} - \alpha_{ij} P \delta_{ij} \tag{9-27}$$

Where,

σ_{ij}—— the total stress tensor;

E_{ijkl}—— the elastic tensor of the solid phase;

ε_{kl}—— the strain tensor;

α_{ij}—— a positive constant which is equal to 1 when individual grains are much more incompressible than the grain skeleton;

P —— the hydraulic pressure;

δ_{ij}—— the Kronecher delta function.

9.4.4 Coordinate Transformation

Generally, planes of joints are inclined at an angle to the major principal stress direction as shown in Fig. 9-53. In establishing these equations, the x, y and 1, 3 axes are taken to have the same $z(2)$ axis, and the angel θ is measured from the x to the 1 axis.

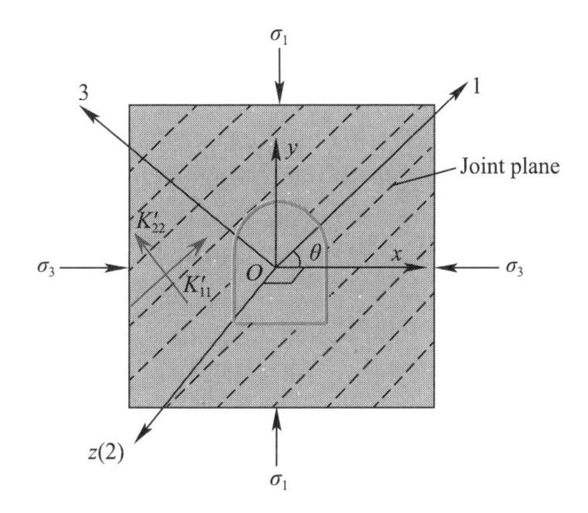

Fig. 9-53 The relationship between planes of joints and the principal stress direction

Considering the hydraulic pressure needs no coordinate transformation, Eq. (9-27) will be expressed as Eq. (9-28).

$$\begin{bmatrix} \sigma_{11} \\ \sigma_{22} \\ \tau_{12} \end{bmatrix} = [T_\sigma]^{-1}[E][T_\varepsilon]\begin{bmatrix} \varepsilon_{11} \\ \varepsilon_{22} \\ \gamma_{12} \end{bmatrix} - \alpha_{ij}\begin{bmatrix} P \\ P \\ 0 \end{bmatrix}\delta_{ij} \tag{9-28}$$

Where,

$[T_\sigma]^{-1}$—— the reverse matrix for stress coordinates transformation;

$[T_\varepsilon]$ —— the strain co ordinates transformation matrix.

They can be expressed as follows:

$$[T_\sigma]^{-1} = \begin{bmatrix} \cos^2\theta & \sin^2\theta & -2\cos\theta\sin\theta \\ \sin^2\theta & \cos^2\theta & 2\sin\theta\cos\theta \\ \sin\theta\cos\theta & -\sin\theta\cos\theta & \cos^2\theta - \sin^2\theta \end{bmatrix} \tag{9-29}$$

$$[T_\varepsilon]=\begin{bmatrix}\cos^2\theta & \sin^2\theta & \cos\theta\sin\theta\\ \sin^2\theta & \cos^2\theta & -\sin\theta\cos\theta\\ -2\sin\theta\cos\theta & 2\sin\theta\cos\theta & \cos^2\theta-\sin^2\theta\end{bmatrix}\tag{9-30}$$

Similarly, the coordinate transformation form can be expressed as Eq. (9-31).

$$\begin{cases}K_{11}=\dfrac{K'_{11}+K'_{22}}{2}+\dfrac{K'_{11}-K'_{22}}{2}\cos2\theta\\ K_{12}=K_{21}=-\dfrac{K'_{11}-K'_{22}}{2}\sin2\theta\\ K_{22}=\dfrac{K'_{11}+K'_{22}}{2}-\dfrac{K'_{11}-K'_{22}}{2}\cos2\theta\end{cases}\tag{9-31}$$

Where,

K'_{11} —— the hydraulic conductivity coefficient along the distribution direction of joint planes;

K'_{22} —— the hydraulic conductivity coefficient perpendicular to the distribution direction;

θ —— the angle between the optimal direction of permeability and the coordinate system x.

9.4.5 A Case Study

(1) Description of the area under study

The model described above, is applied to evaluate the seepage field and stress field of jointed rock roadway in the Heishan metal mine (Fig. 9-54). The mine is located in Chengde city, Hebei Province, in northern China. Heishan metal mine has transferred from open pit mining to underground mining since 2009. The deformation and stability of the roadways for mining the hanging-wall ore become the key technique issue. The elevation of the crest of the slope is 920m. The roadway in the research is in 674m level of the high northern slope. The rock mass mainly composes of anorthosite and norite in the northern slope area. The anisotropic properties of seepage and stress field will be discussed in detail.

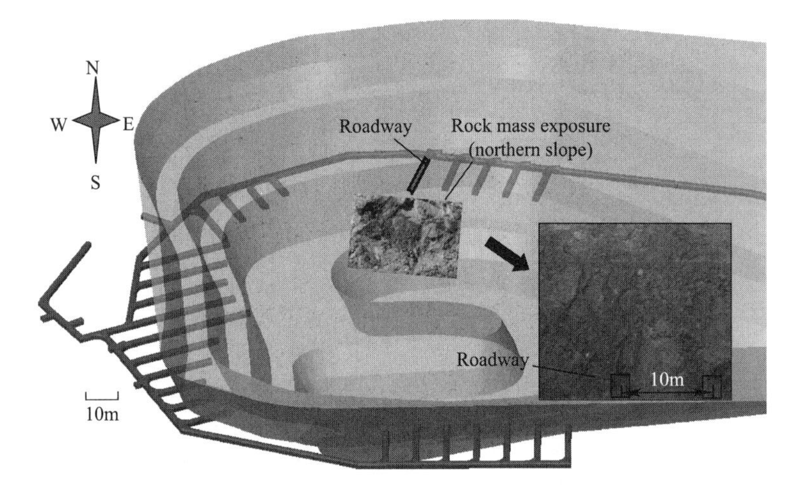

Fig. 9-54 The rock mass exposure of northern slope in Heishan open pit mine

(2) Capture of joint network

Depending on the 3D contact-free measuring system, discontinuities on the exposure can be easily captured. The collection process for the geology information of northern slope in Heishan metal mine has been discussed above. A 3D fracture network of rock mass can be obtained using the Monte Carlo method by analyzing the statistical parameters (Fig. 9-55a). Finally, the fracture network is generated and expanded to the roadway and the fracture network perpendicular to the center line of roadway could be easily captured (Fig. 9-55b).

(3) Permeability investigation

According to the symmetry of geometry, the method for solving the hydraulic conductivity coefficient in different directions of fracture network is presented to save computational time. Rotated every 15° clockwise, exerted certain water pressure on the boundary shown in Fig. 9-56, the hydraulic conductivity coefficient of vertical direction $K_{[90]}$, 75° direction $K_{[75]}$, 60° direction $K_{[60]}$, 45° direction $K_{[45]}$, 30° direction $K_{[30]}$ and 15° direction $K_{[15]}$ can be acquired, respectively. Then the permeability tensors by fracture parameters can be determined. By enlarging the geometric size of the fracture network, the permeability scale effect can be investigated finally.

When the edge length is 7m, the fracture network in the study area is shown in Fig. 9-57. Based on the previous algorithm, the size effect of rock mass in different sizes of statistics window is studied. The region of fracture network is enlarged from 1m to 9m with different step lengths until the equivalent parameters in different

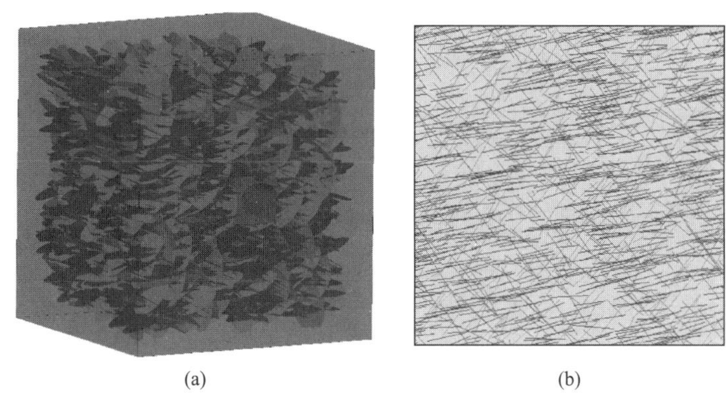

Fig. 9-55 Joint network based on the statistical data using Monte Carlo method
(a) 3D joint network in rock mass; (b) Profile with 10m edge length perpendicular to the roadway

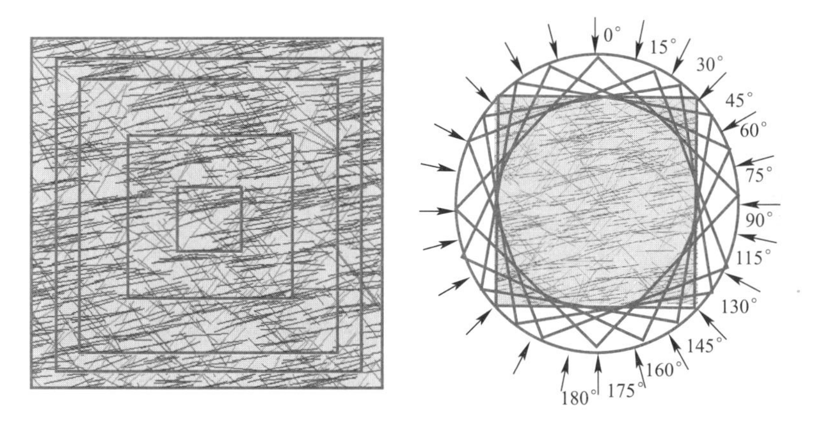

Fig. 9-56 Generated fracture network by Monte Carlo method (side length of internal squares are 1m, 4m, 7m, 8m, 9 m, respectively)

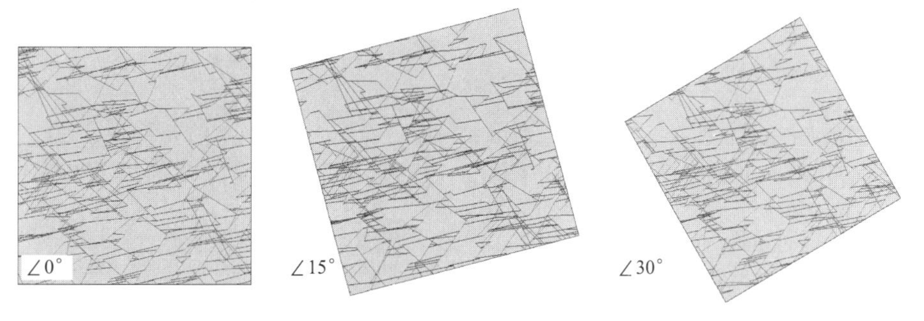

Fig. 9-57 Fracture network in the study area with edge length of 7m (one)

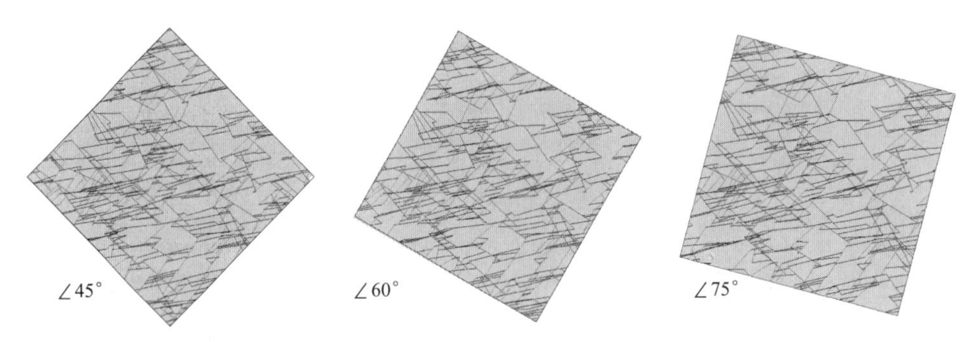

Fig. 9-57 Fracture network in the study area with edge length of 7m (two)

directions achieve a stable value. With a fixed the center of this region, the equivalent permeability is calculated rotating every 15° clockwise as shown in Fig. 9-58.

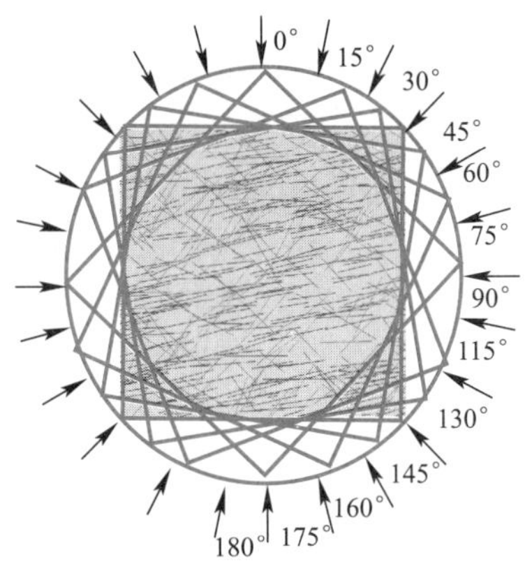

Fig. 9-58 Generated fracture network by Monte Carlo method (side length of internal squares are 1m, 4m, 7m, 8m, 9 m, respectively)

The variation of hydraulic conductivity coefficients with the increase of direction angles and sample size are depicted in Fig. 9-59. The main direction angle is approximate to 15°. It can be seen that the anisotropy of seepage property is apparent with the distribution of joints. Seven hydraulic conductivity coefficients of different sizes (1m, 3m, 4m, 5m, 6m, 8m, 9m) are chosen to be compared with that of the sample whose size is 7m. The coefficient deviation between one particular size and 7m will be found and presented in Fig. 9-60. Values of permeability decrease with the increase of sample size and tend to stabilize when sample size comes to 7m.

The principal values of permeability in maximum and minimum in a rock sample with joint plane angle is 15° are listed in Table 9-14. The principal permeability values decrease from 4.23×10^{-6} to 2.66×10^{-6} m/s as the sample size increases from 1 to 9m along the main direction. According to the results discussed above, the REV is 7m×7m for this fracture sample and the related hydraulic conductivity is 2.77×10^{-6} m/s

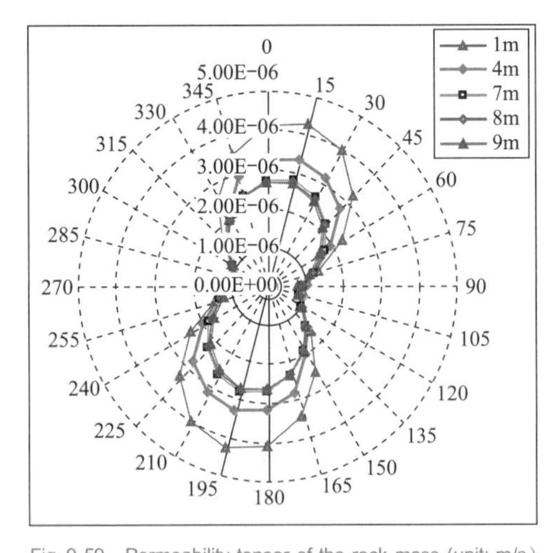

Fig. 9-59 Permeability tensor of the rock mass (unit: m/s)

for maximum and 0.87×10^{-6}m/s for minimum. The ratio for the maximum to minimum hydraulic conductivity value is 3.18.

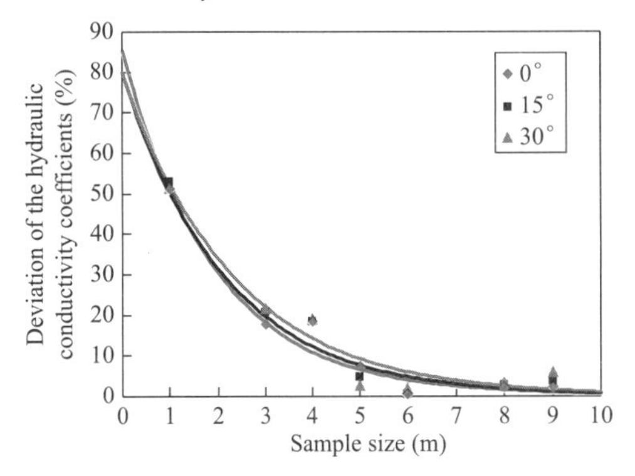

Fig. 9-60 Deviation of the permeability values to that of 7m sample under different directions of samples in different sizes

(4) Damage tensor

Similar to the previous algorithm, the size effect of sample damage in different sizes of statistics windows is studied. The region of fracture network is enlarged from 3 to 12m with different step lengths until the damage values in different directions stabilize. With a fixed center of this region, the principal damage values are calculated rotating every 15° clockwise. Fig. 9-61 shows the sample damage in different sizes and directions. Similarly to the deviation analysis of permeability, the damage values also fluctuatingly tend to stability according to the deviation of the damage tensor in different sizes and directions shown in Fig. 9-62. In this study, the size of represent element volume is also 7m in length and the initial damage tensor of REV of jointed rock mass can also be obtained. The principal damage values D_1 and D_2 for fracture sample is 0.17 for minimum and 0.50 for maximum and the principal direction angle θ of damage tensor is approximate 105° which is perpendicular to the direction of principal permeability. The related rock parameters, including undamaged Young's modulus E_0, Poisson's ratio v_0, uniaxial tensile strength, et al., are listed in Table 9-14. It should be noted that, the elasticity and strength of rock can be determined in the laboratory tests. Finally, the constitutive relation can be found.

Principal values $K_{[15]}$ of fracture hydraulic conductivity at different sizes **Table 9-14**

Sample size (m)		1	4	7	8	9
Principal value of hydraulic conductivity (×10^{-6}m/s)	Maximum	4.23	3.28	2.77	2.69	2.66
	Minimum	0.80	0.97	0.87	0.88	0.85
	Ratio	5.29	3.38	3.18	3.06	3.13

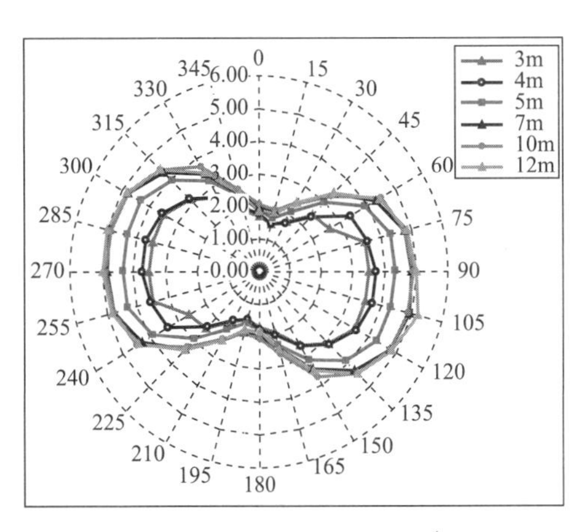

Fig. 9-61 Damage tensor(×10^{-1})

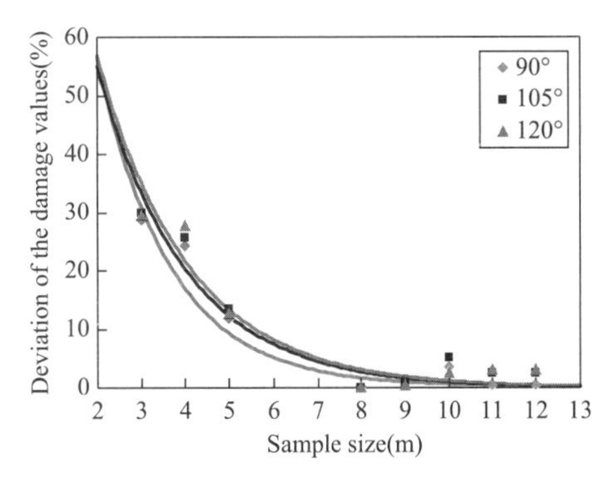

Fig. 9-62 Deviation of the damage tensor to 7m sample under different directions of samples in different sizes

(5) Geometry and boundary conditions

A numerical model of Heishan Metal Mineis established in order to simulate the mechanism of stress and seepage in jointed rock mass, taking into account the anisotropic property, as shown in Table 9-15. Fig. 9-63 shows the basic mechanical properties. Parameters listed in Table 9-16 are the hydraulic conductivity coefficients K_{ij}, damage tensor D_{ii} and Young's modulus E_{ij}, shear modulus G_{ij} of the fracture sample along joint plane direction used in the finite element code. The model contains two roadways with a 3.5m×3m three-centered arch section within a 50m×50m domain. The bottom boundary of the domain is fixed in all directions, and the left and the right boundaries are fixed in the horizontal direction. In this regard, a pressure σ_s of 5.97MPa is applied on the top boundary of

Parameters used in the model to validate the model in simulating the anisotropic properties of stress and seepage **Table 9-15**

Material parameters	Values
Undamaged Young's modulus E_0(GPa)	65
Undamaged Poisson's ratio v_0	0.23
Density (kg/m^{-3})	2700
Principal direction angle of joint plane θ (°)	15
Coupling parameter β (Pa^{-1})	0.5

The hydraulic conductivity coefficients K_{ij}, damage tensor D_{ii} and Young's modulus E_{ij}, shear modulus G_{ij} of the fracture sample along joint plane direction **Table 9-16**

Subscripts ij	K_{ij}(×10^{-6}m/s)	D_{ii}	E_{ij}(GPa)	G_{ij}(GPa)
11	2.77	0.50	18.83	—
22	0.87	0.17	51.91	—
12	—	—	9.34	10.97

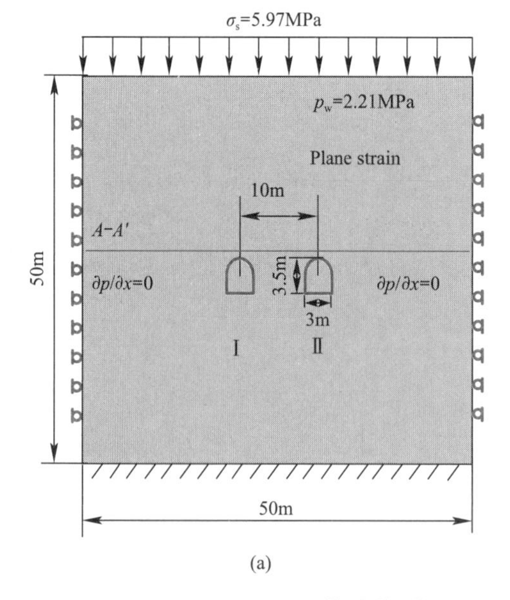

(a)

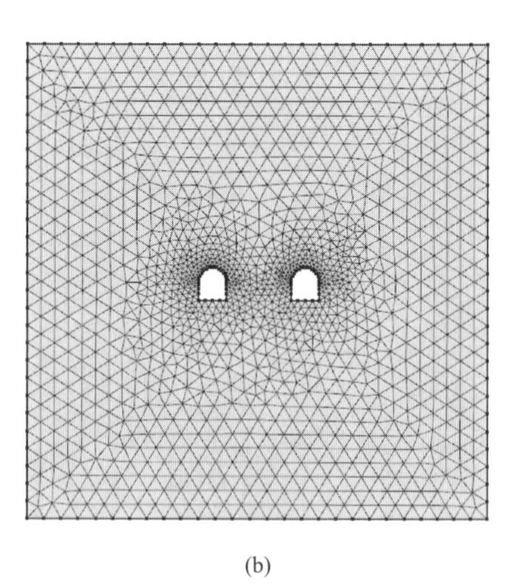

(b)

Fig. 9-63 Plane strain roadway model

(a) Boundary conditions; (b) The finite element mesh

the model to represent the 221-m-deep overburden strata. Under the steady-state groundwater flow condition, a hydrostatic pressure P_w, of 2.21MPa is applied upon top boundary. No-flow conditions are imposed on the three boundaries of the rectangular domain. The initial water pressure on the roadways boundaries is 0MPa. All the governing equations described above are implemented into COMSOL Multiphysics, a powerful PDE-based multiphysics modeling environment. The model is assumed to be in a state of plane strain (with no change in elastic strain in the vertical direction), and static mechanical equilibrium.

9.4.6 Results and Discussion

(1) Stress distribution

Adverse performance of the rock mass in the post-excavation stress field may be caused by either failure of the anisotropic medium or slip on the weakness planes. The elastic stress distribution around the roadways directly influences the deformation of rock and thus determines the design process. Fig. 9-64 shows the contour of first principal stress coupled with the seepage process. The orientation and magnitude of maximum principal stress controlled the distribution of the stress concentration in the heterogeneous media. The simulation result shows that the principal stress concentration zones appear mainly in rock surrounding the roadways. There exist maximum stress concentration areas in the arch foot and floor. Measures should be taken to control the deformation and assure the construction safety.

To characterize the response of the stress to the hydraulic mechanics, a comparison of two scenarios is also presented as shown in Fig. 9-65. The first principal stress in the stress-seepage coupled model along the horizontal section A-A' where $y=27$m is compared with a decoupled model. The result shows that the first principal stress increases when the seepage process is considered. Fig. 9-66 shows a plot of normal stress in joint plane direction and hydraulic conductivity along the horizontal section A-A' where $y=27$m. The coupled and decoupled model can be analyzed using Comsol Multiphysics code. In the coupled case, the governing equations for solid and fluid phase are solved in weakly coupled sense. For the sake of convenient contrast, the normal stress distribution as well as hydraulic conductivity is plotted when no seepage-coupled process is considered. The result shows that the normal stress will be

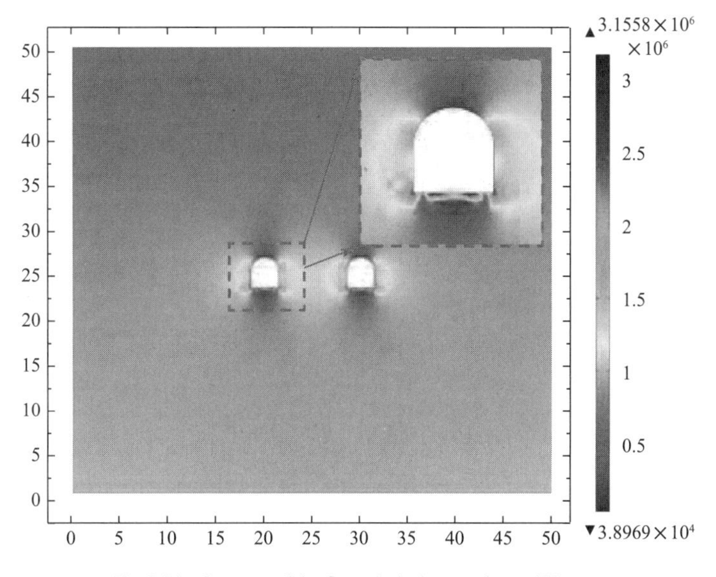

Fig. 9-64 Contours of the first principal stress for $\theta=15°$ case

underestimated when no coupling of seepage process is included. Moreover, the compressive stress leads to the decrease of permeability and the hydraulic conductivity increases when seepage process is included.

(2) Seepage distribution

Flow velocity with the angle of joint plane is 15° is shown in Fig. 9-67. All the process is not considered time-dependent. Therefore, the velocity change is induced by the permeability variation caused by the compressive stress upon the joint plane. In this case study, water flows along the principal direction of joint plane where the permeability is largest in the model. An asymmetric seepage field is observed and the maximum of flow velocity is distributed on the right top of the roadway roof. Moreover, the seepage pressure is also asymmetrically distributed as shown in Fig. 9-68.

(3) Damage zone

For underground mining, two principal engineering properties of the joint planes should be considered. They are low tensile strength in the direction perpendicular to the joint plane and the relatively low shear strength of the surfaces. As discussed above, the fluid pressure and

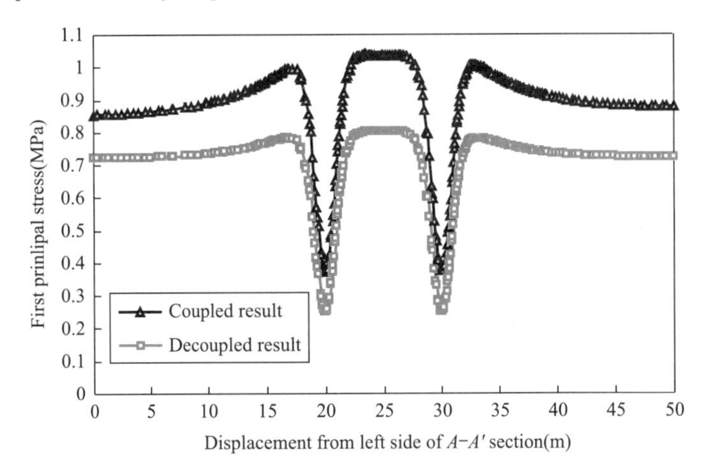

Fig. 9-65 Contrast of first principal stress between coupled and decoupled model

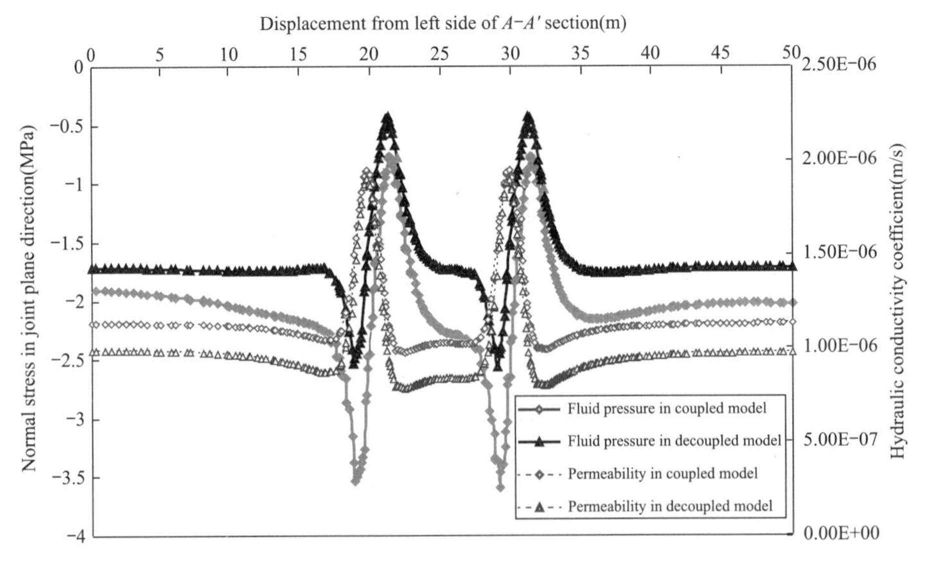

Fig. 9-66 Observed changes of normal stress in joint plane direction and hydraulic conductivity between coupled and decoupled model

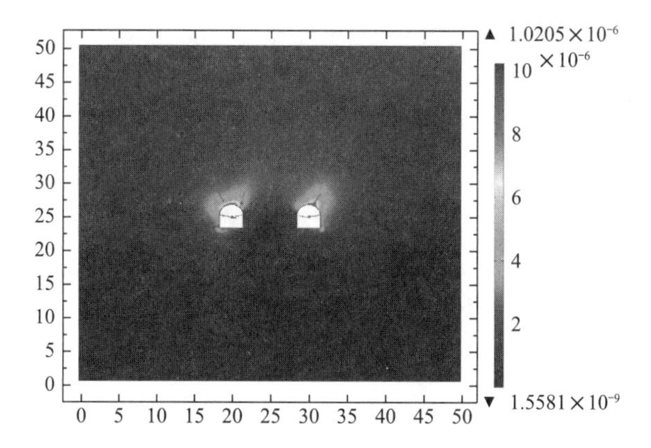

Fig. 9-67 Darcy's velocity magnitude (unit:m/s)

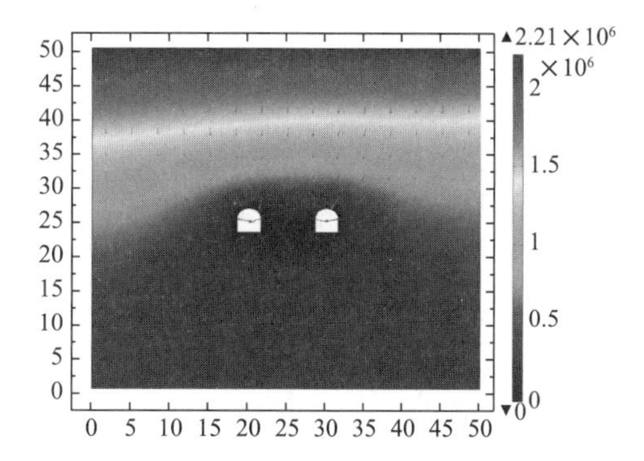

Fig. 9-68 The seepage pressure distribution (unit:Pa)

velocity are sensitive to the joint plane angles θ. In this section, the Hoffman anisotropic strength criterion is used to assess the damage zone in this numerical model as shown in Eq. (9-32).

$$\frac{\sigma_1^2}{X_tX_c}-\frac{\sigma_1\sigma_2}{X_tX_c}+\frac{\sigma_2^2}{Y_tY_c}+\frac{X_c-X_t}{X_tX_c}\sigma_1+\frac{Y_c-Y_t}{Y_tY_c}\sigma_2+\frac{\tau_{12}^2}{S^2}=1 \tag{9-32}$$

The properties X_t and Y_t represent the tensile strength along joint plane direction and perpendicular to joint plane direction, respectively. X_c and Y_c represent the compressive strength along joint plane direction and perpendicular to joint plane direction, respectively. S is the shear strength of the material along the joint direction. σ_1 represents the normal stress along the principle direction of elasticity and σ_2 represents the normal stress perpendicular to the principle direction of elasticity. τ_{12} represents the shear stress. The shear strength of the joints can be described by the simple Coulomb law.

$$S=c+\sigma_2\tan\varphi \tag{9-33}$$

Where,

c —— cohesive strength;

φ —— the effective angle of friction of the joint surfaces.

It should be noted that the mechanical parameters are acquired from laboratory or in-situ tests. However, it is difficult to directly employ the strength parameters for jointed rock mass due to the inaccessibility of the tests for huge rock mass. Based on the in-situ and laboratory tests of Heishan Metal Mine, the tensile, compressive and shear strength parameters are listed in Table 9-17.

Mechanical properties of rock mass in the direction of joint plane

Table 9-17

Tensile strength (MPa)	X_t	16.0
	Y_t	4.0
Compressive strength (MPa)	X_c	120
	Y_c	96
Shear strength (MPa)	c	1.5
	φ	50

The direction of joint plane is 15° and thus the tensile strength in the direction perpendicular to joint plane is relatively low, compared with that in other directions. Fig. 9-69 shows the damage zone in this case study and the failure area is mainly distributed in the direction perpendicular to the weakness plane, which gives an illustration of the roadway failure mode in tabular orebodies. Moreover, the failure of covered rock mass and the rock pillar in and between the two roadways doesn't influence each other in this case study and thus, the choice of roadway's interval is proper from the aspect of the mechanics analysis.

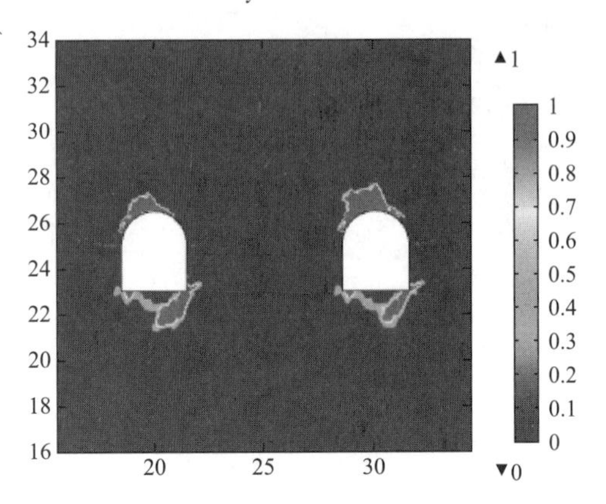

Fig. 9-69 Damage zone of the simulating model (the direction of damage zone is approximately perpendicular to that of joint planes whose angle is 15°)

9.4.7 Further Discussion

These simulations are performed to develop an understanding of the mechanics of joints and influence on stress and seepage fields and to gauge the ability of the proposed transversely anisotropic model to capture the response of jointed rock mass. For this purpose, a total of six scenarios with the joint plane angles θ ranging from 0° to 150° with an interval of 30°, are simulated in order to examine the effect of joint plane directions. And the anisotropic properties of seepage field and damage zones are examined in these simulations.

(1) Seepage field

Fig. 9-70 presents the fluid pressure distribution and flow field arrows with different joint plane directions. It can be seen that the maximum pressure, which is located on the top boundary,

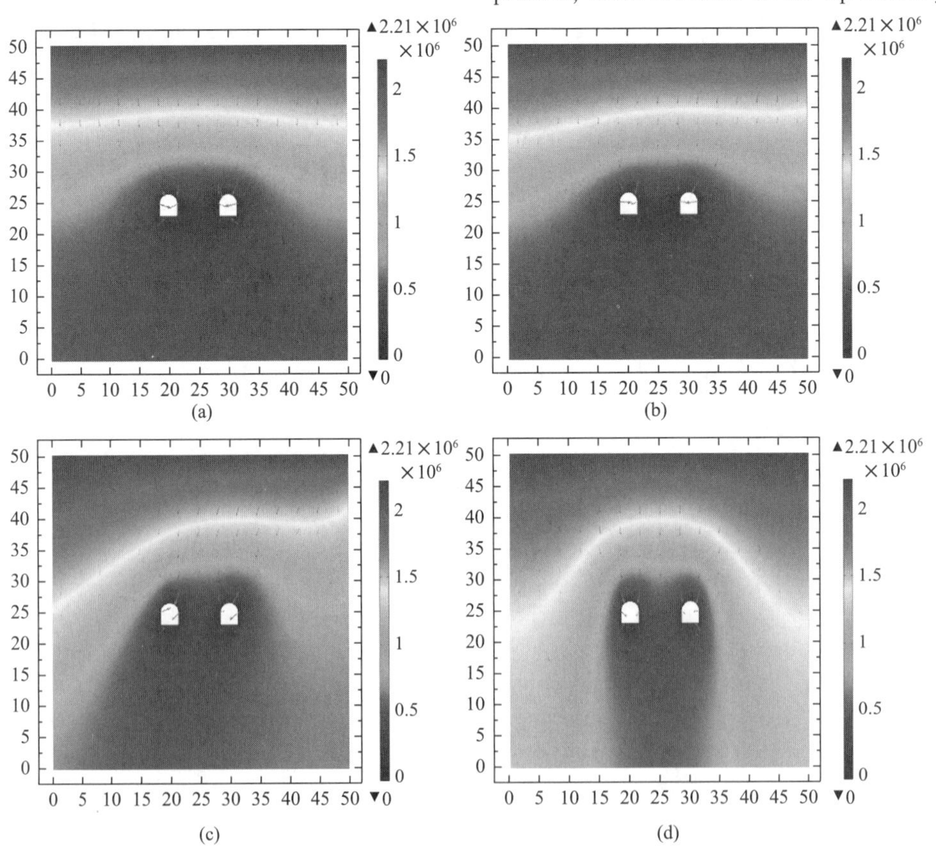

Fig. 9-70 Contour of the fluid pressure and flow velocity vectors in different joint plane directions (one)
(a) Angle of joint plane θ=0°; (b) Angle of joint plane θ=30°; (c) Angle of joint plane θ=60°; (d) Angle of joint plane θ=90°

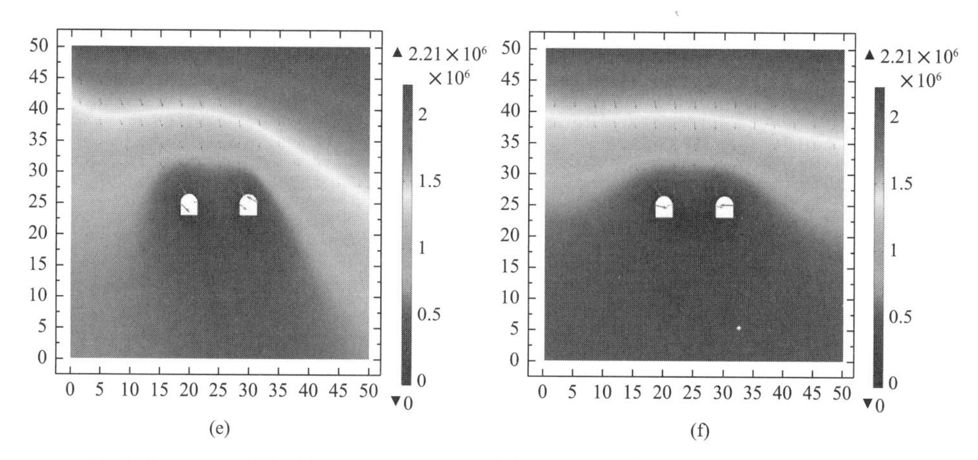

Fig. 9-70 Contour of the fluid pressure and flow velocity vectors in different joint plane directions (two)
(e) Angle of joint plane θ=120°; (f) Angle of joint plane θ=150°

equals the initial fluid pressure (2.21MPa). The fluid pressure distribution on top of the roadways differs with the increase of directions.

When the joint plane is parallel or perpendicular to the floors of roadways, the fluid pressure distributions and flow arrows are all axially symmetric, which agrees with expectations. The fluid pressure in the 30° or 60° case is distributed unsymmetrically as shown in Fig. 9-70 (b) or (c). The flow velocity in y direction along the horizontal section A-A' where y = 27m as shown in Fig. 9-70 (a) curves are plotted in Fig. 9-71, for θ = 0°, 30°, 60°, and 90°, respectively. In all cases, the absolute value of velocity increases approximately exponentially above the roadways. As shown in Fig. 9-71, the flow velocity in the scenario where θ = 0° is in the lowest level. The reason is that the joint plane in this scenario is horizontally distributed and the compressive stress leads to the decrease of permeability. When the angle of joint plane increases to 30°, flow velocities above both the roadways increase and Roadway I has a higher velocity than that of Roadway Ⅱ. The scenario where θ = 60° has the similar performance. However, when the joint plane is vertically distributed, flow velocity above Roadway I is lower than that where θ = 60°. The flow velocity in the case where joints are vertically or horizontally distributed is symmetric.

(2) Damage zone

The effect of joint plane angle on the damage zone is illustrated by using the Hoffman anisotropic strength criterion as shown in Eq. (9-32).

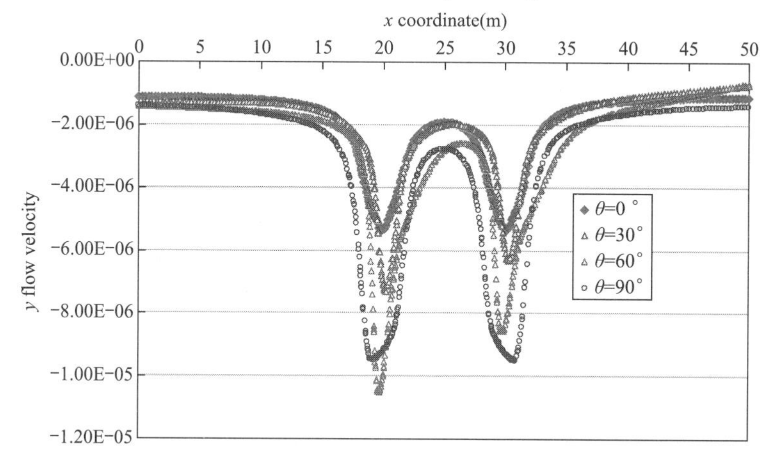

Fig. 9-71 The flow velocity in y direction along the horizontal section A-A' for different numerical models, at θ=0°, 30°, 60°, 90°

The damage zones in different angles of joint planes are shown in Fig. 9-72 and can be an index to visualize the potential failure mode of the roadway. It is clear from Fig. 9-72 (a) — (f) that an increase of the joint plane angle has significantly influenced the shape and size of the damage zone. When the joints are horizontally distributed ($\theta = 0°$), the tensile strength in the direction perpendicular to joint planes is much lower and thus the damage zone mainly concentrates within the roof and bottom of roadways. This failure mainly manifests as roof falling and floor heave. Similarly, when the angle increases to 30°, the damage zones mainly concentrate in the rock mass of the left roof and right floor within the roadways which is also in the direction perpendicular to joint planes. Compared with the result in Fig. 9-72 (b), the direction of the damage zone with $\theta = 60°$ shown in Fig. 9-72 (c) rotated significantly and the area also increases. When the joints are vertically distributed ($\theta = 90°$), the principal direction of the damage zone is nearly horizontal. Lateral rock mass surrounding the roadways stabilize with the increase of

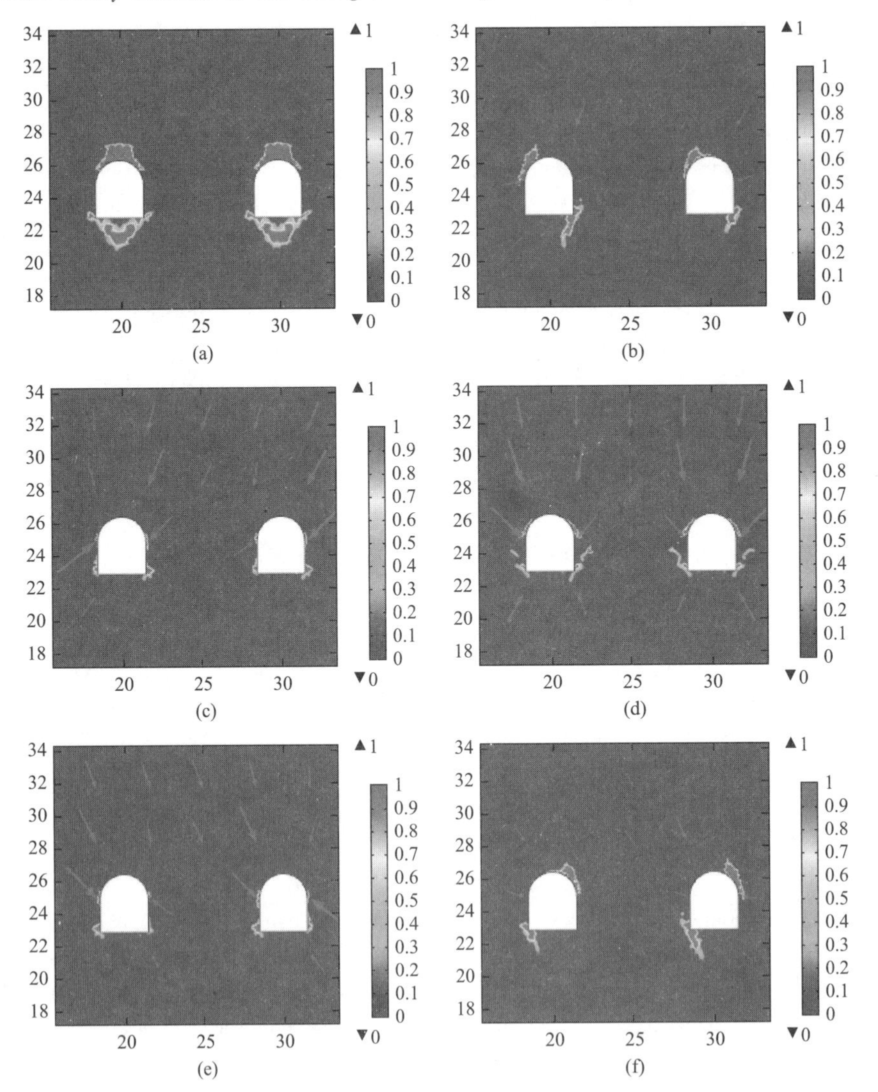

Fig. 9-72 The damaged zone under different principal elastic direction (red arrows represent the flow velocity vector)
(a) Angle of joint plane $\theta = 0°$; (b) Angle of joint plane $\theta = 30°$; (c) Angle of joint plane $\theta = 60°$; (d) Angle of joint plane $\theta = 90°$; (e) Angle of joint plane $\theta = 120°$; (f) Angle of joint plane $\theta = 150°$

joint plane angle in this study and the pillar between two roadways is also stable. The scenarios where $\theta = 120°$ or $\theta = 150°$ have the similar response to the joint plane direction with that of 60° or 30°. The model in this study can to some extent be effectively used to analyze the anisotropic property for jointed rock mass.

References

[1] Barton N. R., Choubey V.. The Shear Strength of Rock Joints in Theory and Practice [J]. Rock Mech anics, 1977, 10(1): 1-54.

[2] Hoek E., Carranza-Torres C., Corkum B.. Hoek-Brown Failure Criterion—2002 edition [C]//Proceedings of the Fifth North American Rock Mechanics Symposium. Toronto, 2002, 1: 267-271.

[3] Fleuty M. J.. The Description of Folds [C]//Proceedings of the Geologists' Association. London, 1964, 75(4): 461-492.

[4] Hu Y., Xu W., Zhan L., et al. Modeling of Solute Transport in A Fracture-matrix System with A Three-dimensional Discrete Fracture Network[J]. Journal of Hydrology, 2022, 605: 127333.

[5] Louis C.. Rock Hydraulics[J]. International Journal of Rock Mechanics, and Mining Science & Geomechanics Abstracts, 1975, 12(4):59.

[6] Oda M.. Permeability Tensor for Discontinuous Rock Masses [J]. Geotechnique, 1985, 35(4): 483-495.

[7] Ramandia H. L., Irtza S., Sirojan T., et al. FracDetect: A Novel Algorithm for 3D Fracture Detection in Digital Fractured Rocks [J]. Journal of Hydrology, 2022, 607: 127482.

[8] Stead D., Wolter A.. A Critical Review of Rock Slope Failure Mechanisms: The Importance of Structural Geology[J]. Journal of Structural Geology, 2015, 74: 1-23.

[9] Wang P. T., Liu C., Qi Z. W., et al. A Rough Discrete Fracture Network Model for Geometrical Modeling of Jointed Rock Masses And The Anisotropic Behaviour [J]. Applied Sciences, 2022, 12(3): 1720.

[10] Wang P. T., Liu Y., Zhang L., et al. Preliminary Experimental Study on Uniaxial Compressive Properties of 3D Printed Fractured Rock Models[J]. Chinese Journal of Rock Mechanics and Engineering, 2018, 37(2): 364-373.

[11] Wang P. T., Ren F. H., Cai M. F.. Influence of Joint Geometry and Roughness on The Multiscale Shear Behavior of Fractured Rock Mass Using Particle Flow Code[J]. Arabian Journal of Geosciences, 2020, 13(4): 165.

[12] Wang X. H., Zheng J., Sun H. Y.. A Method to Identify The Connecting Status of Three-dimensional Fractured Rock Masses Based on Two-dimensional Geometric Information [J]. Journal of Hydrology, 2022, 614: 128640.

[13] Wang X. H., Zheng J., Sun H. Y.. Comparative Study on Interconnectivity Between Three-dimensional and Two-dimensional Discrete Fracture Networks: A Perspective Based on Percolation Theory[J]. Journal of Hydrology. 2022, 609: 127731.

[14] Yan W., Komvopoulos K.. Contact Analysis of Elastic-plastic Fractal Surfaces[J]. Journal of Applied Physics, 1998, 84 (7): 3617-3624.

[15] Zhang H. T., Xu G. Q., Zhan H. B., et al. Simulation of Multi-period Paleotectonic Stress Fields and Distribution Prediction of Natural Ordovician Fractures in The Huainan Coalfield, Northern China[J]. Journal of Hydrology, 2022, 612: 128291.

Appendix A: Terms and Concepts Specific to Geotechnical Engineering

Approximation: a numerical value or method used to estimate or approach the true value of a function or equation.

Atterberg limits: a set of tests used to determine the water content at which a fine-grained soil transitions from solid to plastic and then to liquid states. These tests help assess the consistency and compressibility of clayey soils.

Bearing capacity: the maximum load capable of being sustained by a soil or rock without excessive settlement or failure. It is a critical consideration in the design of foundations for structures.

Boundary element method: a numerical method that discretizes the domain into boundary elements and solves the equations by representing the solution on the boundary.

Computational fluid dynamics (CFD): the study and simulation of fluid flow using numerical methods to solve the governing equations of fluid dynamics.

Condition number: a measure of how sensitive a problem or equation is to changes in the input or initial conditions. A high condition number indicates a problem that is ill-conditioned and prone to numerical instability.

Consolidation: the process by which soils undergo volume reduction due to the expulsion of water under load. Consolidation analysis evaluates the settlement behavior of saturated soils and determines the time rate at which consolidation occurs.

Convergence: the property of a numerical method that describes how close the approximated solution approaches the true solution as the number of iterations or grid points increases.

Deep foundations: structural elements designed to transfer loads from a structure to stable soil or rock layers at significant depths. Examples include piles (concrete, steel, or timber) and drilled shafts.

Differential equation solvers: these methods are used to approximate solutions to ordinary and partial differential equations. Examples include Euler's method, Runge-Kutta method, and finite difference method.

Differentiation: a numerical method used to approximate the derivative of a function at a given point.

Earth pressure: the lateral pressure exerted by soil or backfill against a retaining structure, such as retaining walls or bridge abutments. Understanding earth pressure is crucial for designing structurally sound and stable retaining systems.

Earth retaining structures: structural systems designed to resist lateral forces applied by soil or water, such as retaining walls, sheet piles, or soil nails. They provide stability for excavated slopes or differential level changes.

Earthquake engineering: geotechnical engineering plays a crucial role in analyzing and designing structures to resist earthquake-induced forces. It involves considering dynamic soil properties, liquefaction potential, ground motion amplification, and seismic isolation techniques.

Earthworks: any man-made soil structure or embankment that serves a specific purpose, such as a road or a dam.

Elasticity: the ability of a rock mass to deform under stress and return to its original shape upon stress removal.

Embankment: a soil structure used to support or contain soil, often used in road construction or landfill sites.

Error analysis: methods to quantify and assess the accuracy of numerical solutions. Key concepts include absolute and relative error, truncation error, round-off error, and convergence analysis.

Excavation: the process of digging or removing material from the ground, often used for constructing foundations, tunnels, or basements.

Extrapolation: a numerical method used to estimate the value of a function outside the range of known data points.

Fast fourier transform (FFT): an algorithm that efficiently computes the discrete Fourier transform and is widely used in numerical methods for solving differential equations.

Finite difference method: a numerical method that approximates derivatives by finite differences and solves differential equations by discretizing the domain.

Finite element method: a numerical method that divides the domain into smaller subdomains and approximates the solution using piecewise functions.

Finite volume method: a numerical method that discretizes the domain into control volumes and solves the equations by balancing the fluxes across the control volume boundaries.

Foundation Design: geotechnical engineers design foundations to evenly distribute structure loads to the underlying soil or rock. Different types of foundations include shallow foundations (such as spread footings and mat foundations) and deep foundations (such as piles and caissons).

Foundation: the substructure that supports a structure, typically made of soil or concrete.

Fracture: the process of a rock mass breaking or cracking due to excessive stress.

Geosynthetics: synthetic materials used in geotechnical engineering to improve soil behavior, reinforce slopes or retaining structures, control erosion, or separate different soil layers. Examples include geotextiles, geogrids, and geomembranes.

Geotechnical hazards: natural hazards or potential risks associated with geotechnical conditions. Examples include landslides, slope instability, liquefaction (soil losing strength due to excessive shaking), and ground subsidence.

Geotechnical instrumentation: the use of instruments and sensors to monitor and measure geotechnical parameters, such as groundwater levels, soil pressures, settlement, and slope movements. This data helps validate design assumptions and detect potential issues during construction and operation.

Geotechnical investigation: the process of gathering data about subsurface conditions and geotechnical properties to assess the suitability and stability of a site for construction. This includes techniques such as borehole drilling, sampling, laboratory testing, and in-situ testing.

Geotechnical risk assessment: geotechnical engineers assess and mitigate risks associated with subsurface conditions that may affect project safety and performance. This involves conducting site investigations, analyzing data, and developing risk management strategies.

Geotechnical: relating to the branch of engineering that deals with the behavior of earth materials, such as soil and rock, and their interaction with structures.

Grid spacing: the distance between grid points or nodes used in numerical methods that discretize the domain. Smaller grid spacing generally leads to more accurate results but requires more computational resources.

Ground improvement techniques: these are methods used to improve the engineering properties of soils and rocks. Examples include soil compaction, soil reinforcement with geosynthetics, grouting (injection of stabilizing materials), and deep soil mixing.

Ground improvement: techniques used to modify or enhance the properties of the soil or rock mass to improve its engineering performance. This can involve methods like soil stabilization, compaction, grouting, ground reinforcement, or the use of geosynthetics.

Groundwater control: managing groundwater is critical in geotechnical engineering projects. Techniques such as dewatering (removing or controlling groundwater) and seepage control measures are employed to maintain stable ground conditions during construction.

Groundwater flow: the movement of water within the subsurface, typically within pore spaces of soil and rock formations. Understanding groundwater flow is crucial for analyzing seepage, pore pressure distribution, and slope stability.

Groundwater: water found underground in soil or rock pore spaces or in fractures and serves as an important consideration in geotechnical engineering due to its influence on soil/rock behavior and stability.

Integration methods: numerical integration methods are used to calculate the definite integral of a function over a given interval. Common methods include the trapezoidal rule, Simpson's rule, and Gauss-Legendre quadrature.

Integration: a numerical method used to approximate the definite or indefinite integral of a function.

Interpolation: interpolation refers to the process of estimating the values between two known data points.

Iterative methods: algorithms that use an initial guess to repeatedly refine the solution until a desired level of accuracy is achieved.

Joint: a natural fracture or planar discontinuity within a rock mass that affects its mechanical behavior.

Landslide mitigation: the measures taken to reduce the risk of landslides in areas prone to soil movement or failure.

Linear algebra methods: numerical methods for solving systems of linear equations include Gaussian elimination, LU decomposition, and iterative methods like the Jacobi and Gauss-Seidel methods.

Linear equations: systems of equations where each equation is a linear combination of the variables. Numerical methods are used to solve these equations.

Linear systems: methods for solving systems of linear equations, such as Gaussian elimination, LU decomposition, and the Jacobi method.

Mohr-Coulomb criterion: a mathematical equation that describes the failure of rocks based on their strength properties.

Monte Carlo methods: Monte Carlo methods use random sampling techniques to estimate unknown quantities or simulate complex systems. They are commonly used in areas such as finance, physics, and computer simulations.

Nonlinear equations: systems of equations where at least one equation is not a linear combination of the variables. Numerical methods are used to solve these equations.

Numerical integration: techniques for approximating the definite integral of a function. Methods include the Trapezoidal rule, Simpson's rule, and Monte Carlo integration.

Numerical linear algebra: this field involves algorithms and techniques for solving problems related to matrices, eigenvalues, eigenvectors, singular value decomposition, and matrix factorizations.

Numerical methods: mathematical techniques used to solve problems in mathematics, science, engineering, and other fields using numerical approximations.

Optimization methods: these methods aim to find the maximum or minimum value of a function within a given domain. Examples include gradient descent, Newton's method, and simulated annealing.

Optimization: procedures for finding the maximum or minimum value of a function. Popular optimization algorithms include the Golden Section Search, Gradient Descent, and Genetic Algorithms.

Ordinary differential equations (ODEs): procedures for numerically solving differential equations. Common techniques include Euler's method, Runge-Kutta methods, and finite difference methods.

Partial differential equations (PDEs): approaches to numerically solve partial differential equations. Examples include finite difference methods, finite element methods, and spectral methods. Numerical methods are used to solve these equations.

Permeability: the measure of a soil's ability to transmit fluids (usually water) through its void spaces. It is an essential property for evaluating seepage and drainage characteristics in geotechnical designs.

Pile foundation: a type of foundation that transfers structural load to deep soil or bedrock through a series of piles.

Plasticity: the ability of a rock mass to undergo permanent deformation under stress.

Retaining structures: these structures are designed to retain or support soil or other materials at different elevations. Common types include retaining walls, sheet piles, soldier piles, and anchored bulkheads.

Retaining wall: a structure designed to hold back soil or other materials to create a difference in elevation between two areas, typically used in landscape design or construction.

Rock anchorage: system of installing rock bolts or other reinforcing elements to provide stability and support to a rock mass.

Rock anisotropy: the variation of rock mechanical properties with orientation or direction due to the presence of preferred mineral orientation, foliation, or jointing.

Rock blasting: the use of controlled explosives to fragment and excavate rocks.

Rock bolt design: the design and installation of rock bolts or other reinforcement mechanisms to improve the stability and load-bearing capacity of rock structures.

Rock engineering design: the application of rock mechanics principles and techniques to design safe and efficient engineering structures in rock formations, such as tunnels, dams, foundations, and underground mines.

Rock mass back analysis: the technique of conducting inverse analysis to determine the in-situ rock mass properties or stress conditions based on measured displacements or other observed responses.

Rock mass behavior in high-stress environments: the study of rock mass behavior and failure mechanisms in high-stress environments such as deep underground mines, deep tunnels, or high-pressure geothermal reservoirs.

Rock mass behavior under high confining pressure: the response and mechanical behavior of a rock mass under high confining pressures, such as those encountered at great depths or beneath large overburden.

Rock mass behavior: the response of a rock mass to loading and unloading conditions, including stress-strain relationships, creep, and time-dependent deformations.

Rock mass blast vibration analysis: the evaluation of vibrations and their impact on a rock mass due to blasting operations, including the prediction and control of blast-induced damage and rock fragmentation.

Rock mass characterization: the process of identifying and quantifying the physical and mechanical properties of a rock mass, including rock type, weathering, jointing, and in-situ stress conditions.

Rock mass classification systems: methods used to categorize rock masses based on their structural, geological, and engineering properties.

Rock mass creep behavior: the slow and continuous deformation of a rock mass under constant stress over an extended period of time, which can affect long-term stability and structural integrity.

Rock mass design parameters: the selection and determination of appropriate design parameters, such as rock mass properties, material properties, safety factors, and design methodologies, in rock engineering projects.

Rock mass discontinuity analysis: the analysis and characterization of rock mass discontinuities, including their orientation, persistence, roughness, spacing, and their influence on rock mass behavior.

Rock mass displacement monitoring: the use of geotechnical instrumentation, such as extensometers, inclinometers, crack meters, or GPS, to measure and track the displacements and deformations within a rock mass.

Rock mass displacement: the movement or displacement of rock mass along fractures, joints, or fault surfaces due to applied stress or geological processes.

Rock mass dynamic response: the behavior of a rock mass under dynamic loading conditions, such as earthquakes, blasting, or vibrations from machinery, and the assessment of potential damages and stability issues.

Rock mass empirical methods: simplified and empirical approaches used to assess rock mass behavior and properties, relying on established relationships and correlations based on past experiences and observations.

Rock mass energy dissipation: the ability of a rock mass to dissipate energy under dynamic loading conditions, such as seismic events or rock blasting, to minimize damage and vibrations.

Rock mass excavatability: the evaluation of the ease and efficiency of excavating rock masses using different excavation methods and equipment, considering their geological and geotechnical properties.

Rock mass excavation design: the development of excavation and sequencing plans for rock mass projects, considering factors such as rock mass behavior, support requirements, safety considerations, and project objectives.

Rock mass excavation methods: various methods used to excavate rock masses, including drilling and blasting, mechanical excavation, tunnel boring machines, and the New Austrian tunneling method (NATM).

Rock mass excavation sequence optimization: the optimization of the order and timing of rock mass excavation to minimize ground support requirements, control hazards, and improve overall project efficiency.

Rock mass excavation support design: the design of temporary or permanent support systems, such as shotcrete, rock bolts, or steel arches, to ensure the stability and safety of excavations in rock masses.

Rock mass excavation-induced damage: the evaluation and mitigation of potential damage to a rock mass during the excavation process, including rock deformation, fracturing, spalling, or loosening of support.

Rock mass failure mechanisms: different mechanisms that can lead to rock mass failure, including spalling, wedge failure, toppling, sliding, squeezing, or buckling of rock structures.

Rock mass failure modes: different modes of failure that can occur in a rock mass, including shear failure, tensile failure, compressive failure, and combined failure modes.

Rock mass geohazards: potential geological hazards associated with rock masses, such as landslides, rockfalls, rockslides, or rock bursts, which pose risks to infrastructure, human safety, or environmental stability.

Rock mass geologic structure analysis: the identification, characterization, and analysis of geological structures, such as folds, faults, or bedding planes, in a rock mass, and their impact on its mechanical behavior and stability.

Rock mass geological mapping: the process of geological mapping of a rock mass, including the identification and characterization of different rock types, geological structures, and their spatial distribution.

Rock mass geomechanical design: the integration of geological, geotechnical, and structural considerations to design rock engineering projects that ensure safety, stability, and sustainability.

Rock mass geotechnical mapping: the systematic mapping and characterization of geological and geotechnical features within a rock mass, including joint spacing, orientation, persistence, and roughness.

Rock mass geotechnical risk assessment: the evaluation and quantification of potential hazards, uncertainties, and risks associated with a rock engineering project, considering geological, geotechnical, and structural factors.

Rock mass geothermal energy extraction: the extraction of geothermal energy from a rock mass by utilizing heat transfer mechanisms, assessing the potential for sustainable and efficient energy production.

Rock mass ground improvement techniques: methods used to enhance the engineering properties of a rock mass, such as grouting, compaction, soil reinforcement, or geosynthetic reinforcement, to increase stability and load-bearing capacity.

Rock mass groundwater interaction: the investigation of the interaction between a rock mass and groundwater, including water flow, pore pressure changes, seepage forces, and their influence on rock mass stability.

Rock mass grouting techniques: different techniques employed for grouting in rock masses, including permeation grouting, fracture grouting, curtain grouting, and compaction grouting.

Rock mass grouting: the injection of grout material into rock fractures or voids to fill and strengthen the rock mass, improve ground stability, and control water flow.

Rock mass instrumentation: the use of monitoring systems to measure and analyze the behavior and response of a rock mass over time, including stress, strain, displacement, and groundwater conditions.

Rock mass numerical modeling validation: the process of comparing the results of numerical simulations, using methods like finite element modeling or distinct element modeling, with actual field observations or laboratory tests to validate their accuracy and reliability.

Rock mass numerical modeling: the use of numerical techniques, such as finite element analysis or discrete element method, to simulate and predict the behavior of a rock mass under different conditions.

Rock mass numerical stability analysis: the use of numerical methods, such as finite element analysis, distinct element modeling, or limit equilibrium analysis, to assess the stability of a rock mass under different loading and boundary conditions.

Rock mass permeability: the ability of a rock mass to allow the flow of fluids, such as water or gases, within its pore spaces or fractures.

Rock mass properties: the physical and mechanical characteristics of a rock mass, including strength, density, porosity, and permeability.

Rock mass reinforcement design: the design of reinforcement measures and support systems, such as rock bolts, rock anchors, rock dowels, rock netting, or shotcrete, to enhance the stability and load-bearing capacity of a rock mass.

Rock mass reinforcement: the implementation of engineering measures to stabilize and reinforce a rock mass, including rock bolting, shotcrete, ground anchors, soil nails, and geosynthetics.

Rock mass remote sensing techniques: the use of remote sensing technologies, such as LiDAR (light detection and ranging), photogrammetry, or satellite imagery, to gather data on rock mass characteristics, surface deformations, or hazards.

Rock mass sesponse to stress: the behavior of a rock mass in terms of deformation, strength, and failure under different stress conditions.

Rock mass rheology: the study of the flow and deformation behavior of a rock mass under applied stress, including the identification of viscoelastic and nonlinear properties.

Rock mass rockburst analysis: the study and prediction of rockbursts, which are sudden and violent failure events in a rock mass due to high stress concentrations, and their impact on underground excavations.

Rock mass rockfall protection systems: the design and implementation of systems to protect against rockfall hazards, including catchment fences, rockfall barriers, rockfall nets, or energy-absorbing attenuators.

Rock mass seepage analysis: the assessment of water flow and seepage characteristics within a rock mass, including the evaluation of permeability, pore pressure distribution, and potential for groundwater-related hazards.

Rock mass seismicity: the study of the occurrence, distribution, and characteristics of earthquakes and seismic activity within a rock mass, including induced seismicity from mining or geothermal activities.

Rock mass slope stability analysis: the assessment of the stability of rock slopes or cliffs under the influence of gravity and other external forces, considering factors such as slope geometry, rock properties, water presence, and potential failure mechanisms.

Rock mass stability analysis in underground mining: the assessment of rock mass stability in underground mining operations, considering factors such as mining methods, support systems, and mining-induced stresses.

Rock mass stability analysis: the assessment of the stability and safety of a rock mass against potential failure or collapse, including the calculation of factors of safety and stability charts.

Rock mass stability monitoring: the continuous or periodic monitoring of stability-related parameters, such as displacements, pore pressure, stress changes, or seepage, to assess the ongoing stability of a rock mass structure.

Rock mass strength criteria: various criteria used to evaluate the strength and stability of a rock mass, such as the Hoek-Brown criterion or the Generalized Hoek-Brown criterion.

Rock mass testing and sampling: the collection of representative rock samples and conducting laboratory tests to determine physical and mechanical properties, such as intact strength, porosity, permeability, and elastic modulus.

Rock mass water infiltration: the infiltration of water into a rock mass through joints, fractures, or permeable layers, leading to changes in mechanical properties, erosion, or groundwater-related issues.

Rock mechanics testing: experimental methods used to determine the mechanical properties and behavior of rocks, such as uniaxial compression testing, triaxial testing, and direct shear testing.

Rock mechanics: the study of the mechanical properties and behavior of rocks, including their strength, deformability, and failure mechanisms.

Rock slope stability analysis: the assessment of the stability of rock slopes or cliffs to prevent failures and landslides.

Rock support systems: measures taken to reinforce and stabilize rock masses during excavation, such as shotcrete, rock bolts, mesh, and rock anchors.

Rockfall hazard assessment: the evaluation of the potential for rocks to detach and fall from slopes, cliffs, or excavations, posing a risk to infrastructure or human safety.

Root finding: techniques for finding the value(s) of a function that make it equal to zero. Examples include the bisection method, Newton-Raphson method, and secant method.

Root-finding methods: these methods are used to find the roots (solutions) of equations or functions. Examples include the bisection method, Newton-Raphson method, and secant method.

Round-off error: the error introduced due to the limited precision of numerical calculations and representation of numbers in a computer.

Sensitivity analysis: the study of how changes in the input or initial conditions of a problem affect the output or solution, often used to determine the robustness of numerical methods.

Settlement analysis: the estimation of expected settlement or vertical deformation of a soil or foundation under loading. Settlement analysis helps to ensure that structures will not settle excessively, leading to structural damage or functional issues.

Settlement control: settlement refers to the downward movement of the ground due to the applied load. Geotechnical engineers analyze settlement behavior to ensure that structures do not experience excessive or differential settlements. Techniques such as preloading, surcharge, and ground improvement can be employed to control settlement.

Settlement: vertical compression or displacement of the ground surface due to applied loads. Settlement analysis helps ensure that structures will not experience excessive or differential settlements that could lead to structural damage.

Shear strength parameters: in geotechnical engineering, shear strength is a fundamental property of soils and rocks that determines their resistance to shear forces. The shear strength parameters include cohesion (the internal bonding forces of the particles) and angle of internal friction (the resistance caused by particle interlocking and surface roughness).

Shear strength: the ability of a soil or rock material to resist shear forces or sliding. It is influenced by factors such as cohesion (the internal attraction between particles) and frictional resistance (resulting from interlocking or roughness).

Shear strength: the resistance of a rock mass to sliding along a plane or failure in shear.

Slope stability analysis: slope stability analysis involves assessing the stability of natural or man-made slopes. Factors such as slope geometry, soil properties, groundwater conditions, and external loads are considered to determine the potential for slope failure.

Slope stability: the ability of a slope to resist downward movement or failure. It involves evaluating factors such as the slope angle, soil strength, groundwater conditions, and external loading to assess stability and design appropriate slope reinforcement measures.

Soil classification: the categorization of soils based on their particle sizes, composition, and engineering properties. Common soil classifications include cohesive soils (clay), granular soils (sand and gravel), and organic soils (peat).

Soil liquefaction: a phenomenon in which saturated soils temporarily lose their strength and behave like a liquid due to increased pore water pressure during seismic shaking.

Soil mechanics: the study of the mechanical properties and behavior of soils, including their strength, deformation characteristics, and flow properties.

Soil: a natural material composed of particles that can be classified as rock, sand, silt, or clay.

Soil-structure interaction: this refers to the interaction between the soil and the structure that is built upon it. Geotechnical engineers analyze how soil stiffness, settlement, and lateral displacements affect the performance of structures and vice versa.

Spectral methods: numerical methods that approximate functions using a series of basis functions, such as Fourier series or Chebyshev polynomials.

Stability: the property of a numerical method that ensures the solution does not become significantly distorted or divergent due to errors or approximations.

Strain: the deformation experienced by a rock mass in response to applied stress.

Stress: the force applied per unit area within a rock mass.

Truncation error: the error introduced due to approximating an infinite process, such as an infinite series or integral, by a finite process.

Tunneling in rocks: excavation of underground tunnels or channels through rock masses using various methods, including drilling, blasting, and mechanical excavation.

Underground rock excavation: the process of excavating underground openings, such as tunnels, caverns, or shafts, in a rock mass, involving design, support, and excavation methods specific to underground environments.

Appendix B: Key Words of Geotechnical Engineering

A

abrasion action 磨蚀作用
absorbed water 吸[附]水
absorption loss 吸附损失
accident investigation 事故调查
accumulated deformation 累积变形
accuracy 准确度;精度
acidic igneous rock 酸性火成岩
acidity and alkalinity test 酸碱度试验
acoustic emission monitoring 声发射监测
acoustic exploration 声波勘探
acoustic piezometer 声辐射测压计
acoustic prospecting 声波勘探
active earth pressure [P_a] 主动土压力
active fault 活断层;活动断裂
active period 活动期,活跃期
active pile 主动桩
active state 主动状态
active water (有侵蚀性)活性水
active zone 活动层,膨胀土层
additional stress 附加应力
additive constant 附加常数
adhesion 黏着力;黏着作用;内聚力
adhesive 胶粘剂
admissible load 容许荷载
aerated water 饱气水
aerial photographic survey 航空摄影测量
aggregate (土壤) 团粒;(混凝土) 骨料
allowable bearing pressure 容许承载力
alluvial (clay, silt, sand) 冲积(黏土、粉土、砂)
alluvial plain 冲积平原
alluvial soil 冲积土
alluvial terrace 冲积阶地
alluvium 冲积物;冲积层
all-weather 全天候
alternating strain 交变应变
alternating stress 交变应力
altitude 高程,海拔
ambient pressure 环境压[应]力
ambient temperature 环境[介质]温度
amplitude magnification factor 振幅放大因子
amplitude of vibration 振幅
amplitude ratio 振幅比
anchor (in slope works) (边坡工程) 锚杆,锚;锚杆支撑
anchor bolt 锚栓,螺栓
anchor cable 锚索
anchorage 锚定[固]作用;地锚
anchorage force 锚固力
anchoring 锚固
andesite 安山岩
angle of dilatancy 剪胀角;扩容角
angle of inclination 倾角
angle of internal friction 内摩擦角
angle of repose 休止角
angle of rupture 破裂角
angle of shearing resistance 抗剪角,剪切角
angle of shearing strength 剪切强度角
angle of skin friction 表面摩擦角
angle of slope 斜坡角度
angle of true internal friction 真内摩擦角
angular 棱角状的
anisotropic rock 各向异性岩石
anisotropic soil 各向异性土
anisotropy 各向异性
anticline 背斜
anti-slide pile 抗滑桩
aperture size 缝隙尺寸
aplite 细晶岩
aquiclude 阻水层
aquifer 含水层
arching 拱作用
argillaceous rock 泥质岩;黏土(质)岩
atmospheric water 大气水
attenuation 衰减
attitude (地质) 产状;构造面位置
average stress 平均应力
axial 轴向
axial bearing capacity 轴向承载[压]力
axial compression 轴向压缩
axial extension test 轴向拉伸试验

axial load 轴向荷载;轴压
axial pressure 轴向压力
axial strain 轴向应变
axial stress 轴向应力
axial symmetry 轴对称
azimuth 极方位;极方位角

B

back analysis 反演分析
back-calculation 反算法,反推法
backfill 回填土,回泥;回填
backfilling 回填
background noise level 背景音量
backing concrete 堤背混凝土
backward erosion 向源侵蚀;逆向侵蚀
backwater effect 回水作用
bank (of channel) 堤岸;河畔;(航道)边坡
basalt 玄武岩
basalt dyke 玄武岩岩墙
basic intensity 基本烈度
basin 水槽;港池;盆地;峡
bearing 方位;方向角;支承(座架)
bearing capacity 承载力
bearing capacity factor (地基) 承载力因子,承载力系数
bearing test 载荷试验
bedded rock slope 层状岩石斜坡
bedding 层理
bedding plane 层理面
bedrock 基岩
bending moment 弯矩
bending strength 抗弯强度
bentonite 膨润土,膨土岩
bias 偏差
biaxial state of stress 双轴应力状态
biological weathering 生物风化
Biot's Consolidation Theory 比奥固结理论
biotite 黑云母
Bishop's Simplified Method of Slice 毕肖普简化条分法
blasting site 爆石工地;爆破工地
blasting works 爆石工程;爆破工程
bolt 螺栓
bolting 栓固
bonding 黏结;结合力
borehole 钻孔
borehole camera 钻孔照相机
boring bit 钻头
boring rod 钻杆
boundary condition 边界条件
boundary element 边界元
Boundary Element Method 边界元法
Brazilian Test 巴西试验,劈裂试验
brittle failure 脆性破坏
brittle fracture 脆裂
bulk compressibility 体积压缩性
bulk density 体密度
bulk modulus 体积模量
bulk weight 体重量
bulk volume 扩张体积

C

calibrate/calibration 校准[正];校定
case study 实例研究
cataclastic 碎裂的
cavern 洞穴;岩洞
caving 坍落
cavity (岩溶) 溶洞;空隙;洞穴
Circular Arc Analysis 圆弧分析法
cementing agent 胶结剂
characteristic 特性
claystone 黏土岩
cleavage (岩石) 劈理;(矿物) 解理
chemical stabilization 化学加固

chemical weathering 化学风化
coarse particle 粗粒,粗颗粒
coarse sand 粗砂
coefficient of friction 摩擦系数
coefficient of permeability [k] 渗透系数
coefficient of rigidity 刚度系数
coefficient of shrinkage 收缩系数
coefficient of softening 软化系数
coefficient of stiffness 刚度系数
cohesion [c] 黏聚力,内聚力,凝聚力
collapse 崩塌,坍陷;湿陷
columnar joint 柱状节理
compaction 压实,夯实,夯压
compass 罗盘仪;罗盘,指南针
compressible soil 压缩性土;可压缩土
compression 压缩
Compression Test 压缩试验
compression wave, P-wave 压缩波,P 波
compressive strength 抗压强度
concentration factor (应力) 集中因子
concrete 混凝土
confining pressure [σ_3] (三轴试验) 侧压,围压
consequence 后果;结果;影响
consolidation 固结;加固
consolidation apparatus 固结仪
consolidation settlement 固结沉降
Consolidation Test 固结试验
Constitutive Equation 本构方程
Constitutive Law 本构定律
Constitutive Relation 本构关系
contact pressure 接触压力
continuous medium 连续介质
continuous sampling 连续取样
coordinates 坐标
core 岩芯
core sample 岩芯
core sampler 岩芯取样器
corrosion 腐蚀;侵蚀;锈蚀
corrosion protection 锈蚀防护
Coulomb's Earth Pressure Theory 库仑土压力理论
Coulomb's Law 库仑定律
crack 裂缝
crack deformation 裂缝变形
creep 蠕变,徐变;蠕动
creep limit 蠕变极限
creep rate 蠕变速率
creep rupture 蠕变破坏
creep settlement 蠕变沉降
creep strain 蠕变应变
criteria 准则;标准;尺度
critical state 临界状态
crop 露头
cross section 横切面;横断面;横剖面,剖面
crushing strength 压碎强度
cyclic triaxial test 周期加荷三轴试验
cubical triaxial test apparatus 真三轴试验仪
curing 养护
curtain grout 帷幕灌浆
curtain wall 幕墙
curvature 曲率
curve fitting 曲线拟合

D

dangerous rockmass 危岩体
Darcy's Law 达西定律
data acquisition 数据收集
data processing system 数据处理系统
debris 岩屑;(滑坡) 泥石
debris flow 泥石流;岩屑流
debris flow deposit 泥石流坡积物
deflectometer 挠度计
deformation 变形
density [ρ] 密度
density of collapse 塌陷密度
density probe 密度探测仪
densometer 密度计
deposit 沉积;沉积物;淤积物
depression cone 水位降落漏斗
design 设计
design and construction 设计及施工
design load 设计荷载

design strength 设计强度
detonator 雷管
detrimental settlement 有害沉降;不稳定沉降
detrital sediment 碎屑沉积
detritus 岩屑
deviator stress 偏应力,轴差应力,应力差
deviatoric state of stress 偏应力状态
dike 堤;岩脉,岩墙
dilatancy 膨胀性;剪胀性
dilatation 剪胀;松胀,膨胀,扩大
dilatational wave, P-wave 疏密波,P 波
dilate 膨胀
dilation 剪胀;松胀
dilation rate 膨胀率;剪胀率
dilative soil 剪胀性土;膨胀性土
diorite 闪长岩
dip 倾向;倾角
dip angle 倾角
dip direction 倾(斜方)向;倾向方位角
dipmeter 倾角仪
direct shear apparatus(设备)直剪仪
direct shear test 直剪试验
direct simple shear test [DSS-test] 直接单剪试验
disaster reduction 减灾
disconformeable plane 假整合面
discontinuity (plane)不连续面
discontinuous structure 不连续结构[构造]
discrete element method 不连续单元方法
discrete fracture network 离散裂隙网络
discretization 离散化
disintegration 分解;崩解
disklike rock core 饼状岩芯
displacement (slope)(斜坡)位移
disposal 处理;弃置
distortional wave, S-wave 剪切波,S 波
distributed load 分布荷载
distribution 分布
disturbance 扰动
disturbed soil 扰动土
dolerite 粗晶玄武岩
dolerite dyke 粗晶玄武岩岩脉
dolomite 白云岩,白云石
draft 草图;通风;(船舶)吃水
draft plan 草拟图
drain 排水渠;排水管;排水沟;渠管;水管
drainage 排水
drainage system 排水系统
drill bit 钻头
drilled shaft 钻孔竖井;钻孔桩井
drilling hole 钻孔
dry compaction 干压实
dry density [ρ_d] 干密度
dynamic load 动荷载
dynamic strain 动应变
dynamic stress 动应力
dynamic triaxial test 振动三轴试验
dynamite 炸药

E

earth 土,泥;陆地;地球
earth anchor 土锚,土层锚杆
earth load 土压力,土荷载
earth pressure 土压(力)
earth pressure coefficient 土压力系数
earth science 地球科学
earth stress 地应力
earthflow 泥石流
earthquake 地震
earthquake fault 地震断层
earthquake focus 震源
earthquake hypocenter 震源
earthquake intensity 地震烈度
earthquake magnitude 地震震级
earthquake prediction 地震预报
earthquake scale 地震级
earthquake wave 地震波
effective drainage porosity(排水)有效孔隙率
effective normal stress [σ'] 有效法向应力
effective pressure 有效压力
effective stress parameters 有效应力参数

effective stress path [ESP] 有效应力路径
elastic compression 弹性压缩
elastic constant 弹性常数
elastic deformation 弹性变形
elastic medium 弹性介质
elastic modulus 弹性模量
elastic wave 弹性波
electron microscope 电子显微镜
element 单元;元素;要素;零件
elevation 高程
embankment 堤;路堤
embankment collapse 路堤塌陷
empirical correlation 经验关系
empirical method 经验方法
engineering analysis 工程分析
engineering geological exploration 工程地质勘察
engineering geology 工程地质学
environmental geology 环境地质学
epoxy resin 环氧树脂
equivalent stress 等效应力
equivalent time 等效时间
exploitation 开采
exploration 勘探
extension test 拉伸试验
extensometer 伸延测量表,伸长计,延伸仪;应变计
extrusive rock 喷出岩

F

factor 系数;因素
factor of reduction (桩工) 折减因子
factor of safety against sliding 抗滑安全系数
factor of safety (global, partial) [F] 安全系数(整体,单元)
fail 崩塌,塌毁,毁坏;故障;破坏,损坏
failure 破坏;崩塌;故障;塌毁;毁坏,损坏
failure criterion 破坏准则
failure envelope 破坏包线
failure load 破坏荷载
failure mechanism 导致倒塌的机制;破坏机制
failure plane 破坏平面;崩塌面
failure strain 破坏应变
failure stress 破坏应力
failure surface 破坏面
failure wedge 破坏棱体
fall (锤) 落距;(岩、土) 倒塌
fault 断层;断裂
fault creep 断层蠕动
fault displacement 断层位移
fault zone 断裂带;断层带
feasibility 可行性
feasibility design 可行性设计
feasibility study 可行性研究
field bearing test 现场载荷试验
field data 现场数据
field density test 现场密度试验
field identification 现场鉴定
field inspection 现场视察[检查],实地视察[检查]
field investigation 现场勘察[调查]
field loading test 现场载荷试验
field survey 实地勘察;实地测量
field test 现场试验
fill 填土,填方;填料;填塞
fill material 填料,填土材料
filtration 过滤
fine aggregate 细粒料
fine gravel 细砾
fine sand 细砂
fine-grained soil 细粒土
finite difference method 有限差分法
finite element method [FEM] 有限元法
fissure 裂隙
fissure water 裂隙水
flat jack 扁千斤顶
flow 流动;水流;流水
fold 褶曲,褶皱
foliation 叶[页]理;剥理
forecast 预测;预报
foundation engineering 基础工程;基础工程学

frozen soil 冻土
foundation stability 地基稳定性
functional requirement 功能要求
foundation works 基础工程;地基工程
fracture 断裂,破裂;裂隙
fracture zone 破碎带;破裂带
fragment 碎屑
free vibration 自由振动
free water 自由水
free water elevation 地下水位;自由水位
free water surface 地下水位;自由水位
freeze-thaw test 冻融试验
freezing method 冻结法
frequency 频率
friction angle 摩擦角
friction coefficient 摩擦系数
friction damping 摩擦阻尼
friction resistance 摩擦阻力

G

gabbro 辉长岩
general specification 一般规格
generalized procedure of slices 广义条分法
general shear failure 整体剪切破坏
gently dipping fault 缓倾断层
gently dipping joint 缓倾节理
Geographical Information System [GIS] 地理信息系统
geographical mapping 地理勘察;地理测绘
geo-hazard 地质灾害
geohydrology 水文学地质;地下水水文学
geological condition 地质条件
geological environment 地质环境
geological hazard 地质灾害
geological model 地质模型
geological profile/section 地质剖面(图)
geological structure 地质构造
geological survey 地质调查;地质测绘
geological texture 地质构造
geology 地质学;地质
geomechanics 地质力学
geometric parameter 几何参数
geomorphology 地貌学
geotechnical engineering 岩土工程
geotechnical standard 岩土工程标准
geotechnical study 岩土工程研究[调查]
geotechnique 岩土工程;岩土技术
global factor of safety 综合安全系数
Global Positioning System [GPS] 全球定位系统
global position survey 大地定位测量
gneiss 片麻岩
grain 颗粒
grain size 粒径
grain size distribution 粒径分布
granite 花岗岩,花岗石
granite facing 花岗岩面层;花岗石砌面
granitic rock 花岗质岩石
granitic saprolite 花岗岩残积土
granitoid 似花岗岩状,花岗岩类;人造花岗石面
granodiorite 花岗闪长岩
granular 颗粒状的,粒状
gravel 砾石
gravity 重力
greenhouse effect 温室效应
grid 坐标网;方格网
grit 粗砂;砂砾;粗砂岩
gritstone 粗砂岩
ground collapse 地面塌陷
ground stress 地应力
ground surface 地表
ground survey 地面测量
groundwater elevation 地下水位
groundwater flow direction 地下水流向
groundwater reservoir 地下水库
groundwater resource 地下水资源
groundwater seepage 地下水渗流
groundwater steady flow 地下水稳定流
groundwater surface 地下水位
grout 浆液,注浆
grout curtain 灌浆帷幕
gypsum 石膏

H

hanging rock 危岩
hardening 硬化
hazard 危险,事故;灾害
hazard assessment 危险性评估;灾害评估
hazard effect 灾害效应
hazard geology 灾害地质学
hazard intensity 灾害烈度
hazard model 灾害模式
hazard potential 潜在危险性
hazard risk prediction 灾害风险预测
head (of landslide) (滑坡) 顶端;(山泥倾泻) 源头
heat radiation 热辐射
heat treatment 热处理
heave 隆胀
heave compensation 波浪补偿
heave force 隆胀力
height 高程;高度
height above mean sea level 海拔
heterogeneity 非均质性
heterogeneous 非均质的,不均匀的
heterogeneous material 不均匀物质,非均质物质
heterogeneous soil 非均质土
hidden fault 隐伏断层,潜伏断层
hidden karst 隐伏喀斯特,潜伏喀斯特
high permeability 高透水性;高渗透性
high pressure grouting 高压灌浆
high speed landslide 高速滑坡
high stress 高应力
highly decomposed rock 高度风化岩(石)
homogeneity 均质性
homogeneous 均质的
homogeneous layer 均质层
homogeneous soil 均质土
Hooke's Law 虎克定律
Hookean Body 虎克体
Hookean Model 虎克模型
horizontal load 水平荷载
horizontal plate gauge 板式水平位移计
hornblende 角闪岩
hydration water 吸附水;结合水
hydrauger hole 水冲钻钻孔
hydraulic conductivity 导水性
hydraulic excavation 水力开挖
hydraulic extruder 液压(样本)挤出器
hydraulic failure 水力破坏
hydraulic head 水头
hydraulic modeling 水力模型
hydrogeological investigation 水文地质勘察
hydrogeological parameter 水文地质参数
hydrogeology 水文地质学
hydrology 水文学
hydrostatic load 静水荷载
hydrostatic pressure 静水压力
hydrostatic profile gauge 静水位位移监测计
hydrostatic state of stress 静水应力状态
hygrometer 湿度计
hygroscopic capacity 吸湿容量
hygroscopic water content 吸湿含水量
hyperbolic model 双曲线模型

I

igneous rock 火成岩
illite 伊利石
immature residual soil 新残积土
immediate compression 瞬时压缩
immediate settlement 瞬时沉降
immersion test 浸水试验
impact load 冲击荷载
impact strength index 冲击强度指数
imperial unit 英制
impermeability 不透水性

impermeable boundary 不透水边界
impermeable layer 不透水层
impervious boundary 不透水边界
imposed load 使用荷载
impression packer survey 压印器测试
improvement works 改善工程
in situ 原位,原地,原处;现场;就地
in situ density 原位密度
in situ density test 原位密度试验
in situ soil test 原位土工试验
in situ stress 现场应力;原位应力
in situ test 原位测试;现场试验,实地试验
in situ testing 原位试验
inclined shaft 斜井
inclinometer 测斜仪,倾斜仪
indirect approach 间接法
induced geological hazard 诱发地质灾害
infill material 填充材料
infiltration 渗水;渗入;渗透(作用),渗滤
infiltration capacity 入渗量
infiltration point 入渗点
infiltration rate 入渗速率
infiltration test 渗透试验;渗水试验
infiltration velocity 入渗速度
infiltration well 入渗井
infiltrometer 渗透仪
infrared analysis 红外线分析
inhomogeneity 非均质性
inhomogeneous soil 非均质土
initial compression 初始压缩
initial consolidation 初始固结
initial stress 初始应力
initial tangent modulus 初始切线模量
initial void ratio 初始孔隙比
injection grout 压力灌浆
injection test 注入试验
inspection 检查,视察
instability 不稳定性
instantaneous load 瞬时荷载
instantaneous pore pressure 瞬时孔隙压力
instruction manual 指引手册
instrumentation 仪器;仪器测试,仪器监测
intact rock 完整岩石
intact specimen 原状试件
intensity 强度,烈度
interface 交界面
intermediate principal plane 中主平面
intermediate principal strain 中主应变
intermediate principal stress 中主应力
International Organisation for Standardization [ISO] 国际标准化组织
International System [SI] 国际单位
interparticle force 粒间作用力
interparticle friction 粒间摩擦
interparticle repulsion 粒间斥力
interval 间距
intrusive rock 侵入岩
investigation 勘察;勘测;勘探;调查
investigation assignment 勘测任务;调查任务
isogram 等值线图
isotropic consolidation 各向等压固结
isotropic deposit 各向同性沉积物
isotropic stress 各向等应力
isotropy 各向同性
iteration method 迭代法
iterative 迭代的;反复的

J

jack 千斤顶,起重器
jet boring 喷射钻探
joint (岩石)节理;(建筑物料)接头,接缝
joint filler 接缝填料;节理充填(物)
joint grouting 接缝灌[注]浆
joint opening 缝隙
joint plane 节理面
joint system 节理系统
jointing (in rock) (岩石)节理
jointing 填缝;接合
Joosten Process 乔斯登硅化加固法
jumper 击钻杆
Jurassic (年代) 侏罗纪;(地层) 侏罗系

K

kaolin 高岭土
kaolinite 高岭石
kaolinitic clay 高岭石黏土
kaolinized zone 高岭石化带
karst 岩溶,喀斯特
karst aquifer 岩溶含水层,喀斯特含水层
karst base level 岩溶(侵蚀)基准,喀斯特(侵蚀)基准
karst water inflow 岩溶涌水
karstic 岩溶的,喀斯特的
kataclasite 碎裂岩,破碎岩
kataclastic 破碎的,压碎的,碎裂的
kataclastic structure 破碎构造
Kelvin Model (流变) 开尔文模型
key observation well 基准观测井
key pile 主桩
key wall 齿墙
kinematic hardening 运动硬化
kinematic viscosity 运动黏滞性
kinetic friction 动摩擦
kneading compaction 搓揉压实

L

laboratory 试验室;试验所
laboratory soil test 室内土工试验,试验室土工试验
laboratory test/testing 室内试验,试验室试验
laboratory test method 试验室试验方法
lamina 层状体
laminar flow 层流
land collapse 地面塌陷
land deformation 地面变形
land subsidence 地面下沉,地面沉降
land upheaval 地面隆起
land uplifting 地面隆起,地面上升
landfill 堆填区
landslide 滑坡,山泥倾泻,地滑,山崩,塌方
landslide body 滑坡体
landslide control 滑坡防治
landslide hazard 滑坡[山泥倾泻]危险
landslide mass 滑坡体
landslide surface 滑坡[山泥倾泻]面
lateral deformation 侧向变形
lateral pressure 侧向压力
lateral strain 侧向应变
lateral stress 侧向应力
layer 层,岩层
layered aquifer 层状含水层
layered strata 层状地层
layout 布置图;规划设计;规划图;分布;流程图
layout plan 布置图;分布图;平面图
leakage 渗漏;漏出量
length height ratio 长高比
level (测量) 高程;水平
likelihood of failure 崩塌[破坏]的可能性
lime pile 石灰桩
limestone 石灰岩,灰岩
limit 极限;限度;范围
limit analysis 极限分析
limit equilibrium (slope stability analysis)(斜坡稳定性分析)极限平衡
limit equilibrium method 极限平衡法
limit pressure 极限压力
limit state 极限状态
limit state design 极限状态设计法
limit state method 极限状态法
line load 线荷载
linear strain [ε] 线应变
linear strain rate 线应变率
linear variable differential transformer

[LVDT] 线性交换差动变压器
liquefaction 液化;液化过程
liquid limit 液限
liquid phase 液相
liquid state 液态
lithic 石质的;岩屑的
lithification 岩化作用
lithofraction 岩裂作用
lithological composition 岩性成分
lithology 岩性;岩性学
lithosphere 岩石圈
load 荷载;负荷;负载量
load factor 荷载因子,荷载系数
loading 加荷,负荷,荷载;输入装填
loading frame 载荷架
loading test 载荷试验
local shear failure 局部剪切破坏
locality 位置,地点
longitudinal crack 纵向裂缝
longitudinal joint 纵向节理
longitudinal profile 纵剖面
longitudinal section 纵剖面;纵切面
longitudinal wave 纵波
long-term load 长期荷载
long-term stability 长期稳定性
long-term strength 长期强度

M

magnetite 磁铁矿
magnitude (地震)震级
major principal plane 主平面
major principal strain 最大主应变
major principal stress 最大主应力
mantle friction 表面摩擦
mantle of soil 表皮土
manual boring 人力钻探
mapping (地图) 绘制;(地质) 填图,测绘
marble 大理岩,云石
mass density 质量密度
mass force 体积力
mass movement/transport (坡面土石) 重力迁移
mass permeability 大体积渗透性
mass ratio 质量比
material damping 材料[物料]阻尼
material parameter 材料[物料]参数
material property 材料[物料]性质
material strength 材料[物料]强度
material testing 材料[物料]试验
mathematical model 数学模型
matrix 母岩;基质
matrix material 基质材料
maximum dry density 最大干密度
maximum particle size 最大粒径
maximum principal effective stress 最大有效主应力
mean diameter 平均粒径
mean effective stress 平均有效应力
mean principal stress 平均主应力
mechanics of landslide 滑坡的力学性质
mechanism of deformation 变形机理
mechanism of deformation and failure 变形破坏机理
mechanism of landslide 滑坡机理
metamorphic rock 变质岩
metamorphism 变质作用
Method of Moments 力矩法
Method of Slices 条分法
mica 云母
micron 微米
microstructure 微观结构
mine tailings 尾矿砂
mining 采矿
mining subsidence 采空坍陷
minor fold 小折皱
minor principal plane 小主平面
minor principal strain [ε_3] 小主应变
minor principal stress [σ_3] 小主应力
mobilization 活动性;活化作用
mode of failure 破坏模式;破坏形式;崩塌形式
mode of movement 位移形式;位移模式
mode of vibration 振型
model 模型;模式

model test 模型试验
modelling 模型试验;模拟
modelling research 模拟研究
modulus of deformation 变形模量
modulus of dilation 膨胀模量
modulus of elasticity [E] 弹性模量
modulus of elasticity in shear [G] 剪切模量
modulus of linear deformation 线变形模量
Mohr's Circle 摩尔圆
Mohr's Envelope 摩尔包线;强度包线
Mohr's Theory 摩尔理论
Mohr-Coulomb Criterion 摩尔-库伦准则
Mohr-Coulomb Envelope 摩尔-库伦包络线
moisture content [ω] 含水量
moisture equivalent 含水当量
molecular cohesion 分子黏聚力;内聚力
moment 力矩;瞬时;瞬间;动量
moment of rupture 破坏力矩;弯折力矩
monitoring 监测
monitoring survey 监测
montmorillonite 蒙脱石;蒙脱土
mortar 砂浆
movement 位移;移动;移动途径
mucky soil 淤泥质土
mud (silt/clay)泥,淤泥,软泥;泥浆
mudstone 泥岩
multiple point extensometer(现场观测)多点应变计
muscovite 白云母
mylonite 糜棱岩

N

natural disaster 自然灾害,天灾
natural exposure 天然露头
natural hillside 山坡
natural moisture content 天然含水量
natural oscillation 固有振动
natural slope 山坡,天然坡
natural soil 天然土
natural strength 天然强度
natural terrain (滑坡) 山坡;坡地
natural vegetation 天然植披
neutral pressure 孔隙水压力
New Austrian Tunneling Method 新奥(地利)隧道(建筑)法
Newtonian Liquid 牛顿液体
node 节点;波节
non-coring drilling 无岩芯钻进,不取芯钻进
nonlinearity 非线性
non-plastic 无塑性
normal fault 正断层
normal force 法向力
normal loading condition 正常荷载条件
normal strain [ε_n] 法向应变,正应变
normal stress [σ_n] 法向应力,正应力
normality law 正交定律
normalization 归一化
numerical analysis 数值分析

O

observation 观测,观察
observation target 测量观标
observed profile 观测剖面;实测剖面
ocean engineering 海洋工程;海洋工程学
offset 偏移;水平错断
offshore 离岸地区;近海地区
open joint 张开节理,裂开节理
open pit(场地勘探) 露天探坑;(场) 露天开采
opening 孔洞,张裂隙出口
optimization 最优化,最佳化;最佳条件选择
orientation 定向;取向;定位;方向
oriented sample 定向样品
outcrop (岩石) 露头
outflow 流出,外溢;流量
outlet 出水孔

overall shear failure 整体剪切崩塌
overall stability 整体稳定性
overburden 覆盖层;表土;积土压力
overburden layer 覆盖层;表土
overburden pressure 覆盖压力;自重压力
overstrain 超限应变
overstress 超限应力

P

packer (permeability) test (钻孔) 压水(渗透)试验
parameter 参数
parametric study 参数研究
parent rock 母岩
partial saturation 不完全饱和
partially weathered rock 部分风化岩石[岩体]
particle 土粒;质点;微粒;颗粒
particle size 粒径,粒度
particle size distribution [PSD] (土壤) 粒径分布,粒径级配
peak strength 峰值强度
pebble 小卵石,圆砾
permanent deformation 永久变形
permeability 渗透性
permeability coefficient 渗透系数
permeability criterion 渗透性准则
permeability test 渗透试验
permeable 透水;透水的
permeable layer 透水层
petrology 岩石学
photogrammetric plot 摄影测量图
photogrammetry 摄影测量学
photographic inclinometer 摄影测斜仪
phreatic line 浸润线,潜水线
physical model 物理模型;实体模型
physical weathering 物理风化
piezometer 测压管;孔隙水压计
pile 桩
planar failure 平面破坏
plane of rupture 破裂面
plane of shear 剪切面
plane of sliding 滑动面
plane of weakness 弱面
plane slide 平面滑动
plane strain 平面应变
plane strain compression 平面应变压缩试验
plane strain extension 平面应变拉伸试验
plane stress 平面应力
plane wave 平面波
plastic deformation 塑性变形
plastic equilibrium 塑性平衡
plastic failure 塑性破坏
plastic strain 塑性应变
plastic zone 塑性区
plasticity 塑性
plasto-elasticity 塑弹性
plate tectonic 板块构造
point load 集中荷载,点荷载,集中载重
point load index (岩石) 点荷载指标
point load index test 点荷载指标试验
point load test 点荷载试验
Poisson's ratio [v] 泊松比
pore 孔隙
pore air pressure 孔隙气压力
pore pressure dissipation 孔隙压力消散
pore pressure distribution 孔隙压力分布
pore pressure parameters 孔隙压力系数
pore pressure ratio 孔隙压力比
pore pressure [u] 孔隙压力,孔压
porous media 多孔介质
precision 精度;精密度
pressure 压力
pressure arch 压力拱
pressure transducer 压力传感器
pressure wave 压力波
primary wave, P-wave 初波,P 波
principal effective stress 有效主应力
principal plane 主平面
principal strain 主应变
principal stress 主应力
principal stress ratio 主应力比

principal stress space 主应力空间
probability 概率;可能性
probe 探头,探测器,探针
processor 加工装置;加工程序
profile 轮廓,外形;剖面
progressive settlement 累进性沉降
pumping test 抽水试验
punch shear test 冲剪试验
punching failure 冲剪破坏
punching shear 冲剪
polyvinylchloride [PVC] 聚乙烯
PVC split liner 开边 PVC 衬管

Q

quarry 采石场,石矿场
quarry stone 毛石
quarrying 采石
quartz 石英
quartz syenite 石英正长岩
quartzite 石英岩
quasi-static 预静态的
quick lime 生石灰,氧化钙
quick shear test 快剪试验
quick soil classification 土的简易分类法
quicklime pile 石灰桩
quicksand 流砂
quick-setting (混凝土) 快速凝固的,早凝的
Q-value (岩体分类法) Q 值

R

radial consolidation 径向固结
radial deformation 径向变形
rainfall 降雨;降雨量
rainfall intensity 降雨强度;雨量强度
rainfall infiltration 雨水渗入
random analysis 随机分析
random error 随机误差
rapid drawdown 水位骤降
rate of consolidation 固结速率
rate of creep 蠕变率
rate of loading 加荷速率
rate of settlement 沉降速率
rate process theory 速率过程理论
Rayleigh wave, R-wave 瑞利波,R 波
reaction force 反应力,反力
reclamation 填海;围垦;回收
recompression 再压缩
recrystallization 重结晶作用
reduction factor 折减系数
reflection 反射
relative density 相对密度
relative humidity 相对湿度
relaxation (桩承载力) 松弛;歇后降低
reliability analysis 可靠性分析
remote sensing 遥感
remote sensing image 遥感图像
remote sensing technique 遥感技术
residual angle of internal friction 残余内摩擦角
residual cohesion intercept 残余黏聚力[内聚力]
residual deformation 残余变形
residual strength 残余强度
residual stress 残余应力
reverse fault 逆断层
rheological model 流变学模型
rheological test curve 流变试验曲线
rheology 流变学
rhyolite 流纹岩
rigid 刚性的;坚硬的
very high risk 非常高风险
high risk 高风险
moderate to high risk 中等至高风险
moderate risk 中等风险

low to moderate risk 低至中等风险
low risk 低风险
negligible risk 微风险
rock 岩石,石头
rock anchor 岩层锚杆
rock blasting 爆石
rock bolt 岩层锚杆,岩层螺栓
rock deformation parameter 岩石变形参数
rock fall 滚石;岩石滚下,岩石滚落
rock fall fence 防石栏
rock joint shear strength test 岩石节理剪力试验
rock mass 岩体
rock mass quality 岩体质量,岩体质量
rock mass rating(岩体分类)岩体等级
rock mechanics 岩石力学,岩体力学
rock outcrop 外露岩石;岩石露头
rock permeability 岩石渗透性
rock pressure 岩体压力
Rock Quality Designation [RQD] 岩石质量指标
rock quality index 岩石质量指数
rock slope 岩石坡
rock strength parameter 岩石强度参数
rock structure 岩石构造
rock texture 岩石结构
rock weathering 岩石风化;岩石风化作用
rock weathering index 岩石风化系数
rockburst 岩爆
rupture 破裂,断裂
rockfill dam 堆石坝
rupture circle 破裂圆
rockfill slope 填石坡,堆石坡
rough discrete fracture network 粗糙离散裂隙网络
roughness 粗糙度

S

safety factor 安全系数,保险系数
sample 试样,样本,样品
sampling 取样;抽样
sand 砂,砂粒;砂地;砂层
sandstone 砂岩
sandy clay 砂质黏土
sandy clay loam 砂质黏壤土
sandy gravel 砂砾石
sandy silt 砂质粉土,砂质粉砂
sandy soil 砂质土
saturated sample 饱和样本
saturation 饱和
scale 比例;比例尺
scale effect 尺度效应
scanning electron microscope 扫描电子显微镜
scheme 计划,方案;图表,图解
schist 片岩
schistosity 片理
section 部分;剖面,断面
sedimentary discontinuity 沉积结构面
sedimentary rock 沉积岩
sedimentary soil 沉积土
sedimentation 沉积作用
seepage 渗流;渗漏,渗水,渗出
seepage analysis 渗漏分析
seepage failure 渗透破坏
seepage flow 渗流
seepage line 渗漏线
seepage path 渗透路径
seepage pressure 渗透压力
seepage velocity 渗流速度
seismic engineering 地震工程
seismic focus 震源
seismic force 地震作用
seismic intensity 地震烈度
seismic load 地震荷载
seismic strain energy 地震应变能
seismic survey 震波勘测;地震勘探
seismic wave 地震波
sensitivity 灵敏度
sensitivity analysis 灵敏度分析;不确定性分析
sensor 传感器;探测设备;敏感组件
spectrum analysis 光谱分析
shear box 剪切盒

shear box clearance 剪切盒开缝
shear box test 剪切盒试验
shear coefficient of viscosity 剪切黏滞系数
shear crack 剪切裂缝
shear failure 剪切破坏
shear force 剪切力,剪力
shear modulus 剪切模量
shear plane 剪切面
shear strain 剪应变
shear strain rate 剪应变速率
shear strength 抗剪强度
shear strength envelope 抗剪强度包线
shear strength parameter 抗剪强度参数
shear strength test 抗剪强度试验能极限状态
shear stress [τ] 剪应力
shear wave, S-wave 剪切波,S 波
servo-accelerometer 伺服加速仪[器]
shear zone 剪切带
shearing resistance 抗剪强度,抗剪阻力
settlement 沉降,地陷;沉降量;沉积物
shaft 竖井,桩身
shaft-sinking (for caisson work) (沉箱工程) 沉井
shale 页岩
shallow failure 浅层崩塌
shotcrete 喷浆,喷射水泥砂浆,喷射混凝土
shotcrete lining 喷射混凝土衬砌
shrinkage 收缩;收缩量
side slope 边坡
silt 粉土,粉砂
siltstone 粉砂岩
silty clay 粉质黏土
silty clay loam 粉质黏壤土
silty fine sand 粉质细砂
site 现场;地盘,工地,场地;事发地点
site investigation 场地勘察;实地勘测;现场勘察
site survey 工地[地盘]测量
size effect 尺寸效应
size grading 粒径分级
skarn 硅卡岩
sketch plan 草图
slab 板,平板
slate 板岩;石板
slide/sliding 滑动
sliding boundary 滑坡边界
sliding failure 滑动破坏
sliding plane 滑动面
sliding resistance 抗滑阻力
sliding zone 滑动带
slightly weathered 轻度风化
slip surface 滑动面
slope 斜坡,边坡
slope angle 坡角
slope circle 坡面圆
slope crest 坡顶
slope cutting 切坡,削坡
slope failure 斜坡坍塌,山泥倾泻,滑坡
slope stability 边坡稳定性,斜坡稳定性
slope stability analysis 斜坡稳定性分析
slope toe 坡脚
socketed pile 嵌岩桩
soil 土,土壤
soil classification 土的分类
soil slope 土坡
solidification 固化作用
space distribution 空间分布
spacing 间距;闲隙
specific heat 比热
specimen 试件,试样,样品
specimen size 试样尺寸
split test 劈裂试验
spread foundation 扩展基础;独立基础
spring constant 弹簧常数
spring shock absorber 弹簧减振器
stability 稳定性
stability analysis 稳定性分析
stability factor 稳定性因子
stabilization 加固,固化
state of limit equilibrium 极限平衡状态
stereometry 立体测量学;立体几何
stereoplot 立体测图;赤平极射投影
stereoscope 立体仪
stereoscopic monitoring 立体监测
stiffness 刚度,劲度
strain 应变;变形;拉紧,加压
strain control 应变控制
strain deviator 应变偏量
strain gauge 应变计
strain invariant 应变不变量
strain path 应变路径

strain rate 应变速率
strain softening 应变软化
strain tensor 应变张量
strain-hardening 应变硬化
stratified soil 层状土
strength 强度
strength characteristics 强度特性
strength envelope 强度包线
strength parameters 强度参数
stress 应力
stress concentration 应力集中
stress deviator 应力偏量
stress difference 应力差
stress field 应力场
stress invariant 应力不变量
stress level 应力水平
stress path 应力路径
stress ratio 应力比
stress redistribution 应力重分布
stress tensor 应力张量
stress-strain behavior 应力-应变性状
stress-strain curve 应力-应变曲线
stress-strain relationship 应力-应变关系
strike（地质）走向
structural geology 构造地质学
subsidence 下沉;坍陷,沉陷,下陷,沉降,塌陷
subsidence rate 沉降速度
subsurface 地下;下层面
subsurface excavation 地下开挖
subsurface exploration 地下勘探
subsurface seepage flow 地下渗流
surface flow 表流,地表径流
surface roughness 表面粗糙度
surface texture 表面纹理;表面结构
survey 测量;查勘;调查
surrounding rock（洞室）围岩
swelling clay 膨胀土
swelling pressure 膨胀压力
swelling soil 膨胀土
swelling test 膨胀试验
syenite 正长岩
syncline 向斜
synthetical strength 综合强度

T

tailings 尾矿,尾矿砂
tectonic discontinuity 构造结构面;构造不连续面
tectonic earthquake 构造地震
tectonic joint 构造节理
tectonic stress field 构造应力场
telemetry 遥测
television borehole camera 钻孔电视摄影机
temperature effect 温度效应
temperature gradient 温度梯度
tensile crack 张性破裂,张裂隙
tensile failure 拉伸破坏
tensile strength 抗拉强度
tensile stress 拉应力;张应力
tensile test 抗拉试验;拉力试验
tensiometer 张力计
tension 张力;拉力
tension crack 张裂;张力裂缝
test data 试验数据
textural porosity 结构孔隙度
thrust 推;上冲;冲断层,逆断层
topographic map 地形图
topographic survey 地形测量
topography 地形;地势;地貌
topple 倾覆;坍塌
toppling 崩塌;倾倒
toppling mass 倾倒体;崩塌物
topsoil 表土
total normal stress [σ] 法向总应力
total settlement 总沉降量
total stress [σ] 总应力
true angle of internal friction 真内摩擦角
true-triaxial apparatus 真三轴仪
tunnel 隧道;平洞,平巷
tunnelling 隧道工程
transverse extensometer 横向应变计
turbulence 湍流;紊流;乱流
transverse wave 横波

turbulent flow 紊流
Tresca Criterion 特莱斯卡准则
trial 试验;试用,尝试
trial pit 探井,试井,泥井
triaxial apparatus 三轴仪
triaxial cell 三轴压力室
triaxial compression test 三轴压缩试验
triaxial shear test 三轴剪切试验
triaxial state of stress 三轴应力状态
triaxial test 三轴试验

U

ultimate bearing capacity 极限承载力
ultimate limit state 极限状态
ultimate load 极限荷载
ultimate strength value 强度终值
ultimate stress state 极限应力状态
uncertainty 不确定性,不可靠性
unconfined compressive strength 无侧限[无围压]抗压强度
unconfined compression test 无侧限[无围压]压缩试验
unconformity 不整合
underground engineering 地下工程
underground erosion 地下侵蚀
underground exploration 地下勘探
underground mining 地下采矿
uniaxial state of stress 单轴应力状态
uniaxial tensile strength 单轴抗拉强度
uniaxial tensile strength test 单轴抗拉强度试验
uniaxial tension test 单轴拉伸试验
uniformity 均匀度
unit weight [γ] 单位重量
universal testing machine 万用试验机

V

vacuum pump 真空泵
vacuum sampling tube 真空取样器
variance 变量;方差
vein (in a rock)纹理;岩脉
velocity field 速度场
verification 确定;确认;核实;验证
vibrating wire cell 振弦式压力盒
vibrating wire extensometer 振弦式应变计
vibrating wire strain gauge 振动钢弦式应变计
viscoelastic behavior 黏弹性
viscoelasticity 黏弹性
viscosity 黏滞性;黏度
viscous damping 黏滞阻尼
viscous flow 黏滞流
void(s)孔隙,空隙;空的
void water 孔隙水
void ratio [e] 孔隙率;孔隙比
volcanic ash 火山灰
volcanic hazard 火山灾害
volcanic rock 火山岩
volcanic saprolite 火山岩风化土
volcanics 火山岩,近地表火成岩
volcano 火山
volumetric strain [ε_v] 体积应变
Von Mises Criterion 冯米塞斯准则

W

wall foundation 墙式基础
wall friction 墙摩擦力
warning system 预警系统
waste disposal site 废料处理场

waste reception facility 废物接收设施
water absorption 吸水性
water acidification 水体酸化
water contamination 水污染
water content ratio 含水比
water content/moisture content 含水量,含水率
water distribution system (供)水分布系统
water head 水压;水头
water injection test hole 注水试验孔
water jet 射水器;水柱
water pressure 水压[力]
water resources 水资源
water sample 水样
water sampler 取水样器
water table 地下水位
water-pressure test 压水试验
waterproofing measures 防水措施
waterproofing system 防水系统
watertight screen 防渗帷幕
watertightness test 水密性试验
wave force 波浪力
wave front 波前;波锋
wave length 波长
wave number 波数
weak plane 软弱结构面,脆弱面
weak rock 软质岩石,脆弱岩石
weathered zone 风化带
weathering 风化(作用)
weathering grade (岩层)风化度
weathering profile (岩层)风化剖面,风化壳
wedge 楔形体,楔;楔入,格入
wedge aquifer 楔形含水层
wedge theory 楔体理论
Weibull Distribution 威布尔分布
weighting 发重;衡量;加重
wet density [ρ] 湿密度
wharf 顺岸码头
wind load 风荷载
wire gage 电阻应变片

X

X-ray diffraction analysis X 射线衍射分析
X-ray identification X 射线鉴定
X-ray photograph X 射线照片
X-ray powdered crystal diagram method X 射线粉晶照相法
X-ray powdered crystal diffraction method X 射线粉晶衍射法

Y

yield criterion 屈服准则
yield locus 屈服轨迹
yield point 屈服点
yield stress 屈服应力
yield stress model 屈服应力模型
yield surface 屈服面
yield value 屈服值;产量
Young's modulus [E] 杨氏模量,弹性模量

Z

zero air void curve 饱和曲线
zero air void ratio 饱和气隙比
zone of aeration 充气带;包气带
zone of capillarity 毛细管水带
zone of capillary saturation 饱和毛细管带
zone of saturation 饱和带